AF353033

A Themed Issue in Honor of Professor Raphael Mechoulam: The Father of Cannabinoid and Endocannabinoid Research

A Themed Issue in Honor of Professor Raphael Mechoulam: The Father of Cannabinoid and Endocannabinoid Research

Editor

Mauro Maccarrone

MDPI • Basel • Beijing • Wuhan • Barcelona • Belgrade • Manchester • Tokyo • Cluj • Tianjin

Editor
Mauro Maccarrone
University of L'Aquila
Italy

Editorial Office
MDPI
St. Alban-Anlage 66
4052 Basel, Switzerland

This is a reprint of articles from the Special Issue published online in the open access journal *Molecules* (ISSN 1420-3049) (available at: https://www.mdpi.com/journal/molecules/special_issues/Professor_Raphael_Mechoulam).

For citation purposes, cite each article independently as indicated on the article page online and as indicated below:

LastName, A.A.; LastName, B.B.; LastName, C.C. Article Title. *Journal Name* **Year**, *Volume Number*, Page Range.

ISBN 978-3-0365-3006-2 (Hbk)
ISBN 978-3-0365-3007-9 (PDF)

Contents

About the Editor

Mauro Maccarrone, Dr. Enzymology and Bio-Organic Chemistry, is Professor and Chair of Biochemistry at the Department of Biotechnological and Applied Clinical Sciences, University of L'Aquila (Italy). He is also Head of the Lipid Neurochemistry Unit at the European Center for Brain Research—IRCCS Santa Lucia Foundation, Rome. For his research activity, he has received the "4th Royan International Research Award for Reproductive Biomedicine" (2003), the "2007 IACM (International Association for Cannabis as Medicine) Award for Basic Research", the "2016 Mechoulam Award", the "2020 Tu Youyou Award", and the "2020 International Space Station Research and Development Award" by the American Astronautical Society with NASA and CASIS. He has been included by Stanford University in the "World Top 2% Scientists' List" of the most cited authors, and is also listed among the "Top Italian Scientists".

Preface to "A Themed Issue in Honor of Professor Raphael Mechoulam: The Father of Cannabinoid and Endocannabinoid Research"

This honorary issue of *Molecules* showcases contributions by scientists who received the Mechoulam Award in the last 20 years for major contributions to the understanding of how plant-derived (cannabinoid) and endogenous (endocannabinoid) compounds exert their manifold actions on human health and disease. Both fields of investigation were started by Professor Raphael Mechoulam, who has just celebrated his 90th birthday and continues to illuminate our field of research with his inspiring new ideas.

This book also represents the inaugural issue of a new section of *Molecules* on "Bioactive Lipids". The Guest Editor serves as the Editor-in-Chief for this new section.

Mauro Maccarrone
Editor

Editorial

Tribute to Professor Raphael Mechoulam, The Founder of Cannabinoid and Endocannabinoid Research

Mauro Maccarrone [1,2]

[1] Department of Biotechnological and Applied Clinical Sciences, University of L'Aquila, 67100 L'Aquila, Italy; mauro.maccarrone@univaq.it

[2] European Center for Brain Research, Santa Lucia Foundation IRCCS, 00143 Rome, Italy

Citation: Maccarrone, M. Tribute to Professor Raphael Mechoulam, The Founder of Cannabinoid and Endocannabinoid Research. *Molecules* **2022**, *27*, 323. https://doi.org/10.3390/molecules27010323

Received: 21 December 2021
Accepted: 4 January 2022
Published: 5 January 2022

Publisher's Note: MDPI stays neutral with regard to jurisdictional claims in published maps and institutional affiliations.

During the last 60 years the relevance for human health and disease of cannabis (*Cannabis sativa* or *Cannabis indica*) ingredients, like the psychoactive compound Δ^9-tetrahydrocannabinol (THC), cannabidiol, 120+ cannabinoids and 440+ non-cannabinoid compounds, has become apparent [1]. THC was identified in 1964, and approximately 30 years later (in 1992), the molecular reasons for the biological activity of cannabis extracts were made clearer by the discovery of anandamide (*N*-arachidonoylethanolamine). The latter is the first member of a new family of bioactive lipids collectively termed "endocannabinoids", that are able to bind to the same receptors activated by THC. In addition to endocannabinoids (that include several *N*-acylethanolamines and acylesters), a complex array of receptors, metabolic enzymes, transporters (transmembrane, intracellular and extracellular carriers) were discovered, and altogether they form a so-called "endocannabinoid system" that finely tunes the manifold biological activities of endocannabinoids themselves [2].

Both plant-derived THC and the first endocannabinoids were discovered in Israel by the laboratory led by Professor Raphael Mechoulam, who has just celebrated his 90th birthday and clearly stood out as a giant of modern science.

I met Professor Mechoulam (Raphi) back in 1999, when I attended my first International Cannabinoid Research Society (ICRS) meeting in Acapulco (Mexico) as a newcomer in the field. Although already acclaimed as the founder of a new research area, Raphi was extremely friendly to me, and curious about the implications of my data on the anandamide-degrading fatty acid amide hydrolase in the wider context of human heath. After quite some years, I can say that Raphi still represents an inspiration for young scientists, and a solid reference for more experienced colleagues who are interested in any aspect of cannabinoid and endocannabinoid research. It is indeed rather difficult to summarize the many seminal discoveries and the huge impact that Raphi has had over the last 60 years, in particular on advancing therapeutic drug discovery. Just to give a few examples, he was the first to elucidate in 1964 the complete structure of THC [3]. Then, he identified many additional phytocannabinoids in 1965–1972 (reviewed in ref. [1]), and later on discovered also the endogenous counterparts of THC: anandamide in 1992 [4], and 2-arachidonoylglycerol in 1995 [5], the latter at the same time and independently of Sugiura and colleagues in Japan [6]. Then, Raphi identified arachidonoyl-serine, an endogenous vasodilator, in 2006, and oleoyl-serine, an endogenous regulator of bone mass, in 2010 (reviewed in refs [7,8]). These are just some of the milestones in Raphi's (endo)cannabinoid investigations that have boosted intense research on the proteins that bind to and metaboilze these substances, leading to the definition of an entirely new signal transduction system based on bioactive lipids. Such a system, along with plant-derived cannabinoids themselves, is now widely recognized for its therapeutic potential in almost all human diseases, as suggested also by the ever-growing number of investigations that can be retrieved from a PubMed search (Table 1).

The many implications of the seminal work of Raphi for chemistry, biochemistry, biology, pharmacology and medicine are reflected in this special issue by contributions made by Raphi himself and by the selected group of scientists who over the last 20 years received from the ICRS the highest recognition in the field of (endo)cannabinoid research: the Mechoulam Award.

Table 1. Results of a PubMed search from 1964 (when THC was discovered) to 2021 with the entries "cannabinoids and disease" and "endocannabinoids and disease". It should be recalled that the first endocannabinoid anandamide was discovered in 1992.

Time Range	Cannabinoids and Disease	Endocannabinoids and Disease
1964–1970	0	-
1971–1975	14	-
1976–1980	19	-
1981–1985	18	-
1986–1990	23	-
1991–1995	37	0
1996–2000	103	16
2001–2005	497	178
2006–2010	1305	665
2011–2015	1608	884
2016–2021	2924	1580

In this issue, Raphael Mechoulam and his collaborators report novel data on cannabigerol derivatives able to reduce inflammation, pain and obesity, conditions where there is a huge unmet need of efficient drugs. Indeed, the interest in cannabigerol has been growing in the past few years and therapeutic expectations are rather high [9].

Allyn Howlett, the first Mechoulam Award recipient in 2000, John Huffman (also awarded in 2006) and Brian Thomas address the "spicy story" of cannabimimetic indoles, reviewing the discovery of aminoalkylindole analgesics, structure-activity relationship studies in search of their common pharmacophore, and their activity as cannabinoid receptor agonists [10].

George Kunos, awarded in 2005, and his colleagues describe novel findings on the effects of a peripherally restricted hybrid inhibitor of type 1 cannabinoid receptor (CB_1) and inducible NO synthase (iNOS) on alcohol drinking behavior and alcohol-induced gut permeability. Of note, they analyze also the relative role of central versus peripheral CB_1 receptors in alcohol drinking behavior, which may have major implications for drug discovery against alcohol dependence [11].

Vincenzo Di Marzo, awarded in 2007, reports new data on liver-expressed antimicrobial peptide-2 (LEAP-2) in the gut, showing that it is regulated by the endocannabinoidome-gut microbiome axis, an emerging and really hot topic in the field [12].

Ken Mackie, recipient of the Mechoulam award in 2008, examines with his colleagues the effects of several "minor" cannabinoids on neuronal function by using two model systems: cultured autaptic hippocampal neurons and dorsal root ganglion neurons. They show that two of these natural compounds (cannabidivarin and Δ^9-tetrahydrocannabivarin) inhibit CB_1 signaling, yet via distinct mechanisms [13].

Cecilia Hillard, who received the Mechoulam Award in 2011, reports that THC-induced catalepsy requires intact adenosine A_2A receptor signaling to occur. She also shows that cannabidiol and its 4-fluoro derivative both can potentiate the cataleptic effect of THC, an effect that also requires A_2A receptor signaling. Collectively, these data could be explained by cannabinoid inhibition of the equilibrative nucleotide transporter, which will raise adenosine concentrations thus resulting in activation of adenosine receptors, particularly A_2A present in the striatum [14].

Beat Lutz, awarded in 2014, and colleagues describe subsynaptic distribution, lipid raft targeting and G protein-dependent signaling of CB_1 in synaptosomes from the mouse hippocampus and frontal cortex. In summary, their results provide an updated view of the functional coupling of CB_1 to $G_{\alpha i/o}$ proteins at excitatory and inhibitory terminals, and substantiate the utility of the CB_1 rescue model in studying endocannabinoid physiology at the subcellular level [15]. Incidentally, CB_1 location within lipid rafts remains an interesting subject of investigation after 15 years from its first discovery [16].

Mary Abood, who received the Mechoulam award in 2015, and her colleague review CB_1 receptor signaling and biased signaling. The latter involves selective activation of a signaling transducer in detriment of another, mainly involving selective activation of G-protein or β-arrestin. However, biased signaling at the CB_1 receptor is poorly understood due to the lack of strongly biased agonists. Mary also uses crystallographic structures of CB_1 and proposed mechanisms of action of biased allosteric modulators to discuss a putative mechanism for CB_1 activation and biased signaling [17].

Andreas Zimmer received the Mechoulam award in 2018, and with his colleagues reports new data on type 2 cannabinoid receptor (CB_2) that is shown to alter social memory and microglial activity in an age-dependent manner. They demonstrate how physiological brain aging is characterized by gradual, substantial changes in cognitive ability, accompanied by chronic activation of the neural immune system, a relevant form of inflammation that is termed "inflammaging" [18].

Natsuo Ueda, 2020 Mechoulam awardee, and his coworkers describe the involvement of the γ-isoform of cytosolic phospholipase A_2 ($cPLA_2$) in the biosynthesis of bioactive *N*-acylethanolamines (NAEs) like *N*-arachidonoylethanolamine (anandamide), *N*-palmitoylethanolamine and *N*-oleoylethanolamine. In mammalian tissues NAEs are produced from glycerophospholipids via *N*-acyl-phosphatidylethanolamine (NAPE), and the ε isoform of $cPLA_2$ functions as an *N*-acyltransferase to form this precursor. Since the $cPLA_2$ family consists of six isoforms (α, β, γ, δ, ε, and ζ), the present study investigates a possible involvement of the isoforms other than ε in NAE biosynthesis. Presented results suggest that indeed cPLA2γ is involved in the biosynthesis of NAEs through its phospholipase A_1/A_2 and lysophospholipase activities [19].

Finally, Javier Fernandez-Ruiz, awarded in 2021, and his coworkers report a preclinical investigation on neuroprotective effects of the orphan G protein coupled receptor (GPR) 55 ligand VCE-006.1 in experimental models of Parkinson's disease (PD) and amyotrophic lateral sclerosis (ALS). They conclude that targeting GPR55 may afford neuroprotection in PD, but not in ALS, thus stressing the differences in the development of cannabinoid-based therapies in neurodegenerative disorders [20].

This honorary issue of *Molecules* showcases contributions by half of the scientists who received the Mechoulam Award over the years. They are listed in Table 2 along with the awardees who unfortunately could not participate in this editorial project. I thank all colleagues for their valuable contributions to this volume, and I especially thank Professor Raphael Mechoulam for continuing to illuminate our field of research with his always inspiring new ideas.

Table 2. Mechoulam Award recipients. Contributors to the present *Honorary Issue* are in italics.

Mechoulam Award Recipient	Year
Allyn Howlett	2000
Billy Martin	2001
Roger Pertwee	2002
Raj Razdan	2003
Murielle Rinaldi-Carmona and Francis Barth	2004
George Kunos	2005

Table 2. *Cont.*

Mechoulam Award Recipient	Year
John Huffman and Alex Makriyannis	2006
Vincenzo Di Marzo	2007
Ken Mackie	2008
Gerard Le Fur	2009
Patti Reggio	2010
Cecilia Hillard	2011
Ben Cravatt	2012
Aron Lichtman	2013
Beat Lutz	2014
Mary Abood	2015
Mauro Maccarrone	2016
Daniele Piomelli	2017
Andreas Zimmer	2018
Daniela Parolaro	2019
Natsuo Ueda	2020
Javier Fernandez-Ruiz	2021

Acknowledgments: This paper was made possible by financial support from Università degli Studi dell'Aquila under intramural competitive grants "RIA 2021" and "Progetti di Ricerca di Ateneo 2021" to MM.

Conflicts of Interest: The author declares no conflict of interest. The funders had no role in the design of the study; in the collection, analyses, or interpretation of data; in the writing of the manuscript, or in the decision to publish the results.

References

1. Elsohly, M.A.; Slade, D. Chemical constituents of marijuana: The complex mixture of natural cannabinoids. *Life Sci.* **2005**, *78*, 539–548. [CrossRef] [PubMed]
2. Maccarrone, M. Missing pieces to the endocannabinoid puzzle. *Trends Mol. Med.* **2020**, *26*, 263–272. [CrossRef] [PubMed]
3. Gaoni, Y.; Mechoulam, R. Isolation, structure and partial synthesis of an active constituent of hashish. *J. Am. Chem. Soc.* **1964**, *86*, 1646–1647. [CrossRef]
4. Devane, W.A.; Hanus, L.; Breuer, A.; Pertwee, R.G.; Stevenson, L.A.; Griffin, G.; Gibson, D.; Mandelbaum, A.; Etinger, A.; Mechoulam, R. Isolation and structure of a brain constituent that binds to the cannabinoid receptor. *Science* **1992**, *258*, 1946–1949. [CrossRef] [PubMed]
5. Mechoulam, R.; Ben-Shabat, S.; Hanus, L.; Ligumsky, M.; Kaminski, N.E.; Schatz, A.R.; Gopher, A.; Almog, S.; Martin, B.R.; Compton, D.R.; et al. Identification of an endogenous 2-monoglyceride, present in canine gut, that binds to cannabinoid receptors. *Biochem. Pharmacol.* **1995**, *50*, 83–90. [CrossRef]
6. Sugiura, T.; Kondo, S.; Sukagawa, A.; Nakane, S.; Shinoda, A.; Itoh, K.; Yamashita, A.; Waku, K. 2-Arachidonoylgylcerol: A possible endogenous cannabinoid receptor ligand in brain. *Biochem. Biophys. Res. Commun.* **1995**, *215*, 89–97. [CrossRef] [PubMed]
7. Mechoulam, R.; Hanuš, L.O.; Pertwee, R.; Howlett, A.C. Early phytocannabinoid chemistry to endocannabinoids and beyond. *Nat. Rev. Neurosci.* **2014**, *15*, 757–764. [CrossRef] [PubMed]
8. Pacher, P.; Kogan, N.M.; Mechoulam, R. Beyond THC and endocannabinoids. *Annu. Rev. Pharmacol. Toxicol.* **2020**, *60*, 637–659. [CrossRef] [PubMed]
9. Kogan, N.M.; Lavi, Y.; Topping, L.M.; Williams, R.O.; McCann, F.E.; Yekhtin, Z.; Feldmann, M.; Gallily, R.; Mechoulam, R. Novel CBG derivatives can reduce inflammation, pain and obesity. *Molecules* **2021**, *26*, 5601. [CrossRef] [PubMed]
10. Howlett, A.C.; Thomas, B.F.; Huffman, J.W. The spicy story of cannabimimetic indoles. *Molecules* **2021**, *26*, 6190. [CrossRef] [PubMed]

11. Santos-Molina, L.; Herrerias, A.; Zawatsky, C.N.; Gunduz-Cinar, O.; Cinar, R.; Iyer, M.R.; Wood, C.M.; Lin, Y.; Gao, B.; Kunos, G.; et al. Effects of a peripherally restricted hybrid inhibitor of CB_1 receptors and iNOS on alcohol drinking behavior and alcohol-induced endotoxemia. *Molecules* **2021**, *26*, 5089. [CrossRef] [PubMed]
12. Shen, M.; Manca, C.; Suriano, F.; Nallabelli, N.; Pechereau, F.; Al-lam-Ndoul, B.; Iannotti, F.A.; Flamand, N.; Veilleux, A.; Cani, P.D.; et al. Three of a kind: Control of the expression of liver-expressed antimicrobial peptide 2 (LEAP2) by the endocannabinoidome and the gut microbiome. *Molecules* **2022**, *27*, 1. [CrossRef]
13. Straiker, A.; Wilson, S.; Corey, W.; Dvorakova, M.; Bosquez, T.; Tracey, J.; Wilkowski, C.; Ho, K.; Wager-Miller, J.; Mackie, K. An evaluation of understudied phytocannabinoids and their effects in two neuronal models. *Molecules* **2021**, *26*, 5352. [CrossRef] [PubMed]
14. Stollenwerk, T.M.; Pollock, S.; Hillard, C.J. Contribution of the adenosine 2A receptor to behavioral effects of tetrahydrocannabinol, cannabidiol and PECS-101. *Molecules* **2021**, *26*, 5354. [CrossRef] [PubMed]
15. Saumell-Esnaola, M.; Barrondo, S.; Caño, G.G.d.; Aranzazu Goicolea, M.; Sallés, J.; Lutz, B.; Monory, K. Subsynaptic distribution, lipid raft targeting and G protein-dependent signaling of the type 1 cannabinoid receptor in synaptosomes from the mouse hippocampus and frontal cortex. *Molecules* **2021**, *26*, 6897. [CrossRef] [PubMed]
16. Bari, M.; Battista, N.; Fezza, F.; Finazzi-Agrò, A.; Maccarrone, M. Lipid rafts control signaling of type-1 cannabinoid receptors in neuronal cells. Implications for anandamide-induced apoptosis. *J. Biol. Chem.* **2005**, *280*, 12212–12220. [CrossRef] [PubMed]
17. Leo, L.M.; Abood, M.E. CB_1 cannabinoid receptor signaling and biased signaling. *Molecules* **2021**, *26*, 5413. [CrossRef] [PubMed]
18. Komorowska-Müller, J.A.; Rana, T.; Olabiyi, B.F.; Zimmer, A.; Schmöle, A.-C. Cannabinoid receptor 2 alters social memory and microglial activity in an age-dependent manner. *Molecules* **2021**, *26*, 5984. [CrossRef] [PubMed]
19. Guo, Y.; Uyama, T.; Rahman, S.M.K.; Sikder, M.M.; Hussain, Z.; Tsuboi, K.; Miyake, M.; Ueda, N. Involvement of the γ isoform of $cPLA_2$ in the biosynthesis of bioactive *N*-acylethanolamines. *Molecules* **2021**, *26*, 5213. [CrossRef] [PubMed]
20. Burgaz, S.; García, C.; Gonzalo-Consuegra, C.; Gómez-Almería, M.; Ruiz-Pino, F.; Unciti, J.D.; Gómez-Cañas, M.; Alcalde, J.; Morales, P.; Jagerovic, N.; et al. Preclinical investigation in neuroprotective effects of the 2 GPR55 ligand VCE-006.1 in experimental models of Parkin-3 son's disease and amyotrophic lateral sclerosis. *Molecules* **2021**, *26*, 7643. [CrossRef] [PubMed]

Article

Novel CBG Derivatives Can Reduce Inflammation, Pain and Obesity

Natalya M. Kogan [1,*,†], Yarden Lavi [1], Louise M. Topping [2], Richard. O. Williams [2], Fiona E. McCann [3], Zhanna Yekhtin [4], Marc Feldmann [2,3], Ruth Gallily [4] and Raphael Mechoulam [1]

[1] Medical Faculty, Institute for Drug Research, Hebrew University, Jerusalem 91120, Israel; yarden.lavi@mail.huji.ac.il (Y.L.); raphaelm@ekmd.huji.ac.il (R.M.)
[2] Kennedy Institute of Rheumatology, University of Oxford, Oxford OX3 7FY, UK; louise.topping@kennedy.ox.ac.uk (L.M.T.); richard.williams@kennedy.ox.ac.uk (R.O.W.); marc.feldmann@kennedy.ox.ac.uk (M.F.)
[3] 180 Life Sciences, Menlo Park, CA 94025, USA; fmccann@180lifesciences.com
[4] Lautenberg Center of Immunology and Cancer Research, The Hebrew University of Jerusalem, Jerusalem 91120, Israel; zhannay@savion.huji.ac.il (Z.Y.); ruthg@ekmd.huji.ac.il (R.G.)
*   Correspondence: natalya.kogan@mail.huji.ac.il
†   These authors contributed equally to this work.

Abstract: Interest in CBG (cannabigerol) has been growing in the past few years, due to its anti-inflammatory properties and other therapeutic benefits. Here we report the synthesis of three new CBG derivatives (HUM-223, HUM-233 and HUM-234) and show them to possess anti-inflammatory and analgesic properties. In addition, unlike CBG, HUM-234 also prevents obesity in mice fed a high-fat diet (HFD). The metabolic state of the treated mice on HFD is significantly better than that of vehicle-treated mice, and their liver slices show significantly less steatosis than untreated HFD or CBG-treated ones from HFD mice. We believe that HUM-223, HUM-233 and HUM-234 have the potential for development as novel drug candidates for the treatment of inflammatory conditions, and in the case of HUM-234, potentially for obesity where there is a huge unmet need.

Keywords: cannabinoid; cannabigerol; anti-inflammatory; obesity

Citation: Kogan, N.M.; Lavi, Y.; Topping, L.M.; Williams, R.O.; McCann, F.E.; Yekhtin, Z.; Feldmann, M.; Gallily, R.; Mechoulam, R. Novel CBG Derivatives Can Reduce Inflammation, Pain and Obesity. *Molecules* **2021**, *26*, 5601. https://doi.org/10.3390/molecules26185601

Academic Editor: Mauro Maccarrone

Received: 30 July 2021
Accepted: 10 September 2021
Published: 15 September 2021

1. Introduction

CBG (cannabigerol) was discovered by Gaoni and Mechoulam in cannabis resin (hashish) in 1964 and was considered a missing link in the biosynthesis of THC (tetrahydrocannabinol) [1]. Cannabinoid biosynthesis begins with the combination of geranyl pyrophosphate and olivetolic acid to form CBGA (Cannabigerolic acid). CBGA serves as the substrate for the synthesis of Δ9-THCA (THC acid) and CBDA (cannabidiolic acid). Decarboxylation of CBGA, Δ9-THCA, and CBDA by heat results in CBG, Δ9-THC, and CBD (cannabidiol), respectively. Because CBGA serves as the substrate for the synthesis of the major cannabinoids, very little is typically found naturally in material from Cannabis sp. [2].

While the cannabis constituents CBD and THC have been thoroughly investigated [3,4], research on CBG has been relatively neglected. Based on established pharmacological properties, there is growing evidence that CBG has therapeutic potential for treating neurological disease, gastrointestinal disease as well as some metabolic disorders [2,5–12]. Notably, having mechanisms both in common and distinct from THC and CBD respectively, it was of interest to explore the anti-inflammatory and analgesic properties of CBG, that may be relevant in the aforementioned disease areas.

Indeed, CBG has been shown to exert anti-inflammatory effects and some derivatives of CBG have been synthesized and tested in both animal models and human patient [5–9]. CBG has also been proved to possess anti-inflammatory properties in neurological models [6,7,13] and confer other therapeutic benefits such as appetite stimulation [14]. In addition, CBG is known to activate α2-adrenoreceptor [15] and to interact with sub-types of TRPV (Transient Receptor Potential Vanilloid) channels, pertinent to signaling associated with gastrointestinal inflammation [16]. Association with the classical cannabinoid receptors CB1 and CB2R has been determined, with binding demonstrated to modulate signaling mediated by receptors and receptor heteromers even at low concentrations of 0.1–1 μM [17]. There have been some attempts to investigate and expand CBG's SAR (Structure-Activity Relationship) and these investigations have led to novel pharmaceutical candidates for Parkinson's disease [6,8].

The immune and inflammatory system has evolved to protect against foreign organisms and substances, antigens, that may cause damage to the normal function of the body. The immune response, when dysregulated, may also lead to various pathological conditions, including tumors [18], autoimmune diseases [19], obesity [20], diabetes [21], cardiovascular diseases [22] and more. Western medicine has introduced an array of drugs aiming at the immune system trying to manipulate its response and reducing side effects of acute inflammation. Among the most common medications used by the public are corticosteroids and Non-steroidal Anti-inflammatory Drugs (NSAIDs). Despite being the universal first choice in treating inflammatory diseases, these drug classes have long been known for their adverse side effects, limiting their dosage and prolonged treatment regimens [23–27].

Here we report the synthesis of three new synthetic CBG derivatives and show them to possess anti-inflammatory and pain-resolving properties in preclinical models. In addition, one of these molecules, HUM-234, has also shown prominent activity in obesity prevention in mice fed a high-fat diet (HFD). Importantly, the metabolic state of the HUM-234 treated HFD mice is significantly closer to healthy levels than that of vehicle-treated HFD mice, and their liver slices show much less steatosis than untreated, or CBG-treated livers from HFD mic, suggesting that HUM-234 and related compounds may have potential to treat metabolic disorders including liver disease.

2. Results

2.1. Chemistry Development and Synthesis of HUM-223, HUM-233 and HUM-234

We sought to explore the effect of a bulky residue at position 5 of CBG in place of the natural pentyl on its anti-inflammatory activity. This is based on similar observations made for THC, among other cannabinoids, in which a bulkier residue such as dimethylheptyl (DMH) resulted in an improved efficacy at the relevant biological model [28]. We have also seen in our previous work with CBD derivatives that bulkier DMH derivatives were more active in the knee arthritis model (unpublished results). We have therefore prepared compound HUM-223 which is a monomethoxy CBG-DMH (Figure 1). During the preparation, the partial reaction with iodomethane resulted in a mixture of fully reacted dimethoxy product, partially reacted monomethoxy and an unchanged CBG-DMH. This mixture was hard to isolate on its own and we, therefore, decided to add an acetylation step to the preparation. The crude mixture of the iodomethane reaction was reacted with an excess of acetic anhydride. This made the chromatography easier and allowed us to isolate the pure monomethoxylated product (1). The acetate protecting group can then easily be removed using LiAlH₄ (Lithium Aluminium Hydride) to produce the final HUM-223.

Another alteration deemed to be useful, namely the use of a morpholine propionate ester at one of the phenol positions of CBG. This concept is based on observations made by our group for a similar modification of CBD, which produced an improvement in the anti-inflammatory properties [29]. Coupling of monomethoxy CBG (2) with morpholinopropionic acid yielded therefore compound HUM-234 (Figure 2).

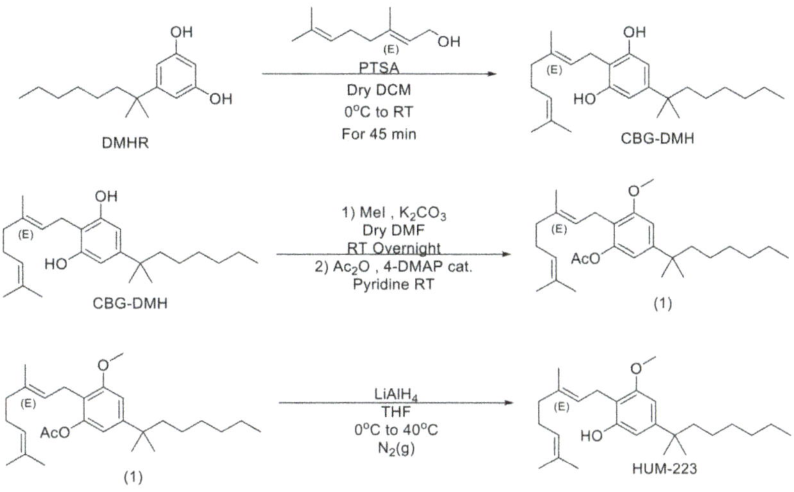

Figure 1. The preparation of HUM-223.

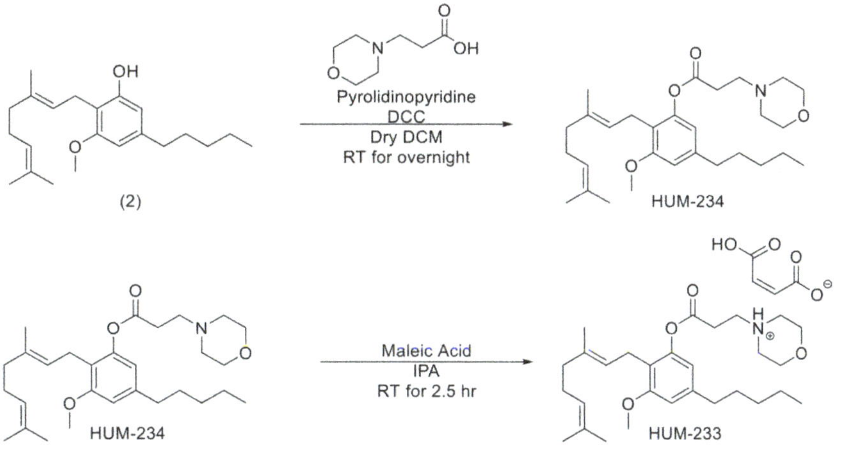

Figure 2. The synthesis of HUM-234 and its maleate salt HUM-233 from monomethoxy CBG.

To further improve its bioavailability, HUM-234 was also turned into a maleic acid salt named HUM-233 (Figure 2). This was based on a similar modification previously carried out by our group with CBD [29]. Since maleic acid has no inherent anti-inflammatory activity, it serves only to increase the compound's bioavailability. As HUM-233 is solid at room temperature, whereas HUM-234 is an oil, we believe that this could offer an additional benefit should this compound prove efficacious since it is easier to handle solid compounds when preparing pharmaceuticals, making the measurement of dose more accurate.

2.2. Evaluation of Anti-Inflammatory Activity

HUM-223 was compared to CBG and to a vehicle control group in three inflammation assays: paw swelling, pain sensation in the paw and circulating TNF-α (Figure 3). CBG itself did not show a consistent pattern of efficacy in all three assays, unfortunately. However, when comparing HUM-223 to CBG we observed significant improvement in the paw swelling assay at a dose of 10 mg/kg.

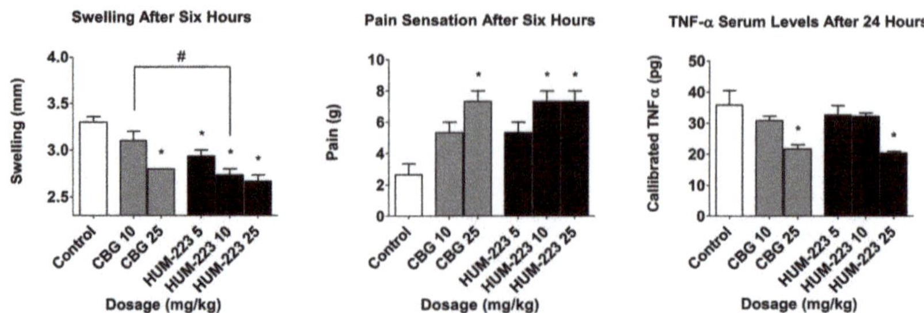

Figure 3. Anti-inflammatory and analgesic evaluation of HUM-223 (black) compared to CBG (cannabigerol) (grey) and vehicle control (white). Statistical comparison was done by 1-way ANOVA (p-value = 0.0002) and post hoc analysis by Tukey's test. * $p < 0.05$ comparing to control group. # p-value < 0.05 in the indicated comparison.

The anti-inflammatory ability of HUM-223 was also assessed in zymosan-induced arthritis (ZIA), a model of acute inflammation in the knee (Figure 4), in comparison to dexamethasone. This molecule has been chosen for knee arthritis assay as previously CBD-DMH derivatives have shown better results in this model than pentyl-chain derivatives. HUM-223 significantly reduced neutrophil elastase levels at 10 mg/kg, in a manner comparable to dexamethasone, when compared to vehicle only treated mice, as measured by in vivo imaging of a fluorescent reporter (IVIS). Additionally, gene analysis of inflamed knee joints demonstrated the ability of HUM-223 to significantly reduce expression of pro-inflammatory genes which encode for enzymes that degrade matrix proteins important for structural integrity, namely adamts4 (A disintegrin and metalloproteinase with thrombospondin motifs 4), neutrophil elastase (Elane) and myeloperoxidase (Mpo), with 5 mg/kg having the greatest effect.

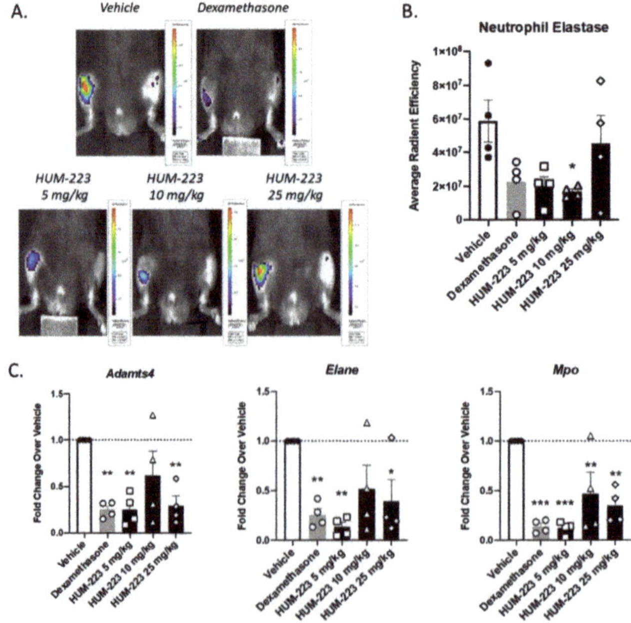

Figure 4. Anti-inflammatory effect of HUM-223 in zymosan-induced arthritis. Mice were treated with vehicle (white bars), dexamethasone (2 m/kg, grey bars) or HUM-223 (black bars). (**A**) Representative

images of neutrophil elastase IVIS fluorescence imaging of knees of ZIA mice treated with vehicle control or HUM-223. (**B**) Quantification of neutrophil elastase average radiant efficiency in the inflamed knee of ZIA mice. (**C**) Gene expression analysis of inflamed ZIA knee joints. Statistical analysis by one-way ANOVA (p-value 0.0392 for Neutrophil elastase IVIS, 0.0080 for Elane, 0.0037 for Adams4 and 0.0002 for MPO) with Tukey's post-hoc test. * $p < 0.05$, ** $p < 0.001$, *** $p < 0.0001$ compared to vehicle-treated mice.

The biological results revealed that both HUM-233 and HUM-234 have anti-inflammatory activity (Figure 5). Both compounds showed improvement of swelling and pain sensation which is comparable to CBG in all tested doses. HUM-233 was able to reduce TNF-α levels in doses of 25 and 50 mg/kg and presented a distinct increase of biological reaction when the doses were increased. HUM-234 showed an opposite trend at elevated doses. HUM-233 and HUM-234 were comparable but were not significantly better than CBG in these assays.

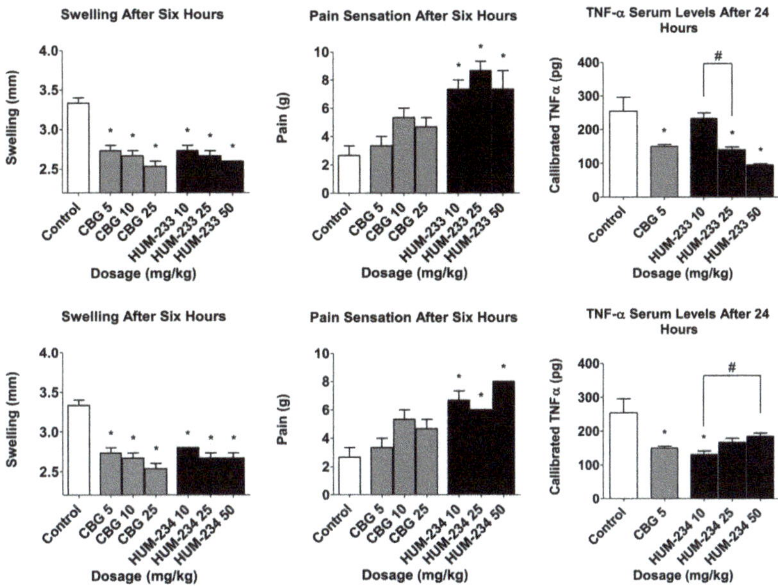

Figure 5. Anti-inflammatory and analgesic evaluation of HUM-233 and HUM-234 (black) compared to CBG (grey) and vehicle control (white). Statistical comparison was carried out by one-way ANOVA (p-value < 0.0001) and post hoc analysis by Tukey's test. * $p < 0.05$ comparing to control group. # $p < 0.05$ in the indicated comparison.

2.3. Evaluation of Effects on Weight Gain

Despite having comparable anti-inflammatory activity, HUM-234 is much more active than CBG in the prevention of obesity. Adult female mice fed an HFD (high-fat diet) gradually gain weight, much faster than STD (standard diet) fed ones. CBG (15 mg/kg, the dose which caused the least weight gain from the preliminary experiments) does not limit weight gain in our model. On the contrary, the weight gain is at some time points even higher for this group than for the vehicle-treated HFD group. However, HUM-234-treated mice (25 mg/kg) gain weight much slower than HFD or HFD + CBG groups (Figure 6). The ALS (Alanine Transaminase) and AST (Aspartate Transaminase) enzymes levels are elevated in the HFD group comparing to STD; HUM-234 significantly reduces their levels (Figure 7), significantly better than CBG. Liver slices of the HFD mice show liver steatosis (liver cells are not dense with large white fat areas between them), while slices of HUM-234-treated mice livers show almost no steatosis (liver cells are dense with almost no white fat areas between them), comparable to healthy livers from mice on the standard diet (STD group, Figure 8); the livers of CBG-treated mice show almost as much steatosis as those of HFD mice.

It is of great interest that while the anti-inflammatory and pain-relieving activity of CBG and HUM-234 are similar, HUM-234 ameliorates weight gain, in contrast to CBG, suggesting this compound has the potential for development as an anti-obesity drug.

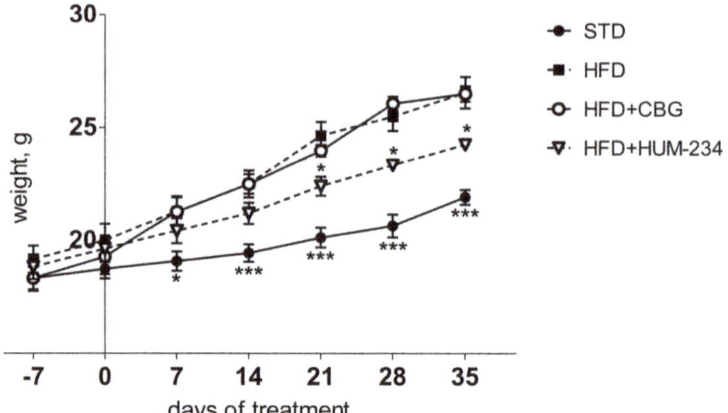

Figure 6. Anti-obesity evaluation of HUM-234 (25 mg/kg) in comparison to CBG (15 mg/kg), HFD (high fat diet) and STD (standard food diet). Statistical analysis by two-way ANOVA matching treatment groups in different days (p-value < 0.0001). Post-hoc analysis by Bonferroni test. * $p < 0.05$, *** $p < 0.001$ comparing to HFD group.

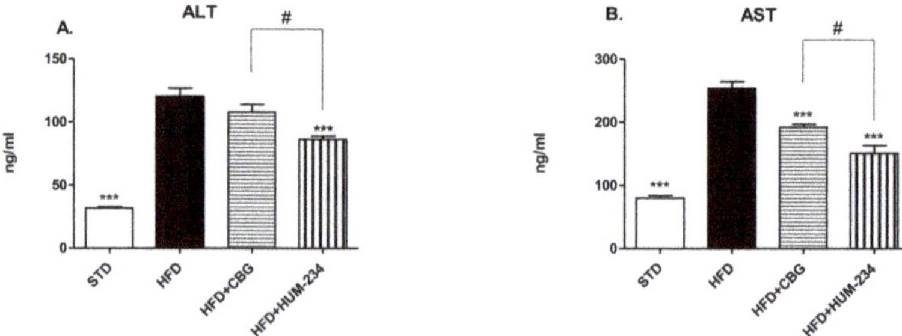

Figure 7. Liver enzymes evaluation of HUM-234 (25 mg/kg) in comparison to CBG (15 mg/kg), HFD (high fat diet) and STD (standard food diet). (**A**) ALT (Alanine transaminase) (**B**) AST (Aspartate transaminase). Statistical analysis by one-way ANOVA (p-value < 0.0001 for both ALT and AST) and post hoc analysis by Tukey's test. *** $p < 0.001$ comparing to HFD group. # p-value < 0.05 in the indicated comparison.

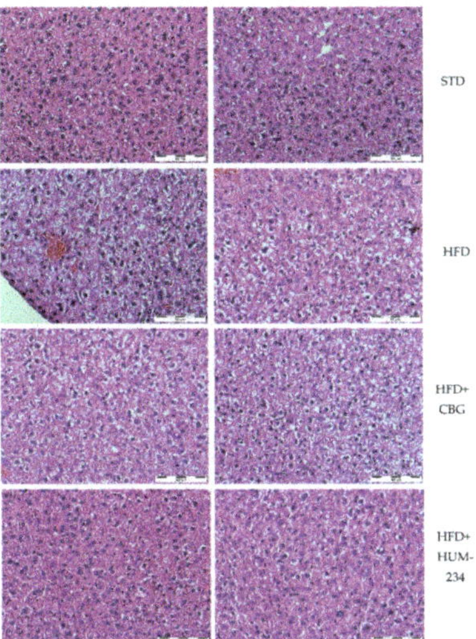

Figure 8. Liver steatosis in HUM-234 (25 mg/kg) in comparison to CBG (15 mg/kg), HFD (high-fat diet) and STD (standard food diet). Magnification 40×, bar length 50 μm.

3. Discussion

Here we report the synthesis of, and the evaluation in-vivo of the anti-inflammatory, analgesic and anti-obesity of three novel CBG derivatives.

Inflammation contributes to the pathogenesis of many diseases. Chronic inflammation can lead to cardiovascular diseases, gastrointestinal diseases, obesity, asthma, arthritis, neurodegenerative diseases, cancer and more.

Currently, the drugs used to treat inflammation are most often NSAIDs and steroids. Monoclonal antibodies to cytokines, TNF, IL6 and IL12/23 are used if these do not work. NSAIDs work by inhibiting the activity of cyclooxygenase enzymes (COX-1 or COX-2). In cells, these enzymes are involved in the synthesis of prostaglandins, which are associated with inflammation, and thromboxanes, which are involved in blood clotting [30]. Most NSAIDs are non-selective and inhibit the activity of both COX-1 and COX-2. These NSAIDs, while reducing inflammation, also inhibit platelet aggregation and increase the risk of gastrointestinal ulcers/bleeds [31]. Side effects can include an increased risk of gastrointestinal ulcers and bleeds, heart attack, and kidney disease [32,33]. COX-2 selective inhibitors have fewer gastrointestinal side effects but promote thrombosis and some of these agents substantially increase the risk of heart attack [31]. By inhibiting physiological COX activity, all NSAIDs increase the risk of kidney disease [34] and through a related mechanism, heart attack. In addition, NSAIDs can blunt the production of erythropoietin resulting in anemia, since hemoglobin synthesis depends on this hormone [35].

Of equal if not greater concern are side effects associated with long-term use of steroidal drugs. Although highly effective in the elimination of inflammation, they can cause obesity, growth retardation in children, and even lead to convulsions and psychiatric disturbances, osteoporosis, adrenal suppression, hyperglycemia, dyslipidemia, cardio-vascular disease, Cushing's syndrome, immunosuppression. an increase in the rate of infections [36].

There has been progress in the treatment of many inflammatory diseases, initially rheumatoid arthritis, and Crohn's disease, then many others. TNF (Tumor Necrosis Factor) inhibitors have been an important step forward in the treatment of several chronic inflammatory diseases, especially rheumatoid arthritis, and Crohn's disease. But these drugs are injectable and costly and have a potential for adverse effects, such as reactivation of latent tuberculosis. Furthermore, many patients treated with TNF inhibitors continue to experience chronic pain and so there is a big unmet need for cheaper, orally available drugs that reduce inflammation and pain [37].

The search for easily prepared, small molecular weight compounds, which can be delivered as an oral formulation continues. A hitherto mostly untapped wealth of novel compounds are natural products from plants, known to possess anti-inflammatory properties. Indeed, Cannabis sativa preparations and pure cannabinoids have been used to relieve symptoms of pain associated with several clinical disease conditions and ailments.

Previously we have reported the inflammatory disease modulating properties of CBD in several in vitro and in vivo systems and have demonstrated a potent ameliorating effect on the clinical signs of arthritis in a model of Rheumatoid Arthritis (RA) [38]. In a related study, our group has also established the potential of HU-320, a synthetic CBD derivative as a promising candidate for use in RA [39]. The anti-inflammatory activity of CBD and HU-320, as previously demonstrated by us in various experimental systems, has led us to explore the potential of CBG derivatives, new, chemically related cannabinoid compounds, which are much less studied, with the aim to evaluate therapeutic potential.

The synthesis of HUM-223, HUM-233 and HUM-234 uses CBG as starting material. CBG is formed in the plant (or by heating) from its precursor CBGA [40]. CBG acid is present in young plants in most Cannabis sativa varieties, including hemp, which is used for industrial purposes. Hemp is grown in many parts of the world, including Europe, for the production of textile fibers. Hence, CBG is potentially an inexpensive natural product. The synthetic pathway from CBG to the derivatives is a short, but relatively low yield one. In view of the simplicity of the reactions described herein, it is predicted that the yields can be increased.

The in-vivo assays used in the present study are well established and widely used in inflammation research [41,42]. The results indicate that HUM-223, HUM-233 and HUM-234 are anti-inflammatory candidates (and HUM-234 also anti-obesity), which we believe have the potential to be developed into therapeutic leads.

HUM-223 was able to significantly reduce swelling in all doses and exceed CBG's effect at a dose of 10 mg/kg. In addition, it showed activity in reducing pain responses and reducing TNF-α levels. The latter effects were comparable to CBG.

We chose HUM-223 to be tested on the knee-arthritis model to complement the described paw arthritis model, as previously similar CBD derivatives, possessing DMH (dimethylheptyl) side chain were more active than pentyl chain derivatives in this model (unpublished results). We were able to prove that HUM-223 is indeed efficacious and doses of 5, 10 and 25 mg/kg have a comparable response to dexamethasone. A dose of 5 mg/kg HUM-223 was proved as the most effective dose out of the three doses tested. At this dose, the effect of HUM-223 was similar to dexamethasone, a known highly effective anti-inflammatory steroidal drug. Furthermore, HUM-223 was shown to reduce local gene expression of Elane, Mpo and Adamts4 at the endpoint of the experiment; enzymes upregulated in inflammatory arthritis, shown to play a central role in oxidant production, cartilage degradation and joint damage [43–45]. In line with our findings, CBG has previously been shown to decrease myeloperoxidase in the context of inflammatory bowel disorder [9], however to our knowledge there have been no previous reports of CBG/CBG analogs reducing neutrophil elastase or Adamts4.

Regarding HUM-233 and HUM-234, both showed anti-inflammatory activity comparable to CBG in all three assays. They were all able to reduce swelling, decrease evoked pain responses upon application of localized pressure on the inflamed paw. We observed an inverse dose-response trend upon increasing the dose of the compounds (Figure 5). The salt HUM-233 at a dose of 10 mg/kg does not equal the same dose of the free base. The inverse linear dose-response behavior of HUM-233 and HUM-234 could be therefore explained as a bell-shaped dose response curve of HUM-234 since the dosage of HUM-233 given represents a smaller dosage of the free amine. A bell-shaped dose-response for CBD has been documented, but for CBG or its derivatives, it has rarely been documented [15].

Of notable interest is the effect on weight gain in an established standard model of obesity. We chose HUM-234 to be assayed on diet-induced obesity, as we previously established a similar derivative of CBD, HU-435, could prevent weight gain in mice [46]. Indeed, HUM-234 prevented the high-fat diet-induced weight gain at a dose of 25 mg/kg. Obesity leads to metabolic abnormalities, so we examined the effect of HUM-234 on the serum levels of liver enzymes ALT, AST and liver damage. We found that serum levels of both ALT and AST were significantly lowered and observed significant amelioration of histological liver damage, with a healthy liver almost fully restored with the administration of 25 mg/kg HUM-234. Importantly, liver sparing was not observed with administration of CBG, suggesting that this novel derivative offers improved benefits over the parent compound for the treatment of obesity and metabolic disease.

The effect of HUM-234 on diet-induced obesity is of great interest to us. Obesity is an "epidemic" of the developed world [47]. It causes abnormal physiological metabolism, which leads to a series of physiological, psychological, and social problems. Additionally, obesity is an important risk factor for diseases such as hypertension [48], hyperlipemia [49], diabetes [13] and even cancer [50], and it is closely associated with the emergence of many chronic diseases [51]. The medications currently used for obesity are only approved for patients who are obese (BMI (body mass index) > 30), or overweight (BMI > 27) with one weight-related health issue, as they possess numerous side effects. The quest for an effective and safe treatment for obesity is ongoing.

We are aware of the limitations posed by the compounds' biological behavior. All three compounds showed some variability in the concentrations most active in the swelling, pain sensation and TNF-α assays as did CBG. This is a known and well-documented trend in cannabis studies [52]. Moreover, the observed bell-shaped dose-response is unfavored when looking for new therapeutics. These challenges should be addressed in further development. However, the activity of HUM-234 in diet-induced obesity assay and in liver enzymes suggests that this may be a candidate therapeutic for obesity.

We believe that CBG derivatives HUM-223, HUM-233 and HUM-234 have the potential to be further developed as novel drug candidates for use in inflammatory conditions, and potentially also as anti-obesity treatment, where there is a huge unmet need.

4. Materials and Methods

4.1. NMR Spectroscopy

NMR data were collected on Varian Unity Inova 300 MHz spectrometer using the standard pulse sequences and processed with Agilent software.

4.2. Mass Spectrometry

The samples were analyzed by GC-MS in a Hewlett-Packard G1800 A GCD system with HP-5971 gas chromatograph with electron ionization detector. Ultra-low-bleed 5%-phenyl capillary column (28 mm $\times$ 0.25 mm (i.d.) $\times$ 0.25 μm film thickness) based on diphenyl methylsiloxane chemistry (HP-5MS; Agilent Technologies, Santa Clara, CA, USA) was used. Experimental conditions were: inlet, 250 °C; detector, 280 °C; splitless injection time; initial temperature, 90 °C; initial time, 3.00 min; rate, 25 °C/min; final temperature, 280 °C; helium flow rate, 1.0 mL/min. The software used was GCD Plus ChemStation.

LC-ESI-MS was done with Waters LC e2695 Separation Module equipped with reversed phase C18 column (Xselect® CSH, 2.1 $\times$ 100 mm, 2.5 um, Waters (TC) Israel Ltd., Petah Tikva, Israel) connected to Waters 2489 UV/visible Detector and Waters QDa Detector. The Software used was MassLynx V4.2. Several gradients of Acetonitrile and water both containing 0.1% FA were developed for analysis starting from 0% to 100% Acetonitrile or 50% to 100% Acetonitrile. Eventually, the addition of Methanol to the gradient was proved effective for the analysis of our compounds and the final method of analysis was as follows: column temperature: 45 °C, UV detector: 225 nm, sampling rate 20 points/s; QDa detector: ES (+) m/z between 100 to 1000, cone voltage 2V, ES ($-$) m/z between 100 to 1000, cone voltage 2 V. Gradient: starting point: 15% acetonitrile (0.1% FA), 15% methanol

(0.1% FA) and 70% water (5% acetonitrile and 0.1% FA) from 0 to 15 min. Then, 40% acetonitrile (0.1% FA), 40% methanol (0.1% FA) and 20% water (5% acetonitrile and 0.1% FA) from 15 to 22 min. Equilibration to starting conditions from 22 to 27 min. Flow rate: 0.25 mL/min. Probe temperature: 600 °C, source temperature: 120 °C, turbo temperature: 49 °C.

4.3. Chemical Synthesis

All the chemicals and solvents used were purchased from well-established commercial sources and used without any further purification procedures.

Newly synthesized cannabinoids and intermediate compounds were characterized by ^{1}H NMR, ^{13}C NMR and either GCMS or LC-ESI-MS.

4.3.1. 1,1-Dimethylheptyl Cannabigerol (CBG-DMH)

1.84 gr (7.8 mmol) of DMHR are dissolved in 3.3 mL of dry DCM with 0.13 gr (0.78 mmol) of p-toluenesulfonic acid (PTSA). This solution is then cooled to 0 °C. Separately, 1.35 mL (7.8 mmol) of geraniol are dissolved in 2.6 mL of dry DCM and then cooled to 0 °C as well. The cold geraniol solution is then added dropwise with a high stir to the cold DMHR solution. The reaction is then stirred at RT for 45 min and quenched by the addition of sat. $NaHCO_3$ solution. The water phase is then separated from the organic phase and the former is further extracted with DCM three times. The combined organic phase is washed with brine, dried over $MgSO_4$ and evaporated. The crude product is then purified by silica gel column chromatography (TLC 20% EtOAc: Pet Ether). CBG-DMH is purified by repeated chromatography. CBG-DMH is obtained as a pale-yellow oil. Yield: 0.09 gr (30%). Analytical characteristics were in accordance with previously published literature [53].

O-Methyl-*O*-Acetoxy-CBG-DMH. (1) is prepared from CBG-DMH in two steps without purification in between. 1.67 gr (1.8 mmol) of CBG-DMH are dissolved in 10 mL of dry DMF under a nitrogen atmosphere. 0.37 gr (2.71 mmol) of potassium carbonate is added and the suspension is then allowed to stir at room temperature for 5 min. To this suspension, 60 µL (0.9 mmol) of methyl iodide is added. The reaction is allowed to stir overnight at room temperature. The reaction is then diluted with 10% w/v HCl to pH 1 and extracted three times with Et_2O. The organic phase is washed with sat. $NaHCO_3$ to pH 10 and with brine to neutral pH. It is then dried over $MgSO_4$ and evaporated. Since chromatography of the crude did not efficiently separate the mono-methoxylated CBG-DMH from the di-methoxylated CBG-DMH, we decided to acetylate the free phenol that remained on the mono-methoxylated CBG-DMH. This method, in our hands, makes the chromatography much simpler and improves the overall yield. The 0.7 gr of crude methoxylation product are therefore carried to the next step without further purification.

The crude is dissolved in 10 mL of pyridine under nitrogen atmosphere with a catalytic amount of 4-dimethylaminopyridine (4-DMAP). 2 mL (18.2 mmol) of acetic anhydride are added slowly and the reaction is stirred at room temperature, monitoring by TLC. The reaction is worked up by diluting with EtOAc and washing the organic phase with 10% w/v HCl to pH 1. The aqueous phase is then extracted three times with EtOAc. The combined organic phase is then washed with sat. $NaHCO_3$ to pH 8 and brine to neutral pH. The organic phase is dried over $MgSO_4$ and evaporated. The crude product is purified by silica gel column chromatography (TLC 10% EtOAc: Pet Ether). Yield after two steps: 0.23 gr (30%). Compound (**1**) is obtained as yellow oil. 1H NMR (500 MHz, $CDCl_3$) δ 6.75 (s, 1H), 6.63 (s, 1H), 5.18 (t, *J* = 6.9 Hz, 1H), 5.11 (t, *J* = 6.9 Hz, 1H), 3.85 (s, 3H), 3.27 (d, *J* = 7.0 Hz, 2H), 2.31 (s, 3H), 2.09 (t, 2H), 2.01 (t, 2H), 1.78 (s, 3H), 1.68 (s, 3H), 1.61 (s, 3H), 1.31 (s, 6H), 1.25 (q, *J* = 9.8, 7.9 Hz, 6H), 1.15 (d, *J* = 13.1 Hz, 2H), 0.89 (t, *J* = 6.9 Hz, 3H). 13C NMR (126 MHz, $CDCl_3$) δ 169.36, 157.94, 149.38, 149.20, 134.76, 131.15, 124.40, 122.26, 119.52, 112.54, 106.22, 77.39, 77.14, 55.74, 44.59, 39.76, 37.86, 31.80, 30.06, 28.87, 26.72, 25.68, 24.64, 22.98, 22.73, 20.95, 17.67, 16.06, 14.12. GCMS: *m/z* 428 tR:13.67 min.

4.3.2. 1″,1″-Dimethylheptyl-monomethoxycannabigerol (HUM-223)

0.52 gr (1.2 mmol) of (1) is dissolved in 15 mL of dry THF under a nitrogen atmosphere and cooled to 0 °C. Then, 0.49 gr (12.8 mmol) of LiAlH$_4$ is added, and the reaction is heated to reflux and monitored by TLC. The reaction is cooled to room temperature and the LiAlH$_4$ is neutralized first by dropwise addition of EtOAc followed by ice and 10% *w/v* HCl. The water phase is extracted three times with EtOAc. The combined organic phase is washed with sat. NaHCO$_3$ and brine, dried over MgSO$_4$ and evaporated. The crude product is purified by medium pressure liquid chromatography (TLC 10% EtOAc:Pet Ether). HUM-223 is obtained as a colorless oil. Yield: 0.25 gr (54%). 1H NMR (300 MHz, CDCl$_3$) δ 6.48 (1H, s), 6.46 (1H, s), 5.31–5.29 (1H, m), 5.08–5.06 (1H, m), 3.83 (3H, s), 3.43 (2H, d, *J* = 6.9 Hz), 2.13–2.07 (2H, m), 1.83 (3H, s), 1.75–1.66 (1H, m), 1.69 (3H, s), 1.61 (3H, s), 1.59–1.54 (1H, m), 1.3 (12H, s), 1.22–1.09 (2H, m), 0.87 (3H, t, *J* = 6.9 Hz). 13C NMR (75 MHz, CDCl$_3$) δ 157.45, 155.13, 149.65, 137.93, 131.85, 123.95, 122.24, 112.01, 106.89, 101.15, 55.78, 44.56, 39.75, 37.75, 31.83, 30.09, 28.98, 26.48, 25.73, 24.69, 22.74, 22.11, 17.72, 16.16. 14.15. GCMS: *m/z* 386 t$_R$: 13.7 min.

4.3.3. Monomethoxycannabigeroyl-3-morpholinopropraonate Maleate (HUM-233)

0.343 gr (0.72 mmol) of HUM-234 are dissolved in 40 mL of isopropyl alcohol (IPA). Then, 83.5 mg (0.72 mmol) of maleic acid are added and the reaction is stirred at room temperature for 2.5 h. IPA is evaporated and HUM-233 is recrystallized from EtOAc and ether. HUM-233 is obtained as crystalline white solid. Yield: 0.32 gr (75%). 1H NMR (300 MHz, CDCl$_3$) δ 6.59 (s, 1H), 6.46 (s, 1H), 6.33 (s, 2H), 5.09–4.96 (m, 2H), 4.04–3.89 (m, 4H), 3.82 (s, 3H), 3.40 (t, *J* = 7.0 Hz, 2H), 3.26–3.05 (m, 4H), 2.56 (t, *J* = 7.8 Hz, 2H), 2.10–1.96 (m, 2H), 1.96–1.85 (m, 2H), 1.70 (s, 3H), 1.64 (s, 3H), 1.56 (s, 3H), 1.40–1.24 (m, 4H), 0.89 (t, *J* = 6.8 Hz, 3H). LCMS:ES (+) *m/z* 472 [M + H], 494 [M + Na] tR: 7.26 min, ES (−) *m/z* 115 [M − H], 231 [2M − H] tR: 1.91 min.

4.3.4. Monomethoxycannabigeroyl-3-morpholinoproprionate (HUM-234)

0.432 mL of methyl morpholinoproprionate are dissolved in 4 mL of 1,4-dioxane, 2.16 mL of water and 1.08 mL of 3N aqueous sodium hydroxide. The reaction is stirred at room temperature for 4.5 h and then the solvent is evaporated. The solids are suspended in 40 mL of dry DCM under nitrogen atmosphere and small amount of MgSO$_4$ is added to insure the absence of water in the reaction flask. Then, 0.889 gr of O-monomethyl cannabigerol (2) are added, followed by 0.04 gr of pyrrolidinopyridine. The reaction is then stirred at room temperature for 5 min. 0.56 gr of *N,N*-dicyclohexylcarbodiimide (DCC) are added and the reaction is stirred at room temperature overnight. The solids are filtered, and the solution is evaporated. The crude product is purified by silica gel column chromatography (TLC 20% EtOAc:Pet Ether). HUM-234 is obtained as yellow oil. Yield: 0.25 gr (20%). ^{1}H NMR (300 MHz, CDCl$_3$) δ 6.57 (s, 1H), 6.50 (s, 1H), 5.19–4.99 (m, 2H), 3.80 (s, 3H), 3.72 (t, *J* = 4.6 Hz, 4H), 3.25 (d, *J* = 6.9 Hz, 2H), 2.88–2.66 (m, 4H), 2.55–2.48 (m, 5H), 2.11–2.00 (m, 2H), 1.99–1.89 (m, 2H), 1.73 (s, 3H), 1.65 (s, 3H), 1.57 (s, 3H), 1.39–1.24 (m, 4H), 0.91 (d, *J* = 6.0 Hz, 3H). ^{13}C NMR (75 MHz, CDCl$_3$) δ 170.89, 158.12, 149.31, 145.98, 141.99, 134.79, 131.18, 124.32, 122.26, 119.72, 114.34, 108.53, 66.91, 55.70, 54.03, 53.42, 39.75, 35.96, 32.30, 31.57, 30.92, 26.69, 25.71, 22.85, 22.57, 17.69, 16.11, d14.08. LCMS: ES (+) *m/z* 472 [M + H], 494 [M + Na] t$_R$: 7.26 min.

4.4. Biological Evaluation

4.4.1. Animals

Female Sabra mice (for the swelling, pain and TNF-α experiments) and C57Bl6 (for obesity experiments, a line that was shown to gain much weight under HFD conditions), 7–8 weeks old, were maintained in the specific-pathogen-free unit of the Hadassah Medical School, Hebrew University, Jerusalem, Israel. The experimental protocols were approved by the Institutional Animal Care Ethics Committee (permission # MD-20-16042-5). The animals were maintained at a constant temperature (20–21 °C) and a 12-h light/12-h dark cycle and were provided a standard pellet diet with water *ad libitum*. The mice were acclimatized in the animal facility for at least 2 weeks before the experiments. The data presented in Figures are representatives of 2 separate experiments.

4.4.2. Induction and Treatment of Paw Inflammation (Paw ZIA)

Inflammation was induced by injection of 40 uL of a suspension of 1.5% *w/v* zymosan A (Sigma-Aldrich Israel Ltd., Rehovot, Israel) in saline into the subplantar surface of the right hind paw of the mice. This was followed immediately by an intraperitoneal injection of the test compound. For injection, the compounds were dissolved in a vehicle containing ethanol:Cremophore:saline at a ratio of 1:1:18. Paw swelling and pain perception were assessed after 2, 6 and 24 h. Blood was collected after 24 h for analysis of TNFα serum levels.

4.4.3. Evaluation of Edema

Calibrated calipers were used to measure paw swelling (thickness) 2, 6 and 24 h after injection of zymosan.

4.4.4. Pain Assay

Pain at 2, 6 and 24 h after zymosan injection was assessed by the von Frey nociceptive filament assay, where 1.4–60 g filaments, corresponding to 4.17–5.88 log of force, was used to test the sensitivity of the swollen paw. The untreated hind paw served as a control. The measurements were performed in a quiet room and the animals were handled for 10 s before the test. A trained investigator then applied the filament, poking the middle of the hind paw to provoke a flexion reflex, followed by a clear finch response after paw withdrawal. Filaments of increasing size were each applied for about 3–4 s. The mechanical threshold force in grams was defined as the lowest force required to obtain a paw retraction response.

4.4.5. Measurement of TNFα

Blood was collected 24 h after zymosan injection, and the sera were assayed for TNFα using a mouse TNFα ELISA kit (R&D Systems, Minneapolis, MN, USA), according to the manufacturer's instructions.

4.4.6. Zymosan Induced Arthritis of the Knee

All experimental procedures were approved by the Ethical Review Process Committee and the UK Home Office, in accordance with the 1986 Animals (Scientific Procedures) Act (permission # 30/3441). Male C57BL/6J mice aged 8–10 weeks were used. Mice were housed in ventilated cages, maintained at 21 °C $\pm$ 2 °C and a 12-h light/12-h dark cycle, with food and water available ad libitum. Mice were anesthetized with 2% isoflurane and both knees were shaved. ZIA was induced by intra-articular injection of 180 µg of Zymosan A (Sigma) suspended in PBS as previously described [9]. Left knee joints received a vehicle control injection. For IVIS imaging, mice received an intravenous injection of 4 nmol Neutrophil Elastase 680 FAST imaging probe (Perkin Elmer, Waltham, MA, USA) and were imaged 4 h post intravenous injection using the IVIS Spectrum (Perkin Elmer) [54]. Images were analyzed using Living Image 4.7 software (Perkin Elmer) to obtain the average fluorescence intensities of a circular region of interest encompassing the knee joint. Mice were humanely culled 8 h post zymosan administration and knee joints were snap-frozen for gene expression analysis. RNA was extracted using the TRIzol method, as previously described [54]. Reverse transcription of 1 ug of total RNA was conducted using a High-Capacity cDNA Reverse Transcription Kit (Thermo Fisher Scientific) with random primers and following the manufacturer's protocol. Real-time PCR was carried out using the power SYBR Green Master Mix in a real-time PCR system (Thermo Fisher Scientific, Waltham, MA, USA). Data are expressed as relative units calculated by $2-\Delta\Delta Ct$ by normalization relative to RPL32 and to fold change over vehicle-treated control samples.

4.4.7. Diet-Induced Obesity

Mice were fed with a standard diet (STD) or high-fat diet (HFD) for 7 days before the beginning of the injections. On day 7 they were divided into the groups: STD, HFD, HFD+CBG and HFD+HUM-234, and treated for 35 days. The mice were weighed every week.

4.4.8. Liver Injury

At the end of the diet-induced obesity experiment, the mice were sacrificed. The livers were removed, fixed with buffer formalin and stained with haematoxylin and eosin, for microscope evaluation. Paraffin sections of 4–5 μm thickness, were stained with hematoxylin and eosin. For microscopic evaluation, sections were examined ($\times 40$).

4.4.9. Determination of ALT and AST Levels

The levels of two aminotransferases, alanine aminotransferase (ALT) and aspartate aminotransferase (AST), were assayed in the sera of mice at the end of diet-induced obesity experiment with or without HUM-234 treatment, by ALT and AST strips respectively (Refloram-Mannheim GmbH, Mannheim, Germany) and quantitated by an automated analyzer (Reflotran Plus, Roche, Basel, Switzerland).

4.5. Statistical Analyses

Statistical analysis was performed with GraphPad Prism software. Statistical analysis details are listed under each figure. The results are presented as value $\pm$ SE (standard error). In rare cases where all the measurements give the same values, no SE bar is presented, as no error can be measured. * p-value < 0.05, ** p-value < 0.01 and *** p-value < 0.001 when comparing to a control group. # p-value < 0.05 when comparing between the different dosages of the tested compounds, or when comparing them to CBG, the exact comparisons are listed on Figures' legends.

Author Contributions: Conceptualization, N.M.K., R.G. and R.M.; methodology, N.M.K., L.M.T., R.G. and R.M.; investigation, N.M.K., Y.L., L.M.T. and Z.Y.; writing—original draft preparation, N.M.K. and Y.L.; writing—review and editing, N.M.K., L.M.T., R.O.W., F.E.M., M.F. and R.M.; visualization, N.M.K.; supervision, R.M. and M.F. All authors have read and agreed to the published version of the manuscript.

Funding: This work was funded by 180 Life Sciences, Menlo Park, CA 94025, USA.

Institutional Review Board Statement: The study was conducted according to the guidelines of the Declaration of Helsinki, and approved by the (or Ethics Committee of the Hebrew University of Jerusalem (protocol code MD-20-16042-5, date of approval 15 January 2020) and by UK Home Office and the University of Oxford Clinical Medicine Animal Welfare and Ethical Review Board (protocol code 30/3441, date of approval 22 September 2016)).

Informed Consent Statement: Not applicable.

Data Availability Statement: The data presented in this study are available on request from the corresponding author.

Acknowledgments: We thank Aviva Breuer for helpful advice.

Conflicts of Interest: M.F. is a shareholder and co-chairman of 180 Life sciences, FM is an employee of 180LS; FM and ROW have share options in 180LS.

Sample Availability: Samples of the compounds are not available from the authors.

References

1. Gaoni, Y.; Mechoulam, R. The structure and synthesis of cannabigerol, a new hashish constituent. *Proc. Chem. Soc.* **1964**, 82. [CrossRef]
2. Nachnani, R.; Raup-Konsavage, W.M.; Vrana, K.E. The Pharmacological Case for Cannabigerol. *J. Pharmacol. Exp. Ther.* **2021**, *376*, 204–212. [CrossRef] [PubMed]
3. Hill, K.P.; Palastro, M.D.; Johnson, B.; Ditre, J.W. Cannabis and Pain: A Clinical Review. *Cannabis Cannabinoid Res.* **2017**, *2*, 96–104. [CrossRef]
4. Nichols, J.M.; Kaplan, B.L.F. Immune Responses Regulated by Cannabidiol. *Cannabis Cannabinoid Res.* **2020**, *5*, 12–31. [CrossRef]
5. Gugliandolo, A.; Pollastro, F.; Grassi, G.; Bramanti, P.; Mazzon, E. In Vitro Model of Neuroinflammation: Efficacy of Cannabigerol, a Non-Psychoactive Cannabinoid. *Int. J. Mol. Sci.* **2018**, *19*, 1992. [CrossRef]
6. Burgaz, S.; García, C.; Gómez-Cañas, M.; Muñoz, E.; Fernández-Ruiz, J. Development of An Oral Treatment with the PPAR-γ-Acting Cannabinoid VCE-003.2 against the Inflammation-Driven Neuronal Deterioration in Experimental Parkinson's Disease. *Molecules* **2019**, *24*, 2702. [CrossRef] [PubMed]
7. di Giacomo, V.; Chiavaroli, A.; Orlando, G.; Cataldi, A.; Rapino, M.; Di Valerio, V.; Leone, S.; Brunetti, L.; Menghini, L.; Recinella, L.; et al. Neuroprotective and Neuromodulatory Effects Induced by Cannabidiol and Cannabigerol in Rat Hypo-E22 cells and Isolated Hypothalamus. *Antioxidants* **2020**, *9*, 71. [CrossRef] [PubMed]
8. Díaz-Alonso, J.; Paraíso-Luna, J.; Navarrete, C.; del Río, C.; Cantarero, I.; Palomares, B.; Aguareles, J.; Fernández-Ruiz, J.; Bellido, M.L.; Pollastro, F.; et al. VCE-003.2, a novel cannabigerol derivative, enhances neuronal progenitor cell survival and alleviates symptomatology in murine models of Huntington's disease. *Sci. Rep.* **2016**, *6*, 29789. [CrossRef]
9. Borrelli, F.; Fasolino, I.; Romano, B.; Capasso, R.; Maiello, F.; Coppola, D.; Orlando, P.; Battista, G.; Pagano, E.; Di Marzo, V.; et al. Beneficial effect of the non-psychotropic plant cannabinoid cannabigerol on experimental inflammatory bowel disease. *Biochem. Pharmacol.* **2013**, *85*, 1306–1316. [CrossRef] [PubMed]
10. Mammana, S.; Cavalli, E.; Gugliandolo, A.; Silvestro, S.; Pollastro, F.; Bramanti, P.; Mazzon, E. Could the Combination of Two Non-Psychotropic Cannabinoids Counteract Neuroinflammation? Effectiveness of Cannabidiol Associated with Cannabigerol. *Medicina* **2019**, *55*, 747. [CrossRef]
11. Henley, D.; Lightman, S.; Carrell, R. Cortisol and CBG—Getting cortisol to the right place at the right time. *Pharmacol. Ther.* **2016**, *166*, 128–135. [CrossRef]
12. Ruhaak, L.R.; Felth, J.; Karlsson, P.C.; Rafter, J.J.; Verpoorte, R.; Bohlin, L. Evaluation of the Cyclooxygenase Inhibiting Effects of Six Major Cannabinoids Isolated from Cannabis sativa. *Biol. Pharm. Bull.* **2011**, *34*, 774–778. [CrossRef]
13. Bjerregaard, L.G.; Jensen, B.W.; Ängquist, L.; Osler, M.; Sørensen, T.I.A.; Baker, J.L. Change in Overweight from Childhood to Early Adulthood and Risk of Type 2 Diabetes. *N. Engl. J. Med.* **2018**, *378*, 1302–1312. [CrossRef] [PubMed]
14. Brierley, D.I.; Samuels, J.; Duncan, M.; Whalley, B.J.; Williams, C.M. Cannabigerol is a novel, well-tolerated appetite stimulant in pre-satiated rats. *Psychopharmacology* **2016**, *233*, 3603–3613. [CrossRef]
15. Cascio, M.; Gauson, L.; Stevenson, L.; Ross, R.; Pertwee, R. Evidence that the plant cannabinoid cannabigerol is a highly potent α2-adrenoceptor agonist and moderately potent 5HT1A receptor antagonist. *Br. J. Pharmacol.* **2010**, *159*, 129–141. [CrossRef]
16. de Petrocellis, L.; Orlando, P.; Moriello, A.S.; Aviello, G.; Stott, C.; Izzo, A.A.; di Marzo, V. Cannabinoid actions at TRPV channels: Effects on TRPV3 and TRPV4 and their potential relevance to gastrointestinal inflammation. *Acta Physiol.* **2012**, *204*, 255–266. [CrossRef]
17. Navarro, G.; Varani, K.; Reyes-Resina, I.; Sánchez de Medina, V.; Rivas-Santisteban, R.; Sánchez-Carnerero Callado, C.; Vincenzi, F.; Casano, S.; Ferreiro-Vera, C.; Canela, E.I.; et al. Cannabigerol Action at Cannabinoid CB1 and CB2 Receptors and at CB1–CB2 Heteroreceptor Complexes. *Front. Pharmacol.* **2018**, *9*, 632. [CrossRef]
18. Greene, E.R.; Huang, S.; Serhan, C.N.; Panigrahy, D. Regulation of inflammation in cancer by eicosanoids. *Prostagland. Other Lipid Mediat.* **2011**, *96*, 27–36. [CrossRef] [PubMed]
19. Shachar, I.; Karin, N. The dual roles of inflammatory cytokines and chemokines in the regulation of autoimmune diseases and their clinical implications. *J. Leukoc. Biol.* **2013**, *93*, 51–61. [CrossRef] [PubMed]
20. Ertunc, M.E.; Hotamisligil, G.S. Lipid signaling and lipotoxicity in metaflammation: Indications for metabolic disease pathogenesis and treatment. *J. Lipid Res.* **2016**, *57*, 2099–2114. [CrossRef]
21. Lackey, D.E.; Olefsky, J.M. Regulation of metabolism by the innate immune system. *Nat. Rev. Endocrinol.* **2015**, *12*, 15–28. [CrossRef]
22. Yang, Y.; Lv, J.; Jiang, S.; Ma, Z.; Wang, D.; Hu, W.; Deng, C.; Fan, C.; Di, S.; Sun, Y.; et al. The emerging role of Toll-like receptor 4 in myocardial inflammation. *Cell Death Dis.* **2016**, *7*, e2234. [CrossRef]
23. Kim, S.Y.; Solomon, D.H. Comparative safety of nonsteroidal anti-inflammatory drugs. *Nat. Rev. Cardiol.* **2011**, *8*, 193–195. [CrossRef]
24. Wehling, M. Non-steroidal anti-inflammatory drug use in chronic pain conditions with special emphasis on the elderly and patients with relevant comorbidities: Management and mitigation of risks and adverse effects. *Eur. J. Clin. Pharmacol.* **2014**, *70*, 1159–1172. [CrossRef] [PubMed]
25. Heffler, E.; Madeira, L.N.G.; Ferrando, M.; Puggioni, F.; Racca, F.; Malvezzi, L.; Passalacqua, G.; Canonica, G.W. Inhaled Corticosteroids Safety and Adverse Effects in Patients with Asthma. *J. Allergy Clin. Immunol. Pract.* **2018**, *6*, 776–781. [CrossRef]

26. Sarnes, E.; Crofford, L.; Watson, M.; Dennis, G.; Kan, H.; Bass, D. Incidence and US Costs of Corticosteroid-Associated Adverse Events: A Systematic Literature Review. *Clin. Ther.* **2011**, *33*, 1413–1432. [CrossRef] [PubMed]

27. Krensky, A.M.; Vicenti, F.; Bennett, W.M. Immunosuppressants, tolerogens, and immunostimulants. In *Goodman & Gilman's the Pharmacological Basis of Therapeutics*; Brunton, L.L., Lazo, J.S., Parker, K.L., Eds.; McGraw-Hill: New York, NY, USA, 2005; pp. 1405–1431, ISBN 0071422803.

28. Bow, E.W.; Rimoldi, J.M. The Structure–Function Relationships of Classical Cannabinoids: CB1/CB2 Modulation. *Perspect. Medicin. Chem.* **2016**, *8*, PMC.S32171. [CrossRef]

29. Mechoulam, R.; Kogan, N.; Gallily, R.; Breuer, A. Novel Cannabidiol Derivatives and Their Use as Anti-Inflammatory Agents. WO Patent 107879, 12 September 2008.

30. Smyth, E.M.; Burke, A.; FitzGerald, G.A. Lipid-derived autacoids: Eicosanoids and platelet-activating factor. In *Goodman & Gilman's the Pharmacological Basis of Therapeutics*; Brunton, L.L., Lazo, J.S., Parker, K.L., Eds.; McGraw-Hill: New York, NY, USA, 2005; pp. 653–670, ISBN 0071422803.

31. Day, R.O.; Graham, G.G. The vascular effects of COX-2 selective inhibitors. *Aust. Prescr.* **2004**, *27*, 142–145. [CrossRef]

32. Bally, M.; Dendukuri, N.; Rich, B.; Nadeau, L.; Helin-Salmivaara, A.; Garbe, E.; Brophy, J.M. Risk of acute myocardial infarction with NSAIDs in real world use: Bayesian meta-analysis of individual patient data. *BMJ* **2017**, *357*, j1909. [CrossRef]

33. Lanas, A.; Chan, F.K.L. Peptic ulcer disease. *Lancet* **2017**, *390*, 613–624. [CrossRef]

34. Brater, D.C.; Harris, C.; Redfern, J.S.; Gertz, B.J. Renal Effects of COX-2-Selective Inhibitors. *Am. J. Nephrol.* **2001**, *21*, 1–15. [CrossRef]

35. Bleumink, G.S.; Feenstra, J.; Sturkenboom, M.C.J.M.; Stricker, B.H.C. Nonsteroidal Anti-Inflammatory Drugs and Heart Failure. *Drugs* **2003**, *63*, 525–534. [CrossRef] [PubMed]

36. Liu, D.; Ahmet, A.; Ward, L.; Krishnamoorthy, P.; Mandelcorn, E.D.; Leigh, R.; Brown, J.P.; Cohen, A.; Kim, H. A practical guide to the monitoring and management of the complications of systemic corticosteroid therapy. *Allergy Asthma Clin. Immunol.* **2013**, *9*, 30. [CrossRef] [PubMed]

37. Feldmann, M.; Maini, R.N. TNF defined as a therapeutic target for rheumatoid arthritis and other autoimmune diseases. *Nat. Med.* **2003**, *9*, 1245–1250. [CrossRef]

38. Mechoulam, R.; Sumariwalla, P.F.; Feldmann, M.; Gallily, R. Cannabinoids in Models of Chronic Inflammatory Conditions. *Phytochem. Rev.* **2005**, *4*, 11–18. [CrossRef]

39. Sumariwalla, P.F.; Gallily, R.; Tchilibon, S.; Fride, E.; Mechoulam, R.; Feldmann, M. A novel synthetic, nonpsychoactive cannabinoid acid (HU-320) with antiinflammatory properties in murine collagen-induced arthritis. *Arthritis Rheum.* **2004**, *50*, 985–998. [CrossRef]

40. Pellati, F.; Borgonetti, V.; Brighenti, V.; Biagi, M.; Benvenuti, S.; Corsi, L. Cannabis sativa L. and Nonpsychoactive Cannabinoids: Their Chemistry and Role against Oxidative Stress, Inflammation, and Cancer. *Biomed Res. Int.* **2018**, *2018*, 1691428. [CrossRef]

41. Malfait, A.M.; Gallily, R.; Sumariwalla, P.F.; Malik, A.S.; Andreakos, E.; Mechoulam, R.; Feldmann, M. The nonpsychoactive cannabis constituent cannabidiol is an oral anti-arthritic therapeutic in murine collagen-induced arthritis. *Proc. Natl. Acad. Sci. USA* **2000**, *97*, 9561–9566. [CrossRef]

42. Gallily, R.; Yekhtin, Z.; Hanuš, L.O. Overcoming the Bell-Shaped Dose-Response of Cannabidiol by Using Cannabis Extract Enriched in Cannabidiol. *Pharmacol. Pharm.* **2015**, *6*, 75–85. [CrossRef]

43. Carmona-Rivera, C.; Carlucci, P.M.; Goel, R.R.; James, E.; Brooks, S.R.; Rims, C.; Hoffmann, V.; Fox, D.A.; Buckner, J.H.; Kaplan, M.J. Neutrophil extracellular traps mediate articular cartilage damage and enhance cartilage component immunogenicity in rheumatoid arthritis. *JCI Insight* **2020**, *5*, e139388. [CrossRef]

44. Rose, K.W.J.; Taye, N.; Karoulias, S.Z.; Hubmacher, D. Regulation of ADAMTS Proteases. *Front. Mol. Biosci.* **2021**, *8*, 701959. [CrossRef] [PubMed]

45. Odobasic, D.; Yang, Y.; Muljadi, R.C.M.; O'Sullivan, K.M.; Kao, W.; Smith, M.; Morand, E.F.; Holdsworth, S.R. Endogenous Myeloperoxidase Is a Mediator of Joint Inflammation and Damage in Experimental Arthritis. *Arthritis Rheumatol.* **2014**, *66*, 907–917. [CrossRef] [PubMed]

46. Gallily, R.; Breuer, A.; Mechoulm, R. Cyclohexenyl Compounds, Compositions Comprising Them and Uses Thereof. U.S. Patent 10,239,848, 26 March 2019.

47. Meldrum, D.R.; Morris, M.A.; Gambone, J.C. Obesity pandemic: Causes, consequences, and solutions—But do we have the will? *Fertil. Steril.* **2017**, *107*, 833–839. [CrossRef]

48. Teixeira, L.G.; Leonel, A.J.; Aguilar, E.C.; Batista, N.V.; Alves, A.C.; Coimbra, C.C.; Ferreira, A.V.M.; De Faria, A.M.C.; Cara, D.C.; Alvarez Leite, J.I. The combination of high-fat diet-induced obesity and chronic ulcerative colitis reciprocally exacerbates adipose tissue and colon inflammation. *Lipids Health Dis.* **2011**, *10*, 204. [CrossRef]

49. Doll, S.; Paccaud, F.; Bovet, P.; Burnier, M.; Wietlisbach, V. Body mass index, abdominal adiposity and blood pressure: Consistency of their association across developing and developed countries. *Int. J. Obes.* **2002**, *26*, 48–57. [CrossRef]

50. Latino-Martel, P.; Cottet, V.; Druesne-Pecollo, N.; Pierre, F.H.F.; Touillaud, M.; Touvier, M.; Vasson, M.-P.; Deschasaux, M.; Le Merdy, J.; Barrandon, E.; et al. Alcoholic beverages, obesity, physical activity and other nutritional factors, and cancer risk: A review of the evidence. *Crit. Rev. Oncol. Hematol.* **2016**, *99*, 308–323. [CrossRef]

51. Suárez-Carmona, W.; Sánchez-Oliver, A.J.; González-Jurado, J.A. Pathophysiology of obesity: Current view. *Rev. Chil. Nutr.* **2017**, *44*, 226–233. [CrossRef]

52. Lima, M.G.; Tardelli, V.S.; Brietzke, E.; Fidalgo, T.M. Cannabis and Inflammatory Mediators. *Eur. Addict. Res.* **2021**, *27*, 16–24. [CrossRef]
53. Baek, S.H.; Yook, C.N.; Han, D.S. Boron-Trifluoride Etherate on Alumina-A Modified Lewis-Acid Reagent(V) A Convenient Single-Step Synthesis of Cannabinoids. *Bull. Korean Chem. Soc.* **1995**, *16*, 293–296.
54. Vermeij, E.A.; Koenders, M.I.; Blom, A.B.; Arntz, O.J.; Bennink, M.B.; van den Berg, W.B.; van Lent, P.L.E.M.; van de Loo, F.A.J. In vivo molecular imaging of cathepsin and matrix metalloproteinase activity discriminates between arthritic and osteoarthritic processes in mice. *Mol. Imaging* **2014**, *13*, 1–10. [CrossRef] [PubMed]

Review

The Spicy Story of Cannabimimetic Indoles

Allyn C. Howlett [1,*], Brian F. Thomas [2] and John W. Huffman [3]

[1]   Department of Physiology and Pharmacology, Wake Forest School of Medicine, Winston-Salem, NC 27157, USA
[2]   Department of Analytical Sciences, The Cronos Group, Toronto, ON M5V 2H1, Canada; brian.thomas@thecronosgroup.com
[3]   Department of Chemistry, Clemson University, Clemson, SC 29634, USA; huffman@clemson.edu
*   Correspondence: ahowlett@wakehealth.edu; Tel.: +1-336-716-8545

Abstract: The Sterling Research Group identified pravadoline as an aminoalkylindole (AAI) non-steroidal anti-inflammatory pain reliever. As drug design progressed, the ability of AAI analogs to block prostaglandin synthesis diminished, and antinociceptive activity was found to result from action at the CB_1 cannabinoid receptor, a G-protein-coupled receptor (GPCR) abundant in the brain. Several laboratories applied computational chemistry methods to ultimately conclude that AAI and cannabinoid ligands could overlap within a common binding pocket but that WIN55212-2 primarily utilized steric interactions via aromatic stacking, whereas cannabinoid ligands required some electrostatic interactions, particularly involving the CB_1 helix-3 lysine. The Huffman laboratory identified strategies to establish CB_2 receptor selectivity among cannabimimetic indoles to avoid their CB_1-related adverse effects, thereby stimulating preclinical studies to explore their use as anti-hyperalgesic and anti-allodynic pharmacotherapies. Some AAI analogs activate novel GPCRs referred to as "Alkyl Indole" receptors, and some AAI analogs act at the colchicine-binding site on microtubules. The AAI compounds having the greatest potency to interact with the CB_1 receptor have found their way into the market as "Spice" or "K2". The sale of these alleged "herbal products" evades FDA consumer protections for proper labeling and safety as a medicine, as well as DEA scheduling as compounds having no currently accepted medical use and a high potential for abuse. The distribution to the public of potent alkyl indole synthetic cannabimimetic chemicals without regard for consumer safety contrasts with the adherence to regulatory requirements for demonstration of safety that are routinely observed by ethical pharmaceutical companies that market medicines.

Keywords: aminoalkylindole; allodynia; antinociception; cannabinoid receptor; CP55940; JWH-018; K2; pravadoline; spice; WIN55212-2

Citation: Howlett, A.C.; Thomas, B.F.; Huffman, J.W. The Spicy Story of Cannabimimetic Indoles. *Molecules* **2021**, *26*, 6190. https://doi.org/10.3390/molecules26206190

Academic Editor: Mauro Maccarrone

Received: 11 August 2021
Accepted: 8 October 2021
Published: 14 October 2021

Publisher's Note: MDPI stays neutral with regard to jurisdictional claims in published maps and institutional affiliations.

1. Introduction: Pravadoline and the Discovery of Aminoalkylindole Analgesics

The Howlett laboratory entered the cannabinoid field from the investigation of analgesic compounds that chemists at Pfizer Central Research had developed [1–3] in their quest to introduce a non-opioid, non-aspirin-like analgesic based upon the structure of the active 11-hydroxylated metabolite of Δ^9-tetrahydrocannabinol (THC) [4,5] (Figure 1). Pfizer discontinued the cannabinoid analgesic program after early clinical trials with levonantradol (Figure 1) [5–9] but left a legacy of promoting cannabinoid therapeutics within the scientific research community (see symposium covering chemistry, biochemistry, pharmacokinetics, pharmacotherapeutic uses, government regulations, and philosophical considerations [10]). Investigations in the Howlett laboratory identified that the antinociceptive activity of the classical and nonclassical cannabinoid ligands was associated with their agonist activity at a G-protein-coupled receptor (GPCR) coupled to Gi that could inhibit cAMP accumulation [1,2,11–13]. These studies led to the development of a radioligand binding assay using [^{3}H]CP55940 (Figure 1) to characterize the cannabinoid receptor in neuronal cells and the brain [12,14–16]. As these studies were being published, Dr. Howlett was contacted by Dr. Susan Ward at Sterling Research Group of Sterling Winthrop, Inc. (a subsidiary of Eastman Kodak), inquiring about whether the Howlett lab would be able to screen analgesic compounds that did not fit the pattern for non-steroidal anti-inflammatory drugs (NSAIDs) or opioid analgesics. A non-disclosure agreement and a library of compounds soon followed.

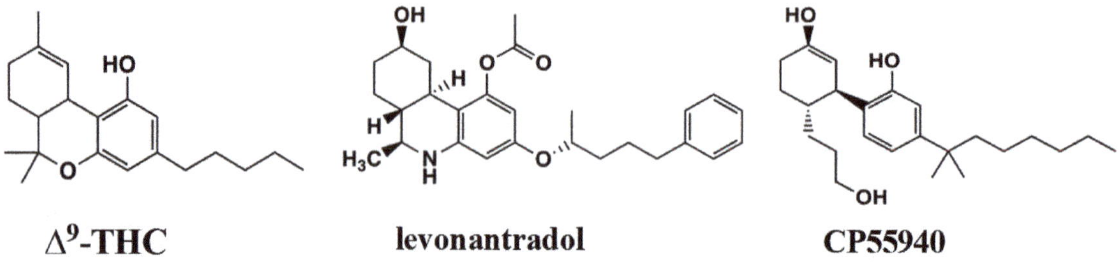

Figure 1. Classical cannabinoids Δ^9-tetrahydrocannabinol (THC) and levonantradol and non-classical A,C-bicyclic cannabinoid CP55940.

Chemists from the Sterling Research Group were exploring non-steroidal anti-inflammatory analgesics, pravadoline, and its analogs, similar in structure to the well-recognized NSAID indomethacin (Figure 2). Pravadoline, comprised of an indole nucleus with an alkylamine substituent extending from the indole N, was a cyclooxygenase inhibitor like NSAIDS and blocked the formation of prostaglandins with a potency comparable to ibuprofen or naproxen, but less than indomethacin and more than acetaminophen [17]. Unlike these NSAIDs, pravadoline was an order of magnitude less potent in acute or chronic anti-inflammatory models and did not promote gastrointestinal ulcers in rodents [17]. Nevertheless, in a battery of seven antinociceptive tests in rodents, pravadoline exhibited potency that was comparable to aspirin and ibuprofen but less than indomethacin or naproxen. Pravadoline was less potent than morphine in these same antinociception tests; however, its effects could not be attributed to an opioid receptor because pravadoline's response in the acetic acid-induced writhing test was not blocked by the opioid antagonist naloxone [17]. Other data not shown indicated that this response was also not due to serotonin receptors, α_1- or α_2-adrenergic receptors, or P_1 or P_2 purinergic receptors [17].

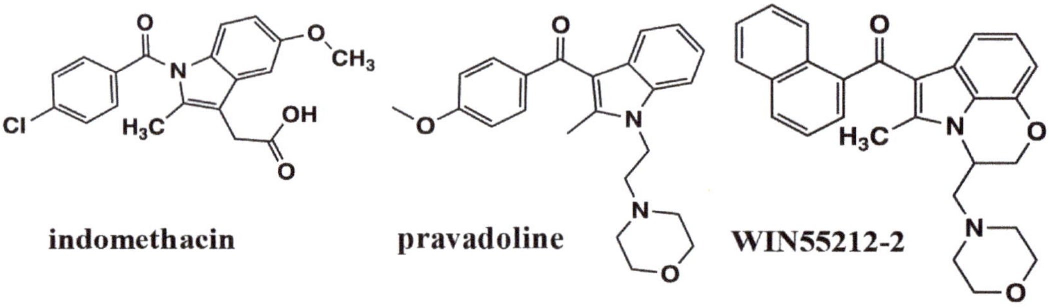

Figure 2. NSAID indomethacin, and aminoalkylindoles pravadoline and WIN55212-2.

To address the mechanism of action, the Sterling Research Group found that pravadoline mimicked the opioid receptor-mediated relaxation of mouse vas deferens contractions, yet this response was not blocked by naloxone [17]. They chose three aminoalkylindole (AAI) analogs that were incapable of cyclooxygenase inhibition to test for their ability to inhibit guinea pig ilium and mouse or rat vas deferens contractions [18,19]. The analogs differed from pravadoline by being devoid of the (R)-α-methyl on the indole or having a naphthoyl group replace the aroyl [18]. The naphthoyl analogs were one to two orders of magnitude more potent than pravadoline at inhibition of electrically contracted vas deferens or guinea pig ileum, whereas prototypical NSAIDs had no effect [19]. These responses to pravadoline and its naphthoyl analog were not reversed by antagonists of mu, delta and kappa opioid, α1-adrenergic, P_1-purinergic, or various serotonergic receptors [19]. Pravadoline and its naphthoyl analog failed to inhibit smooth muscle contractions in response to bradykinin or substance P, suggesting that the AAI effects were on presynaptic neurotransmiiter release. Interestingly, when various other neurotransmitter receptor agonists were tested, delta-opioid agonist peptide DADLE and the cannabinoid analgesic levonantradol were the most potent to inhibit vas deferens and guinea pig ileum contractions [19].

Additional AAI compounds were developed and evaluated using the mouse vas deferens and adenylyl cyclase assays. The naphthoyl AAI evoked inhibition of basal- and forskolin-stimulated adenylyl cyclase in rat brain cerebellar membranes in the presence of a cyclic nucleotide phosphodiesterase inhibitor [20]. For several analogs, the potency to inhibit adenylyl cyclase correlated with their potency to inhibit contractions in the mouse vas deferens [20]. This led to the discovery of a novel conformationally restrained enantiomeric pair in which a morpholinoethyl side chain was closed at position seven on the indole ring. This compound was given the code WIN55212-2 for the active (R) isomer and WIN55212-3 for the inactive (S) isomer (Figure 2) [20,21].

The development of AAI compounds also included an antagonist for the AAI agonists, WIN56098, which was created by the replacement of the C3-naphthoyl with a three-ringed anthracene. WIN56098 evoked competitive antagonism of the mouse vas deferens inhibition by pravadoline, the naphthoyl analog, and WIN55212-2, as well as inhibition of brain adenylyl cyclase by WIN55212-2 [20]. WIN56098 failed to compete in radioligand binding screens for α_1-, α_2-, β_1-, β_2-adrenergic, muscarinic and nicotinic cholinergic, H_1 and H_2 histamine, mu, delta and kappa opioid, 5-HT_{1a-d} and 5-HT_2, NK-1 tachykinin, NMDA, phencyclidine, bombesin, and AngII receptors (Novascreen). Of a number of other neurotransmitter and neuromodulator agonists in the mouse vas deferens assay, the only non-AAI compounds that WIN56098 competitively antagonized were galanin, pargyline, Δ^9-THC, and levonantradol [20]. WIN56098 has not achieved attention from the cannabinoid receptor research community, possibly because its log dose–response curve against WIN55212-2 exhibited a steeper slope than expected for a competitive antagonist [20] (A. Howlett, unpublished data), and it was not able to produce antagonism in vivo in rodent models of cannabinoid activity [22]. The Sterling Research Group also developed the antagonist 6-Br-pravadoline, which antagonized CB_1-mediated inhibition of adenylyl cyclase at very low potency (>1 μM) (A. Howlett, unpublished data).

Thus, armed with the knowledge that antinociceptive AAIs devoid of cyclooxygenase-inhibitory activity could produce in vitro responses resembling those of the cannabinoid agonists, it is not surprising that the Sterling Research Group would engage Dr. Howlett to screen a wide range of AAI compounds in her newly developed [^{3}H]CP55940 radioligand binding assay for cannabinoid receptors. Dr. Howlett reported the final results to the Sterling Research Group in Spring 1990, providing evidence that AAI compounds displaced [^{3}H]CP55940 from rat brain cannabinoid receptors over a wide range of IC_{50} values, with WIN55212-2 being the most potent and pravadoline being the least potent [23].

Simultaneously, the Sterling Research Group developed a radiolabeled [^{3}H]WIN55212-2 for use in binding assays. They demonstrated that the potency of AAI compounds to compete for [^{3}H]WIN55212-2 binding sites in rat cerebellar membranes correlated with inhibition of mouse vas deferens contractions [21,24]. Of the 60 neurotransmitter or neuromodulator agonists tested, none competed for [^{3}H]WIN55212-2 binding except cannabinoid ligands [24]. The final evidence that the AAI analgesic compounds bind to brain cannabinoid receptors came from the development of an irreversibly binding isothiocyanato-desmethyl naphthalene AAI [25]. When this affinity ligand was used to pretreat rat brain membranes, its covalent binding depleted 90% of the [^{3}H]CP55940 binding sites [25].

The greatest density of [³H]WIN55212-2 binding sites occurred in membranes prepared from the cerebellum, hippocampus, and striatum, with very little binding in the midbrain and spinal cord [24]. In studies of [³H]WIN55212-2 autoradiography in rat brain sections, the binding pattern was similar to that reported previously for [³H]CP55940 [26]. Studies in the mouse "tetrad" of cannabinoid-elicited behaviors (hypolocomotion, hypothermia, antinociception, catalepsy-like immobility) indicated that the naphthoyl AAI analogs that could inhibit the mouse vas deferens contractions were able to mimic Δ^9-THC in vivo [22]. In addition, stereospecificity was demonstrated for the WIN55212 enantiomers in the "tetrad" behaviors. Functionally, drug discrimination studies indicated that rats trained to recognize Δ^9-THC were able to identify the naphthoyl AAI analogs and the active enantiomer WIN55212-2 but not the inactive WIN55212-3 [22]. Important considerations in interpreting in vivo investigations include the pharmacokinetics and biotransformation of WIN55212-2. In a study published a decade later, Zhang and colleagues identified up to eight arene oxidative products following incubation with rat liver microsomes [27], which could have influenced biological activity. We can conclude that both common brain anatomic distribution patterns and behavioral similarities in rodent models demonstrate that the analgesic AAI compounds indeed bind to and stimulate the brain cannabinoid receptors.

Sterling Winthrop, Inc. abandoned the AAI analgesic drug discovery project in June 1990 (personal communication S.J. Ward to A.C. Howlett). Some compounds were made available to researchers in collaborative projects, and Sterling Research Group scientists published their research findings to inform the biomedical research community of this novel class of AAI cannabimimetic compounds. Sterling Winthrop, Inc. formed a strategic alliance with the French pharmaceutical company Elf Sanofi in 1991, and the final acquisition of the Sterling Winthrop, Inc. prescription drug component by Elf Sanofi occurred in June 1994 [28].

2. Aminoalkylindoles and Cannabinoids: Structure—Activity Relationship Studies in Search of a Common Pharmacophore

Given the abilities of AAI ligands to displace [³H]CP55940 and cannabinoid ligands to displace [³H]WIN55212-2 in rat brain preparations, an obvious hypothesis to test was that the AAI ligands occupy the same binding pocket of the CB_1 cannabinoid receptor, and further, that AAI ligands share a common pharmacophore with cannabinoid ligands. The common pharmacophore hypothesis was considered by a number of laboratories, each of which proposed models of homologous functionalities overlaying the structures of WIN55212-2 with a cannabinoid ligand. The Structure–Activity Relationship (SAR) studies of AAI compounds was evaluated to test these hypotheses and to establish principles for novel pharmacotherapeutic drug design. The most extensive series of compounds to assess AAI interaction with the CB_1 receptor was developed by the Huffman laboratory. The synthesis and characterization of the Huffman series, as well as AAI compounds from other laboratories, have been comprehensively reviewed [29,30].

In St. Louis, computational chemists Welsh and Shim evaluated the Howlett data for the competition of AAI compounds with [³H]CP55940 binding in rat brain membranes [31,32]. These data became the training set for Comparative Molecular Field Analysis (CoMFA) to develop a 3D Quantitative SAR (QSAR) model based upon the steric and electrostatic fields surrounding the molecules in their protonated or non-protonated states. A parallel analysis was performed using K_i values from the Sterling Research Group, which reported competition of AAI compounds with [³H]WIN55212-2 binding in rat cerebellar membranes [33,34]. The resulting CoMFA models indicate that 80% of the variation in AAI ligand affinities for the CB_1 receptor is based upon steric interactions. The potency of the AAI ligands to compete for the [³H]CP55940 binding site correlated well with their ability to act as agonists to inhibit hormone-stimulated adenylyl cyclase activity, with no evidence in the slope factors to suggest multiple receptors or cooperativity [32]. Based upon both the ligand-binding models and the requirements for agonist activity, it was proposed that the cannabinoid C3 side chain and the AAI C3 aroyl ring moiety both utilize hydrophobic interactions with residues within the CB_1 receptor binding pocket. Further molecular modeling led to an alignment in which these moieties in CP55244 (the most potent and stereo-selective, A,C,D-tricyclic, non-classical cannabinoid of the Pfizer series) and WIN55212-2, respectively, were overlaid (Figure 3A) [31]. However, compelling data also indicated that AAI binding to the CB_1 cannabinoid receptor might not result from the same chemical-binding interactions with receptor residues within a shared or overlapping binding pocket. This prediction was based upon evidence that the affinity of the AAI ligands for the [^{3}H]CP55940 binding site was less than for the [^{3}H]WIN55212-2 binding site (for six of the seven compounds assayed in both binding assays) [32]. Additional evidence was that the correlation was only "moderately strong" (r = 0.73) between the predicted K_i from the [^{3}H]CP55940 binding model and the actual K_i from the [^{3}H]WIN55212-2 binding experimental results [32], which is not supportive of identical ligand–receptor binding mechanisms within the shared binding pocket.

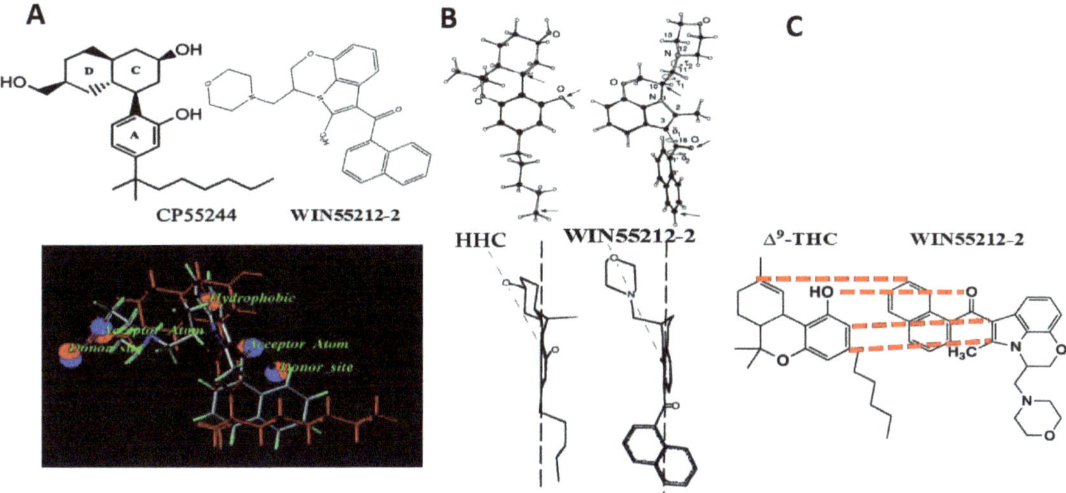

Figure 3. Alignments proposed for the common pharmacophore hypothesis for WIN55212-2 and cannabinoid agonists. (**A**) Alignment with CP55244, Shim and colleagues [31,35]. Reprinted with permission from Shim, J.Y. et al., J. Med. Chem. 45: 1447–1459, copyright 2002, American Chemical Society. (**B**) Alignment with (-)9β-OH-hexahydrocannabinol (HHC), Xie and colleagues [36,37]. Reprinted with permission from Xie, X.Q. et al., Life Sci. 56: 1963–1970, copyright 1995, Elsevier. (**C**) Alignment with Δ⁹-THC, Huffman and colleagues [29,38,39]. Redrawn using WIN55212-2, from Huffman Curr. Med. Chem. 1999 6: 705.

The Makriyannis and Xie laboratory, in collaboration with the Sterling Research Group chemist Eissenstat, used high-resolution 2D NMR with molecular modeling to develop a superimposition of the active enantiomeric structures of WIN55212-2 over the cannabinoid (-)9β-OH-hexahydrocannabinol (HHC) (Figure 3B) [36,37]. In this model, the AAI naphthoyl moiety overlaid the cannabinoid side chain, the WIN55212-2 fixed morpholino group overlaid the HHC cyclohexanol hydroxyl, and the AAI C3-carbonyl overlaid the cannabinoid phenolic hydroxyl [36]. They calculated the minimum energy conformations for the orientation of the naphthyl ring with respect to the carbonyl and of the morpholino group with respect to the C2-methyl in WIN55212-2 [37]. These studies determined a low-energy structure for WIN55212-2.

The Huffman laboratory proposed an alignment of WIN55212-2 with Δ⁹-THC by which the WIN55212-2 fixed morpholino moiety overlaid the cannabinoid C3-alkyl side chain; the AAI 3-carbonyl overlaid the cannabinoid phenolic hydroxyl, and the AAI naphthoyl group overlaid the Δ⁹-THC cyclohexene ring (Figure 3C) [29,38,39]. To test the role of the AAI N-ethylmorpholino of pravadoline and WIN55212-2, the Huffman group developed a series of indole and pyrrole analogs that were substituted at that position with alkyl chains of 1–7 carbon lengths, in the presence or absence of the C2-methyl (Figure 4) [38]. The most potent ligands to bind to the CB_1 receptor [^{3}H]CP55940 site also performed well in the behavioral "tetrad" tests and substituted for CP55940 in the drug discrimination trials. Properties of high potency ligands were: (1) N1-pentyl substituent; (2) no C2-methyl substituent; and (3) an indole rather than a pyrrole nucleus [38,39]. Potencies in all behavioral tests correlated well with the affinity for the displacement of [^{3}H]CP55940 in rat brain membranes. Interestingly, the methyl, ethyl, and propyl pyrroles failed to bind to the cannabinoid receptor [^{3}H]CP55940 site but exhibited behavioral responses of hypolocomotion, hypothermia, and antinociception, albeit with low potency or efficacy (see "What Additional Targets Exist for Aminoalkylindoles?"). Both Δ^9-THC and WIN55212-2, were more potent at decreasing spontaneous activity than antinociceptive or hypothermic responses; however, the difference in potencies with WIN55212-2 was double that of Δ^9-THC.

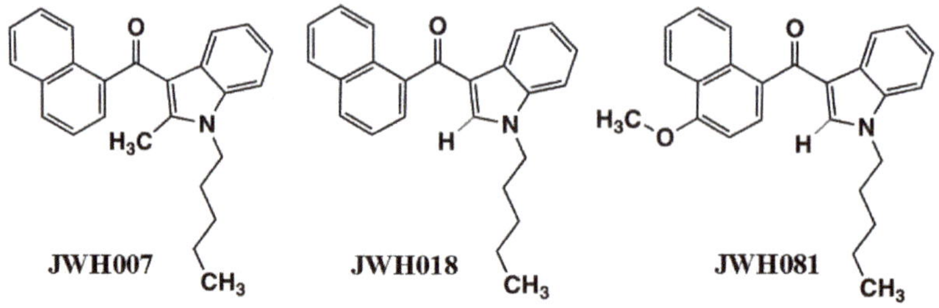

Figure 4. Alkyl indole compounds developed to test the Common Pharmacophore Hypothesis.

In support of the Huffman model, the AAI aminoalkyl group could be replaced with alkyl substituents that resembled the cannabinoid C3 alkyl moiety. In order to assess whether the alignment was correct, the Huffman group synthesized a "hybrid" cannabinoid, JWH-161, in which the structure of Δ^9-THC was fused to an indole nucleus having an N1-pentyl substituent [40]. JWH-161 exhibited potencies for [^{3}H]CP55940 binding to the CB_1 receptor and cannabimimetic "tetrad" tests that were comparable to those of Δ^9-THC. Although this result is consistent with the region of the cannabinoid C3 alkyl side chain interacting with the receptor via hydrophobic interactions, it does not necessarily invoke the necessity of an indole nucleus in this binding domain. The Huffman model aligns the AAI indole carbonyl moiety with the cannabinoid phenolic hydroxyl, which is required for cannabinoid agonist activity at the CB_1 receptor. Removal of the AAI indole carbonyl in naphthylidene indene conformers (E active versus Z inactive) reduced affinity for the [^{3}H]CP55940 binding site [41]. The reduced affinity was calculated to be due to the modification of the linkage angles and orientation of the aryl ring structure [42], which overshadowed the assessment of a potential role for oxygen in hydrogen-bonding interactions.

In a series of N1-ethylmorpholino, 3-naphthyl indoles devoid of the carbonyl oxygen, K_i values for [^{3}H]CP55940 binding displacement were in the 40–42 nM range [42]. For their N1-pentyl analogs, also devoid of carbonyl substituents, K_i values were in the 17–23 nM range [42]. JWH-176, an indene molecule devoid of oxygen or nitrogen atoms, exhibited a $K_i = 26$ nM. These affinities compare favorably with the $K_i = 10$ nM reported for WIN55212-2 in the same data set. These data favor the dominance of aromatic stacking interactions with very little influence of hydrogen bonding for AAI interactions with the CB_1 cannabinoid receptor.

To assess the CB_1 cannabinoid receptor agonist binding requirements, it was known that mutation of a transmembrane helix-3 lysine to alanine in the hCB_1 receptor expressed in HEK293 cells conflicted with competition for [^{3}H]WIN55212-2 by cannabinoid ligands but not by WIN55212-2 [43]. The potency of cannabinoid agonists to inhibit cAMP production was reduced in cells expressing the mutant receptors, but the response to WIN55212-2 was unaffected. These findings suggest that the required phenolic hydroxyl on cannabinoid structures was hydrogen bonding with this lysine but that this hydrogen-bonding interaction was not a factor in the AAI interactions. In contrast, when CB_1 receptor mutants of a highly conserved helix-2 aspartate were expressed in HEK293 cells, cannabinoid agonist displacement of [^{3}H]CP55940 was not affected, but WIN55212-2 binding suffered a 45-fold reduction in affinity when the aspartate was mutated to asparagine, and an 8.5-fold reduction in affinity when mutated to glutamate [44]. These findings suggest that this helix-2 aspartate must be involved in WIN55212-2 but not cannabinoid agonist interactions.

To identify the CB_1 cannabinoid receptor mechanism for AAI ligand binding, the Reggio group developed a homology model based upon the structure of activated rhodopsin [41,42]. The conformation of WIN55212-2 and pravadoline as S-trans (versus inactive S-cis) within the activated cannabinoid receptor binding pocket was predicted by pharmacological results demonstrating the preferred conformation of rigid naphthylidene indene analogs of AAIs to exist as the active "E" (comparable to S-trans) as opposed to the "Z" (comparable to S-cis) conformation [41].

The Reggio group reported that an aromatic cluster of residues in transmembrane helices 3, 4, and 5 are a likely binding pocket to accommodate hydrophobic ligand interactions [45,46]. Using the rhodopsin homology model in the "active state", residues that include helix-3 phenylalanines and helix-4 and helix-5 tryptophans could form an aromatic stack that is energetically favored [46]. A hydrophobic binding pocket of helix-3 valine, isoleucine, and phenylalanine, and helix-6 leucine and isoleucine could accommodate an alkyl chain between three and six carbons in length, and helix-5 and helix-6 tryptophans could allow aromatic stacking interactions with the indole and naphthyl moieties [42]. With this configuration, the binding energy would be due to hydrophobic interactions, although as a minor contribution, a hydrogen bond could exist between N–H of the helix-5 tryptophan and the carbonyl oxygen. This hydrogen bond would not be possible for the indene analogs lacking oxygen and was suggested to be responsible for their reduced potency [42].

Shim and Howlett addressed the mechanism by which WIN55212-2 could trigger a response to activate the CB_1 receptor [47]. Using a homology model based on rhodopsin in the inactive "ground" state, Shim performed Monte Carlo and molecular dynamics simulations to identify the docking conformations exhibiting the lowest ΔE_{bind} values for WIN55212-2 within the CB_1 receptor binding pocket [47]. They correlated the calculated docking ligand–receptor interaction energy with experimental binding affinity data for 37 AAI compounds to compete for [^{3}H]WIN55212-2 binding sites in rat brain membranes under basal conditions (the absence of Na^+ or GTP analogs) [33]. Two conformations having the greatest correlation were identified as having the aroyl groups oriented "up" closest to the extracellular surface of the receptor in the hydrophobic binding space. The interaction energies with amino acids within 3 Å were identified as predominantly van der Waals (steric), with minor contributions of electrostatic (i.e., ionic or hydrogen-bonding) forces, in agreement with previous studies (discussed above). It was hypothesized that the WIN55212-2 structure docked in the ground state would be able to exert a "trigger" to induce one or more micro-conformational changes essential for the process of CB_1 receptor activation. Strain energy is released as the agonist bound to the receptor relaxes to achieve its lowest energy conformation. The energy released from the conformational change in the agonist ligand is the driving force for inducing conformational changes in the receptor that is necessary for transferring the signal to G-proteins. To determine how this might occur, Shim determined the "flexibility" of four torsion angles of the WIN55212-2 molecule to identify intrinsic changes in the agonist's conformations after being bound to the ground state of the CB_1 receptor. In molecular dynamics simulations in the absence of the receptor, a conversion from S-*trans* to S-*cis* could occur as the torsion angle between the carbonyl oxygen and the naphthoyl ring adjusts to reduce the steric repulsion to the indole ring. This allows WIN55212-2 to traverse the lowest possible rotational energy barrier within the allowed conformational space. As the ligand conformation "switches" to release strain energy and attain the lowest possible energy conformation, this "switch" becomes the "steric trigger" to allow WIN55212-2 to force a change in the receptor conformation. If the lowest energy conformation of the agonist creates an unfavorable steric clash with amino acids within the receptor hydrophobic pocket, then the receptor adjusts its conformation. This may occur as series of micro-conformational changes to ultimately achieve the activated state. Conceivably, different ligand-binding conformations for the same binding pocket may initiate diverse types of receptor motions for ligand-specific conformational changes within the receptor. Thus, it is not likely that the AAI [47] and cannabinoid [48,49] agonists utilize the same "mechanism" to trigger micro-conformational changes to activate the CB_1 receptor.

In total, these studies have identified a pharmacophore for AAI ligands to bind within a hydrophobic pocket of the CB_1 receptor. AAI binding overlaps within the binding pocket for cannabinoid ligands. However, the interactions with amino acids and the mechanism for activation of the receptor differ, resulting in subtle conformational differences that could result in selective interactions with their transducers (G proteins, β-arrestins, other associated proteins).

3. The Quest for Selective CB_2 Cannabinoid Receptor Ligands

One of the challenges to cannabinoid pharmacology has been the separation of agonist activities for the CB_2 versus the CB_1 cannabinoid receptors. A highly selective CB_2 agonist would be useful as an anti-hyperalgesic and anti-allodynic agent in neuropathic as well as anti-inflammatory pain [50–52]. The requirements for an ideal CB_2 pharmacotherapeutic agent are (1) to function with high potency and efficacy at the CB_2 receptors, but also (2) to have low affinity for the CB_1 receptors that stimulate untoward central nervous system effects such as sedation and cognitive and memory dysfunction. Evidence based upon the preclinical studies of Huffman and multiple pharmaceutical researchers suggests that the challenge might be met with AAI compounds (reviewed in [53–55]).

3.1. CB_2-Selective Indole Agonists

3.1.1. JWH-015 and Analogs (1-Propyl-2-methyl-3-(1-naphthoyl) indole)

The first observation of cannabimimetic indoles showing CB_2 receptor selectivity was that WIN55212-2 exhibited greater affinity in [^{3}H]CP55940 binding in stably expressing hCB_2-Chinese Hamster Ovary (CHO) fibroblastic cells compared with hCB_1-CHO cells [56]. In an effort to identify additional CB_2-selective ligands, the Abood laboratory examined [^{3}H]CP55940 binding in hCB_2-CHO or hCB_1-CHO cells [57]. WIN55212-2 exhibited a 7-fold selectivity for the CB_2 receptors, but because WIN55212-2 was quite potent at binding to both receptor types, it exhibited potent CB_1-mediated effects in the behavioral "tetrad" tests, which would make it unlikely to serve as a selective CB_2 receptor agonist. The other AAI that showed CB_2-selectivity was JWH-015, which exhibited greater than 25-fold selectivity for binding to the CB_2 receptor [57]. JWH-015, a propyl analog of pravadoline, exhibited very low affinity at the CB_1 receptor and relatively low potency in the behavioral "tetrad" behaviors [57]. It is interesting to note that the slope of the log dose–response [^{3}H]CP55940 binding curve for JWH-015 was shallower than expected for a single binding site, which could indicate either binding to two different receptors, binding to two different affinity states of the CB_2 receptor, or negative allosteric regulation of the CB_2 receptor. This interesting phenomenon has yet to be explained in the research literature.

JWH-015 is a member of a series of C3-naphthoyl indoles in which a propyl substituent was appended to indole N1 (Table 1 and Figure 5). Although the propyl analog reduced the ability to bind to the CB_1 receptor compared with the pentyl analog, it nevertheless retained behavioral "tetrad" activities [38]. Among C3-naphthyl indole analogs lacking the C2 methyl, the N1 alkyl chain length correlated with [^{3}H]CP55940 binding affinity in hCB_2-CHO membranes, increasing nearly 20-fold in going from ethyl to a propyl, whereas the CB_1 receptor binding in rat brain membranes remained at nearly the same poor affinity [58]. The propyl analog, JWH-072, yielded a CB_1/CB_2 selectivity ratio = 6. Both CB_1 and CB_2 receptor binding reached maximal potencies at butyl, pentyl, and hexyl, at which the CB_2/CB_1 selectivity ratio was reduced to ≤3.

Table 1. Cannabimetic indole analogs exhibiting improved CB2/CB1 receptor selectivity.

Name	N1	C2	C3	CB$_2$ K$_i$ (nM)	CB$_2$/CB$_1$ Selectivity Ratio	Reference
JWH-015	Propyl	Methyl	1-naphthoyl	13.8	27	[57]
JWH-046	Propyl	Methyl	7-methyl-1-naphthoyl	16.0	21	[58]
JWH-120	Propyl	H	4-methyl-1-naphthoyl	6.1	170	[57]
JWH-267	Pentyl	H	2-methoxy-1-naphthoyl	7.2	54	[59]
JWH-151	Propyl	Methyl	6-methoxy-1-naphthoyl	30.0	>300	[59]

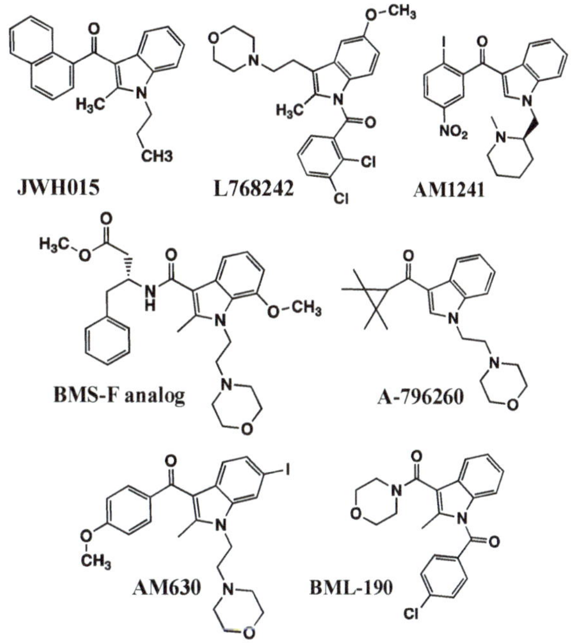

Figure 5. CB$_2$-selective indole ligands.

Because the addition of a C2-methyl reduced affinity for CB$_1$ receptors [38], it was observed that a methyl modification in the propyl analog JWH-015 improved CB$_2$/CB$_1$ selectivity ratio = 24. Selectivity was not improved by adding a C7′-methyl substituent onto the naphthoyl ring system in JWH-046 (CB$_2$/CB$_1$ selectivity ratio = 21) [58], but it was encouraging that for JWH-046, maximal activities in the cannabimimetic "tetrad" tests could not be attained [38]. These compounds (Table 1) were agonists in CHO-CB$_2$ membranes in the [^{35}S]GTPγS binding assay of G protein activation, with JWH-151 showing full efficacy compared with CP55940, and the others having partial agonist activity ranging from 65% to 80% compared with CP55940 [59]. An additional series of halogenated naphthoyl indoles was developed, three of which exhibited optimal high affinity for the CB$_2$ receptor and a good CB$_2$/CB$_1$ selectivity ratio: JWH-423 (1-propyl-3-(4-iodo-1-naphthoyl)indole), JWH-422 (the 2-methyl analog of JWH-423), and JWH-417 (1-pentyl-3-(8-iodo-1-naphthoyl)indole) [60].

Development of the Huffman compounds promoted the recognition by leading drug companies that CB$_2$ receptor selectivity could be achieved. It seemed that nearly half of the participants in the 2005 International Cannabinoid Research Society meeting were pharmaceutical industry scientists. The potential that CB$_2$-selective indole cannabimimetics could be developed as anti-hyperalgesic and anti-allodynic medicines inspired tremendous interest in pharmaceutical companies to engage in preclinical studies (examples follow).

3.1.2. L768242/GW405833 (1-(2,3-Dichlorobenzoyl)-2-methyl-3-(2-[1-morpholine] ethyl)-5-methoxyindole)

The Merck Frosst Centre for Therapeutic Research reported that L768242, also known as GW405833 (Figure 5), exhibited a high affinity for the CB_2 receptor (K_i = 14 nM) and a high CB_2/CB_1 selectivity ratio = 146 [61]. Valenzano and the Purdue Pharma Discovery Research group determined that L768242/GW405833 interacts with human CB_2 receptor [^{3}H]CP55940 sites in hCB_2-CHO cells with high affinity (K_i = 3.9 nM) and a hCB_2/hCB_1 selectivity ratio = 1217 [62]. The affinity was the same for CB_2 binding in rat spleen membranes (K_i = 3.6 nM), and comparison with rat brain membranes yielded a CB_2/CB_1 affinity ratio = 76. In the CB_2-CHO cells, L768242/GW405833 was a partial agonist, exhibiting 50% efficacy compared with CP55940 to inhibit forskolin-stimulated cAMP accumulation [62].

Clayton and colleagues at Glaxo Wellcome Research and Development noted that L768242/GW405833 inhibited carrageenan-induced paw inflammation and hypersensitivity, and these effects were blocked by the CB_2 antagonist SR144528 [63]. Valenzano and colleagues determined that L768242/GW405833 attenuated mechanical hyperalgesia in rat spinal nerve ligation or the rat paw incision tests but had no effect on thermal antinociception (tail-flick or hotplate tests) [62]. In the mouse paw chronic inflammation (Freund's complete adjuvant) model, tactile allodynia was partially reversed, comparable in efficacy to indomethacin. The L768242/GW405833 response was not observed in $CB_2^{-/-}$ mice, but the indomethacin response was not tested (or reported) in the $CB_2^{-/-}$ mice. Beltramo and colleagues at Schering-Plough Research Institute reported that L768242/GW405833 was effective in neuropathic pain tests in rodents in which it attenuated hyperalgesia in the mouse intraplantar formalin model and allodynia in the rat spinal nerve ligation model [64]. Both responses were precluded by pretreatment with the CB_2 antagonist SR144528.

3.1.3. AM1241 ((R-) or (S-) 3-(2-Iodo-5-nitrobenzoyl)-1-(1-methyl-2 piperidinylmethyl)-1H-indole)

AM1241 (Figure 5) displaced [^{3}H]CP55940 with two orders of magnitude greater potency in mouse spleen homogenates (abundant in CB_2 receptors) compared with rat brain synaptosomal membranes (abundant in CB_1 receptors) [65]. Bingham and colleagues at Wyeth Research identified two isomers: R (+) was two orders of magnitude more potent than S (−) to compete for [^{3}H]CP55940 binding to human, rat, and mouse CB_2 compared with CB_1 receptors expressed in CHO cells [66]. Their investigation of forskolin-stimulated rCB_2-CHO cells showed that S-AM1241 inhibited cAMP production, resembling WIN55212-2. In contrast, R-AM1241 augmented forskolin-stimulated rCB_2-CHO cAMP production, resembling SR144528. Enantiomeric response differences between rodent and human CB_2 receptors were complex [66] but might be influenced by the degree of "constitutive" activity in these exogenously expressed systems [67], the serum levels or cellular production of endogenous endocannabinoids, or differential sensitivity to endocannabinoids.

In in vivo models of spinal nerve ligation in rats or mice, AM1241 (ip) dose-dependently attenuated both tactile and thermal hyperalgesia, both of which were antagonized by CB_2-selective AM630 but not by CB_1-selective AM251 [65]. Additional evidence against a CB_1 involvement in the anti-hyperalgesic responses was that AM1241 effects were also observed in $CB_1^{-/-}$ mice. AM1241 attenuated carrageenan-induced inflammatory thermal hyperalgesia when injected directly into the inflamed paw but failed to evoke antinociception in the contralateral control paw [68]. In that model, AM1241 also reversed the local edema, and both edema and hyperalgesia responses to AM1241 were antagonized by AM630 but not AM251.

Beltramo and colleagues showed that AM1241 could attenuate both hyperalgesia in mouse intraplantar formalin and allodynia in the rat spinal nerve ligation tests, and that both responses were inhibited by CB_2-selective SR144528 [64]. S-AM1241 (but not R-AM1241) was as efficacious as indomethacin at prolonging the latency to remove a carrageenin-inflamed paw from a thermal stimulus [66]. The response to S-AM1241 was reversed by CB_2 antagonist AM630, but it was not determined if the response to indomethacin could also be reversed by AM630 or SR144528 [66].

3.1.4. BMS Series and A796260 from 1-Alkyl-3-keto Indole Series

Bristol-Myers Squibb researchers developed a series of compounds based on a substituted indole 3-carboxylic acid nucleus (Figure 5) [69]. Their most promising compound was a phenylalanine-derived amide that exhibited high CB_2 receptor affinity (K_i = 8 nM) and a very high CB_2/CB_1 affinity ratio = 500.

Abbott researchers developed a series of 1-alkyl-3-keto indoles having variations in nitrogen side chains, with saturated cyclic ketones as the C3-aryl substituent. They identified A796260 (Figure 5) having a C3-tetramethylcyclopropyl substituent, as exhibiting extremely high affinity for the CB_2 receptor expressed in CHO cells (K_i = 0.77 nM), an extremely high CB_2/CB_1 selectivity ratio = 2700, and full agonist efficacy in cellular functional assays [70]. A796260 was efficacious in in vivo models of chronic inflammatory pain and chronic neuropathic pain, and its responses were selectively blocked by CB_2 antagonist, but not by CB_1 or μ-opioid antagonists.

In aggregate, these studies identify local, CB_2-dependent, anti-hyperalgesic and anti-allodynic responses in chronic inflammatory and neuropathic pain models that do not require a CB_1 receptor involvement. These promising preclinical experimental results warrant further development in clinical settings. Even so, the cellular and biochemical mechanism of action of these compounds may not be entirely attributable to their actions at the CB_2 receptor. For example, these compounds are analogs of pravadoline, an NSAID exhibiting antinociceptive actions that could be attributed to inhibition of prostaglandin synthesis. Complete understanding of the mechanism of action and potential for untoward side effects will require a more comprehensive investigation of the synthesis of anandamide in the pain process alleviated by these compounds, the contribution of anandamide to the "constitutive" activity of the CB_2 receptor, and the contribution of these CB_2-selective cannabimimetic indoles to the inhibition of COX2 in the inflamed tissue.

3.2. CB_2-Selective Indole Antagonists

3.2.1. AM630 6-Iodo-Pravadoline

AM630 (6-iodo-pravadoline) (Figure 5) appears to respond either as an agonist or as a competitive antagonist and inverse agonist in different types of cell signaling determinations. AM630 was first identified to be a competitive antagonist in the cannabinoid inhibition of mouse vas deferens twitch response, right-shifting the log dose–response curves to Δ^9-THC, CP55940, and WIN55212-2 (K_{inh} values were calculated to be in the 14 nM–36.5 nM range), but not to morphine or clonidine [71]. This report was followed by the determination that AM630 behaved as a low-potency agonist (IC_{50} = 1.9 μM) compared with WIN55212 (IC_{50} = 5.5 nM) to inhibit contractions of the guinea pig ileum [72]. These AM630 log dose–response curves were right-shifted by the CB_1 antagonist SR141716, demonstrating AM630 to be a CB_1 receptor agonist [72]. At high concentrations (100 μM), AM630 behaved as a competitive antagonist to right-shift the WIN55212-2-stimulated [^{35}S]GTPγS binding curves in mouse [73] or guinea pig [74] brain homogenates (assumed to be abundant in CB_1 receptors). In a CB_1-CHO cell [^{35}S]GTPγS binding determination, AM630 behaved as an inverse agonist to inhibit basal by 20% (EC_{50} = 900 nM), under the same conditions that WIN55212-2 behaved as an agonist to stimulate basal activity (EC_{50} = 360 nM) [75].

To clarify the activity of AM630 at the molecular level, Ross, Pertwee, and colleagues used CB_1-CHO and CB_2-CHO cell comparisons to determine affinity to displace [^{3}H]CP55940 and activity for the cannabinoid receptors [76]. As the Pertwee lab had suspected from the studies in tissue preparations, AM630 interacted potently with the CB_2 receptor (K_i = 31 nM) and exhibited a CB_2/CB_1 selectivity ratio = 165. AM630 behaved as a potent (EC_{50} = 76.6 nM) inverse agonist to inhibit basal [^{35}S]GTPγS binding in CB_2-CHO membranes; using the Landsman data in CB_1-CHO cells, this yields a CB_2/CB_1 potency ratio approaching 12. Consistent with these data on G protein activation, AM630 at high concentrations (1 μM) behaved as an inverse agonist in CB_2-CHO cells by augmenting forskolin-stimulated cAMP accumulation. In CH_2-CHO cells, AM630 also behaved as a competitive antagonist for CP55940-Gi-mediated inhibition of forskolin-stimulated cAMP accumulation [76]. In contrast, in CB_1-CHO cells, AM630 at high concentrations (1–10 μM) behaved as an agonist in Gi-mediated inhibition of forskolin-stimulated cAMP accumulation but exerted a tendency to attenuate the Gi-mediated agonist response to CP55940, making AM630 a partial agonist [76]. The Mackie laboratory found that in mCB_2-HEK293 cells, AM630 behaved as an inverse agonist in cAMP production assays but behaved as a low-efficacy agonist in β-arrestin recruitment assays [77].

3.2.2. BML190

BML-190 (Figure 5) has a low affinity for CB_2 receptors exogenously expressed in CHO cells. BML190 appears to be an inverse agonist for the CB_2 receptor, as it augmented forskolin-stimulated cAMP production [61].

3.3. CB_2-Selective WIN55212-2 and AAI Ligand Interactions with the CB_2 Receptor

As described for the CB_1 receptor, the CB_2 receptor engages cannabimimetic indoles via aromatic stacking mechanisms. However, the specific molecular interactions of CB_2-selective AAI ligands with the CB_2 receptor appear to differ from CB_1-selective AAI interactions with the CB_1 receptor.

The importance of amino acids in the CB_2 helix-3 for AAI interactions was reported by Chin and Kendall, who created a chimeric CB_1 receptor possessing the CB_2 helix-3 and expressed the receptors in CHO cells [78]. The affinities for WIN55212-2 (K_d = 4.8 nM) and JWH-018 (K_d = 1.4 nM) were greater for the CB_2-helix 3 chimera than for the CB_1 receptor; however, JWH-015 (K_d = 1 μM) exhibited low affinity but still greater than for the CB_1 receptor [78]. The average CB_2-helix 3 chimera/CB_1 selectivity ratio was 5.6. These affinities paralleled the potencies to inhibit cAMP accumulation in CHO cells expressing these receptors [78]. When individual amino acid differences were investigated by site-directed mutagenesis and expression in CHO cells, it appeared that the serine unique to the CB_2 helix-3 was important for the WIN55212-2 interaction with cannabinoid receptors [78].

The Abood laboratory compared responses of CB_2 to CB_1 receptors expressed in HEK293 cells [79]. For the CB_1 receptor, the helix-3 lysine192 was required for cannabinoid ligand binding but not WIN55212-2 binding. In contrast, when the comparable CB_2 lysine109 was mutated to alanine, there were no differences from wild-type CB_2 in cannabinoid or WIN55212-2 binding or agonist responses to inhibit cAMP accumulation [79]. However, the CB_2 helix-3 serine112 mutation to glycine double mutant with the lysine109 mutation to alanine compromised the cannabinoid agonist but not WIN55212-2 binding [79].

Interestingly, there are two reports of loss of cannabinoid ($[^3H]HU243$ and $[^3H]CP55940$) as well as $[^3H]WIN55212$-2 binding resulting from mutation of the CB_2 receptor helix-3 aspartate that is part of the "DRY" sequence and a coordinating helix-6 alanine [80,81]. Because both amino acids affecting CB_2 receptor binding are located at the intracellular juxtamembrane surface, it is likely that their influence is on rigid helical movement or conformational modifications transmitted along the helices that would affect interactions with the ligands occurring near the extracellular membrane surface.

Several investigations were reported to test the hypothesis that aromatic stacking is important for WIN55212-2 interaction with the CB_2 receptor. Interaction of WIN55212-2 with a phenylalanine in helix-5 unique to the CB_2 receptor was predicted by the Reggio laboratory using in silico docking models [46]. When tested with site-directed mutagenesis and expression in HEK293 cells, the CB_2 receptor mutation of phenylalanine to valine compromised the affinity for WIN55212-2 but did not affect the affinity for cannabinoid ligands HU210 or CP55940 [46]. Parallel changes in the ability to inhibit cAMP accumulation were observed in these cells. A conserved helix-5 tyrosine, important for aromatic stacking in both CB_1 and CB_2 receptors, was necessary for stimulation of signaling by both WIN55212-2 and cannabinoid agonists [45]. Two CB_2 helix-4 tryptophans (or their conservative mutation to phenylalanine) were essential for $[^3H]HU243$ binding and for HU210- or WIN55212-2-mediated inhibition of cAMP production in hCB_2-COS7 cells [82].

Structural interactions between CB_2 receptors and the AAI ligands compared with cannabinoid ligands can lead to functional differences (biased agonism) as demonstrated by the Mackie laboratory for rodent CB receptors expressed in HEK293 cells [77,83]. For example, CP55940 was a full agonist in CB_2-Gi-mediated inhibition of cAMP production, whereas WIN55212-2 had lower efficacy [77]. Both WIN55212-2 and CP55940 recruited β-arrestins to the plasma membrane, whereas classical cannabinoid and most AAI ligands failed [77,83]. CP55940 and cannabinoid ligands promoted the internalization of CB_2 receptors, whereas WIN55212-2 and other AAI ligands did not [83]. The functional selectivity, very likely based upon conformational differences in the structural mechanisms of activation of the receptors by the ligands, can initiate cellular signaling pathways that are uniquely different in target cells. Thus, conflating the cellular responses to cannabimimetic indoles with responses to classical cannabinoids such as Δ^9-THC can lead to misrepresentation of physiological and pharmacological endpoints.

4. What Additional Targets Exist for Aminoalkylindoles?

4.1. Non-CB_1, Non-CB_2 Targets for WIN55212-2

Early in the investigation of WIN55212-2's binding and cellular-signaling properties, Childer's laboratory recognized that displacement of [^{3}H]WIN55212-2 binding by cannabinoid ligand CP55940 differed between rat brain cerebellar membranes (IC_{50} = 1.2 nM) and cultured mouse neuroblastoma–rat glioma hybrid cell NG108-15 membranes (IC_{50} > 5000 nM) [84]. The properties of the binding site in cerebellar membranes were typical of a GPCR in that binding affinity for the agonist [^{3}H]WIN55212-2 was reduced by GTPγS or by Na^+, whereas those binding sites in the hybrid cell were resistant to these regulators. These data suggest that the binding sites were not the same and that only those binding sites in the cerebellar membranes were GPCRs. With the advent of modern molecular biology techniques, the neuroblastoma–glioma hybrid cell line lost its popularity due to its polyploidy, which in fact allows the NG108-15 hybrid cells to express both rat and mouse mRNAs for the CB_1 receptor [85]. The Howlett laboratory determined that the NG108-15 cell line was capable of stimulating a functional inhibition of adenylyl cyclase in membrane preparations, albeit with less response than in membranes from the N18TG2 neuroblastoma parent, and that membranes from the rat C6-glioma parent fail to respond to cannabinoid ligands [13,86]. Thus, although [^{3}H]WIN55212-2 fails to recognize these low-abundance functional CB_1 receptors in the NG108-16 cells, this ligand recognizes an alternative protein target that binds extremely poorly to CP55940 [84] (and perhaps other cannabinoid ligands as well).

If the only target in the brain for WIN55212-2 were the CB_1 receptor, then that target should not be present in the $CB_1^{-/-}$ mouse brain. Breivogel and colleagues performed this test in a study of [^{35}S]GTPγS binding to activated G proteins in brain membranes from the C57Bl/6 $CB_1^{-/-}$ mouse as ablated by Zimmer and colleagues [87]. They demonstrated that the knock-out of CB_1 receptors resulted in a loss of the response to high-efficacy cannabinoid agonists CP55940 and HU210 as well as partial agonist Δ^9-THC [87]. However, anandamide and WIN55212-2 both evoked a response in $CB_1^{-/-}$ mouse brain membranes. Estimates of SR141716-resistant stimulation in wild-type mouse brain membranes suggested that 16% of the anandamide- and 33% of the WIN55212-2-stimulated response might be due to non-CB_1 target(s) [87]. The WIN55212-2-stimulated response in the $CB_1^{-/-}$ mouse brain was localized to regions that in wildtype mice do not express an abundance of CB_1 receptors (brainstem, diencephalon, midbrain, and spinal cord), whereas the WIN55212-2 response was not significantly stimulated in regions expected to express high densities of CB_1 receptors (basal ganglia, cerebellum) [87]. These same findings were reported for the CD1 $CB_1^{-/-}$ mouse ablated by Ledent and colleagues, with some discrepancies in brain regions expressing the response [88]. In their investigation, anandamide and WIN55212-2 were not able to inhibit adenylyl cyclase, suggesting that the novel WIN55212-2-stimulated target does not couple to Gi proteins [88].

Neurophysiological investigations provided additional evidence for a non-CB_1 WIN55212-2 target in the brain. In the mouse hippocampus, which exhibits a well-characterized, CB_1-mediated suppression of neurotransmission at GABAergic presynaptic terminals, Hájos, Ledent and colleagues found that WIN55212-2 compromised neurotransmission at glutamatergic synapses in both wild-type and Ledent CD1 $CB_1^{-/-}$ mice [89,90]. They recognized a high-affinity (nM range), CB_1-mediated reduction in Schaffer collateral-evoked CA1 pyramidal cell excitatory post-synaptic potentials in rat brain slices. However, they also identified a low-affinity suppression of neurotransmission response to WIN55212-2 (µM range) in brain slices pretreated with CB_1 antagonist AM251 [90]. This non-CB_1 response was blocked by pretreatment with Ω-conotoxin GVIA, suggesting that WIN55212-2 might directly target N-type, voltage-gated Ca^{2+} channels or work via a GPCR that targets the N-type channels [90].

WIN55212-2 (µM range) inhibited the frequency of rat nucleus tractus solitarius glutamatergic and GABAergic stimulated postsynaptic currents [91]. This response was not observed with cannabinoid agonist HU210 or CB_1-selective agonist arachidonyl cyclopropylamide. The WIN55212-2 response could not be blocked by CB_1 antagonist AM251, CB_2 antagonist AM630, or TRPV1 blocker AMG9810, suggesting that an alternative target is responsible [91]. Because the nucleus tractus solitarius receives direct inputs from cardiovascular reflex detectors, this novel WIN55212-2 target might disrupt autonomic baroreflex regulation of blood pressure.

4.2. Putative Alkyl Indole Receptors

The Stella laboratory discovered that WIN55212-2 might be acting at brain microglia cell targets via a non-CB_1, non-CB_2 mechanism [92]. In order to characterize the responsible receptor, which they termed the Alkyl Indole (AI) receptor, they developed analogs that could distinguish the novel AI functions [93,94]. ST-11 and ST-48 (Figure 6) are naphthoyl indoles that exhibit high affinity for [^{3}H]WIN55212-2 binding sites (32.6 nM, 23.7 nM, respectively) in membranes from primary cultures of mouse microglia [93,94]. AI receptor stimulation by ST-11 promoted cAMP accumulation and inhibited both basal migration as well as ATP-driven chemokinesis in a Boyden chamber test [93]. ST-11 also inhibited macrophage-colony-stimulating-factor-induced proliferation but did not alter responses to cytokines that direct the determination of microglia to develop M1 (pro-inflammatory) or M2 (anti-inflammatory) phenotypes [93]. However, differentiation to an M2 phenotype was sufficient to attenuate the responses to ST-11, demonstrating that signaling by the AI receptors is subject to modulation by other ongoing cellular signal transduction pathways.

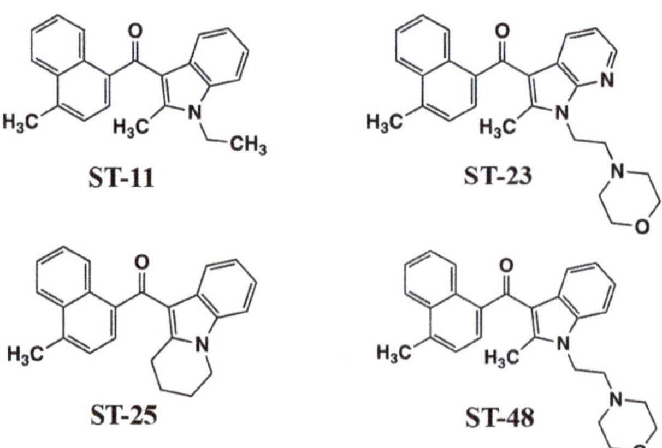

Figure 6. The Stella naphthoyl indoles and analogs.

Previous studies indicated that certain non-CB_1 effects of WIN55212-2 did not appear to involve GPCRs. In the course of investigating ST-11 and its analogs, the Stella laboratory discovered the ability of ST-11 to reversibly interact with the colchicine-binding site of microtubules and attenuate their assembly [95]. In fast-growing glioblastoma tumor cells, this led to disruption of spindle formation, cell cycle arrest in pro-metaphase, and subsequent apoptosis [95]. This response makes ST-11 of great clinical significance as a potential cancer chemotherapeutic agent for glioblastoma. Unlike many mitosis-disrupting cancer drugs, ST-11 avoids multi-drug resistance pumps, and gains access to the brain through the blood–brain barrier when formulated in lipid nanodiscs for efficient delivery [95].

Further drug development to identify the cellular role of AI receptors required a separation of AI activation from microtubule-binding properties in addition to CB_1 and CB_2 cannabinoid receptors. ST-11 fails to bind to CB_1 and CB_2 receptors and exhibits an AI/colchicine binding selectivity ratio = 61.5, which makes it possible to access the brain at concentrations that favor AI receptor-mediated responses [94]. Using a model of DBT cells, which do not express CB_1 or CB_2 mRNA or [^{3}H]CP55940-binding sites, the Stella team demonstrated that the [^{3}H]WIN55212-2 binding site recognized WIN55212-2 (K_i = 6.2 nM) and ST-11, ST-23, ST-25, and ST-48 (K_i's in the 21 nM–52 nM range) (Figure 6), but not CB_1 antagonist SR141716, CB_2 antagonist SR144528, or an inactive indole ST-47 [94]. ST-11, ST-25, and ST-48 were agonists to inhibit basal- and lysophosphatidic acid-mediated chemokinesis, with ST-48 having the greatest potency (EC_{50} = 5 nM). ST-23, ST-25, and ST-48 at high concentrations (3 μM) promoted internalization of HA-mCB$_1$ (but not HA-mCB$_2$) receptors expressed in HEK293 cells. ST-11 and ST-29 at high concentrations (3 μM) competed for [^{3}H]colchicine binding to tubulin preparations [94]. Thus, there is evidence for functional selectivity within this series of AI ligands, with AI receptors regulating cellular signaling at nM concentrations while avoiding off-target properties such as CB_1-binding and tubulin disruption that occur at high concentrations that might not be achievable in vivo.

The subject of non-CB_1, non-CB_2 targets has been comprehensively reviewed recently [92,96]. The Stella review introduces the novel AI receptors for biologically active indole compounds and describes their signal transduction via a Gs-mediated increase in cAMP production [92]. The Reggio review discusses opportunities for overlap in agonist activity among phylogenetically closely related GPCRs, as well as the potential for modifications in pharmacodynamic outcomes based upon heterodimerization or clustering interactions with other GPCRs [96]. The data reviewed herein argue for alternative mechanisms for cannabimimetic indoles to act via orphan or documented GPCRs, or non-GPCR mechanisms, by which AAI analogs could influence behaviors beyond their demonstrated agonist activity at the CB_1 receptors.

5. The Ultimate Diversion of Cannabimimetic Indoles: Spice/K2

For thousands of years, people have experimented with and intentionally consumed or administered novel chemical substances, experienced or observed and recorded their pharmacological effects, and speculated on their mechanisms of action. Preparations of chemicals that produced central nervous system effects such as euphoria, intoxication, stimulation, hallucinations, numbness, analgesia, and anesthesia were often adopted in medical, religious, and recreational practices. Records of preparation methods and pharmacological effects date back to the dawn of writing. With the advent of scientific methods and the disciplines of pharmacology and medicinal chemistry in the nineteenth century, medicinal chemistry data have been preserved in textbooks, scientific literature, patents, and a variety of other archival forms and are often freely available for reference on the internet. The scientific literature and online archives abound with research studies involving new synthetic cannabimimetics being synthesized and tested in in vitro and in vivo experiments, including numerous publications and forensic reports emphasizing the adverse consequences and potential for harm in humans that can be observed with exposure to extremely potent and efficacious synthetic cannabimimetic analogs.

For example, Roger Adam's and colleagues reported their testing of synthetic THC analogs in the 1940s [97–99], including a 1-2-dimethylheptyl analog of Δ^{6a-10a}-tetrahydrocannabinol called pyrahexyl (Figure 7), which was several hundred-fold more potent than the pentyl analog. The potent activity observed after administration (oral consumption) of pyrahexyl did not go unconfirmed by the research scientists or unnoticed by the US Army [100], which included this compound in a development program for incapacitating chemical weapons [101]. The aim of this program was to develop compounds endowed with a "couch lock" or cataleptic effect, that is, non-lethal agents that could be used to incapacitate soldiers. For this reason, pyrahexyl, renamed dimethyl heptylpyran (DMHP) and assigned code number EA-2233 as the mixture of its eight stereoisomers, was included in chemical weapons research that proceeded from 1948 to 1975 at the Edgewood Arsenal in Maryland. In a remarkable effort of resolution and asymmetric synthesis, all eight stereoisomers of DMHP were synthesized, given individual codes EA-2233-1 through EA-2233-8, and investigated for bioactivity. EA-2233-2 was the most potent isomer and could induce confusion, sedation, and hallucinogenic effects at a dosage of 0.5–2.8 µg/kg, corresponding to 35–200 µg for a 70 kg adult. In general, an oral dosage of EA-2233 of 1–2 mg was sufficient to make all human subjects incapable of performing coordinated activities, such as those requested for military action, for as long as 2–3 days. Pyrahexyl was relatively safe, with a therapeutic index of 2000 in laboratory animals, but could occasionally induce severe hypotensive crises, hypothermia, and death, and was not eventually weaponized, in part due to the discovery of more efficacious and safer anticholinergic agents from the quinuclidinyl benzilate series, such as 3-quinuclidyl benzylate) [102].

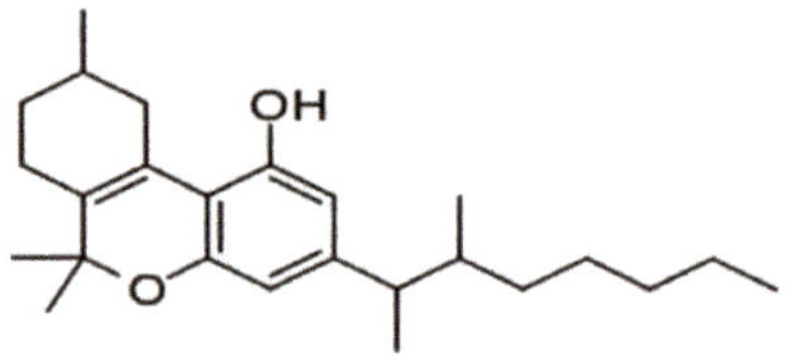

Pyrahexyl; Dimethylheptylpyran (DMHP)

Figure 7. Structure of pyrahexyl (DMHP).

Structure–activity relationships of thousands of opiates and opioids, cannabinoids and synthetic cannabimimetics, dissociative anesthetics, steroids, stimulants, hallucinogens, sedative-hypnotics, and other psychoactive substances of potential abuse and dependence liability, many with synthetic methods and patents published, are readily accessible on-line to the scientific community and the public. Unfortunately, this information is also readily available to clandestine chemists who surreptitiously adopt or extend standard synthetic methods to manufacture and distribute illicit preparations of known psychoactive substances and to develop novel ones to sell on the illicit market as "designer drugs." Based on information available on the internet and in scientific literature published by a wide variety of laboratories and research investigators, potent alkyl indole synthetic cannabimimetic chemicals began to be synthesized in bulk in the early 2000s and were often dissolved in a volatile solvent and sprayed on herbal products that were packaged and made widely available for purchase as "incense" or "spice" and subsequently smoked for their marijuana-like intoxicating properties (Figure 8).

Figure 8. Examples of synthetic cannabimimetic-containing herbal formulations and packaging.

It was during this time that Jenny L. Wiley, a Distinguished Fellow at RTI International with a long history of pharmacological testing of cannabimimetics in laboratory animals, began encountering these illicit herbal products widely available for purchase in convenience stores and gas stations in Virginia and North Carolina. Since they were inappropriately labeled, she and Brian Thomas, the Senior Director of Analytical Chemistry and Pharmaceutics at RTI International, agreed to work together to assist the National Institute on Justice/US Drug Enforcement Agency (DEA) and the National Institute on Drug Abuse (NIDA) in the detection and identification of the synthetic cannabimimetics in these illicit drug products and the characterization of their in vitro cannabinoid receptor affinity and efficacy and in vivo behavioral effects in laboratory animal models of cannabimimetic activity. The results of these investigations, when published in peer-reviewed literature, were intended to facilitate regulation and enforcement, as well as the development of therapeutic treatments for adverse effects, overdose, and substance use disorders.

The spread of bulk synthetic cannabimimetics and synthetic cannabimimetic-containing herbal "spice" blends across international borders occurred rapidly, with products containing JWH-018 accounting for 76% of the 2423 herbal products seized, tested, and reported to the US DEA through the National Forensic Laboratory Information System (NFLIS) in 2010. Even though they were clearly capable of and used to produce profound intoxication, these products were often labeled "not for human consumption" and marketed as "herbal incense" or other misnomers to avoid prosecution by the DEA under the Federal Analogue Act. Unfortunately, the increased availability and use of these potent and efficacious cannabimimetic-containing products led to extreme intoxication, incapacitation, and an increasing number of calls to US Poison Control Centers, which prompted the DEA in March 2011 to use its emergency scheduling authority to temporarily place five of the most commonly encountered synthetic cannabimimetics into the Controlled Substances Act (CSA) as Schedule I; specifically: 1-pentyl-3-(1- naphthoyl)indole (JWH-018), 1-butyl-3-(1-naphthoyl)indole (JWH-073), 1-[2-(4-morpholinyl)ethyl]-3-(1-naphthoyl)indole (JWH-200), 5-(1,1-dimethylheptyl)-2-[(1R,3S)-3-hydroxycyclohexyl]-phenol (CP-47,497), and 5-(1,1-dimethyloctyl)-2-[(1R,3S)-3-hydroxycyclohexyl]-phenol (cannabicyclohexanol; CP-47,497 C8 homolog). This action was deemed necessary by the Administrator of the DEA to avoid an imminent hazard to public safety. As a result, the full effect of the CSA and its implementing regulations, including criminal, civil, and administrative penalties, sanctions, and regulatory controls of Schedule I substances, was brought to bear against the manufacture, distribution, possession, importation, and exportation of these substances and their herbal formulations. The percentage of illicit products containing these five agents seized or otherwise encountered and reported to the DEA decreased from 76% in 2010 to 20% in 2011. However, *a second generation of "legal" synthetic cannabimimetics was already being manufactured and distributed to replace the banned ones, such that during the same timeframe, 2010–2011, the total number of seizures and encounters of illicit products containing positively identified synthetic cannabimimetics increased 10-fold, to over 22,000.* In March of 2012, the DEA used its authority to extend the temporary placement of the five banned agents in Schedule 1 by 6 months. In July of 2012, the FDA Safety and Innovation Act (FDASIA) was passed. It included the Synthetic Drug Abuse Prevention Act that placed several more synthetic cannabimimetic analogs [1-hexyl-3-(1-naphthoyl)indole (JWH-019); 1-pentyl-3-(2-methoxyphenylacetyl)indole (JWH-250); 1-pentyl-3-[1-(4-methoxynaphthoyl)]indole (JWH-081); 1-pentyl-3-(4-methyl-1-naphthoyl)indole (JWH-122); 1-pentyl-3-(4-chloro-1-naphthoyl)indole (JWH-398); 1-(5-fluoropentyl)-3-(1-naphthoyl)indole (AM2201); 1-(5-fluoropentyl)-3-(2-iodobenzoyl)indole (AM694); 1-pentyl-3-[(4-methoxy)-benzoyl]indole (SR-19 and RCS-4); 1-cyclohexylethyl-3-(2-methoxyphenylacetyl)indole (SR-18 and RCS-8); 1-pentyl-3-(2-chlorophenylacetyl)indole (JWH-203), as well as specific synthetic stimulants and hallucinogens, under Schedule 1. It also increased the time that a substance remains in emergency Schedule I status from 1 year to 2, and increased the possible extension period from 6 months to 1 year.

The DEA exercised its emergency scheduling authority again in 2013, 2014, 2015, 2016, 2017, 2019, and in 2021, as it continued to add new cannabimimetic substances under Schedule 1 of the CSA. For example, in 2013, three additional synthetic cannabimimetic analogs [1-pentyl-1H-indol-3-yl)(2,2,3,3-tetramethylcyclopropyl)methanone (UR-144); [1-(5-fluoro-pentyl)-1H-indol-3-yl](2,2,3,3-tetramethylcyclopropyl)methanone (5-fluoro-UR-144, XLR11), and N-(1-adamantyl)-1-pentyl-1H-indazole-3-carboxamide (APINACA, AKB48) were placed under schedule 1 of the CSA. In 2014, the synthetic cannabimimetics quinolin-8-yl 1-pentyl-1H-indole-3-carboxylate (PB-22; QUPIC); quinolin-8-yl 1-(5-fluoropentyl)-1H-indole-3-carboxylate (5-fluoro-PB-22; 5F-PB-22); N-(1-amino-3-methyl-1-oxobutan-2-yl)-1-(4-fluorobenzyl)-1H-indazole-3-carboxamide (AB-FUBINACA); and N-(1-amino-3,3-dimethyl-1-oxobutan-2-yl)-1-pentyl-1H-indazole-3-carboxamide (ADB-PINACA) were added. In 2015, the DEA included the synthetic cannabimimetics N-(1-amino -3-methyl-1-oxobutan-2-yl)-1-(cyclohexylmethyl)-1H-indazole-3-carboxamide (AB-CHMINACA); N-(1-amino-3-methyl-1-oxobutan-2-yl)-1-pentyl-1H-indazole-3-carboxamide (AB-PINACA); [1-(5-fluoropentyl)-1H-indazol-3-yl](naphthalen-1-yl)methanone (THJ-2201), and in 2016 added N-(1-amino-3,3-dimethyl-1-oxobutan-2-yl)-1-(cyclohexylmethyl)-1H-indazole-3-carboxamide (common names MAB-CHMINACA and ADB-CHMINACA), to the rapidly expanding list of Schedule 1 substances. Another DEA scheduling order was published in 2017 for six more synthetic cannabimimetic analogs appearing in illicit products: methyl 2-(1-(5-fluoropentyl)-1H-indazole-3-carboxamido)-3,3-dimethylbutanoate [5F-ADB; 5F-MDMB-PINACA]; methyl 2-(1-(5-fluoropentyl)-1H-indazole-3-carboxamido)-3-methylbutanoate [5F-AMB]; N-(adamantan-1-yl)-1-(5-fluoropentyl)-1H-indazole-3-carboxamide [5F-APINACA, 5F-AKB48]; N-(1-amino-3,3-dimethyl-1-oxobutan-2-yl)-1-(4-fluorobenzyl)-1H-indazole-3-carboxamide [ADB-FUBINACA]; methyl 2-(1-(cyclohexylmethyl)-1H-indole-3-carboxamido)-3,3-dimethylbutanoate [MDMB-CHMICA, MMB-CHMINACA]; and methyl 2-(1-(4-fluorobenzyl)-1H-indazole-3-carboxamido)-3,3-dimethylbutanoate [MDMB-FUBINACA], including their optical, positional, and geometric isomers, salts, and salts of isomers under schedule I. In 2019, ethyl 2-(1-(5-fluoropentyl)-1H-indazole-3-carboxamido)-3,3-dimethylbutanoate (5F-EDMB-PINACA); methyl 2-(1-(5-fluoropentyl)-1H-indole-3-carboxamido)-3,3-dimethylbutanoate (5F-MDMB-PICA); N-(adamantan-1-yl)-1-(4-fluorobenzyl)-1H-indazole-3-carboxamide (common names include FUB-AKB48; FUB-APINACA; AKB48 N-(4-fluorobenzyl)); 1-(5-fluoropentyl)-N-(2-phenylpropan-2-yl)-1H-indazole-3-carboxamide (common names of 5F-CUMYL-PINACA; SGT-25); and (1-(4-fluorobenzyl)-1H-indol-3-yl)(2,2,3,3-tetramethylcyclopropyl) methanone (FUB-144), and their optical, positional, and geometric isomers, salts, and salts of isomers were placed under schedule I; with the addition of these analogs made permanent in March of 2021. Effective as of June, 2021, the DEA has also included naphthalen-1-yl 1-(5-fluoropentyl)-1H-indole-3-carboxylate (NM2201 or CBL2201); N-(1-amino-3-methyl-1-oxobutan-2-yl)-1-(5-fluoropentyl)-1H-indazole-3-carboxamide (5F-AB-PINACA); 1-(4-cyanobutyl)-N-(2-phenylpropan-2-yl)-1H-indazole-3-carboxamide (other names: 4-CN-CUMYL-BUTINACA, 4-cyano-CUMYL-BUTINACA; 4-CN-CUMYL BI-NACA, CUMYL-4CN-BINACA, or SGT-78); methyl 2-(1-(cyclohexylmethyl)-1H-indole-3-carboxamido)-3-methylbutanoate (MMB-CHMICA or AMB-CHMICA); and 1-(5-fluoropentyl)-

N-(2-phenylpropan-2-yl)-1H-pyrrolo[2,3-b]pyridine-3-carboxamide (5F-CUMYL-P7AICA) under Schedule 1 on a permanent basis.

Presently, well over 40 novel synthetic cannabimimetic chemicals have been defined as Schedule 1 controlled substances by the DEA to discourage their further manufacture, distribution, and use (e.g., see Figure 9). However, the illicit drug market persists as new compounds are immediately created to evade detection, regulation, and law enforcement. This iterative cycle of synthesis, use, detection, identification, and banning of chemical substances has had the undesired effect of increasing the chemical diversity of illicit analogs being distributed in these products, thereby exposing users to a wider variety of compounds of unknown pharmacological activity and potential long-term negative consequence, while having a limited positive effect on the aggregate distribution and use.

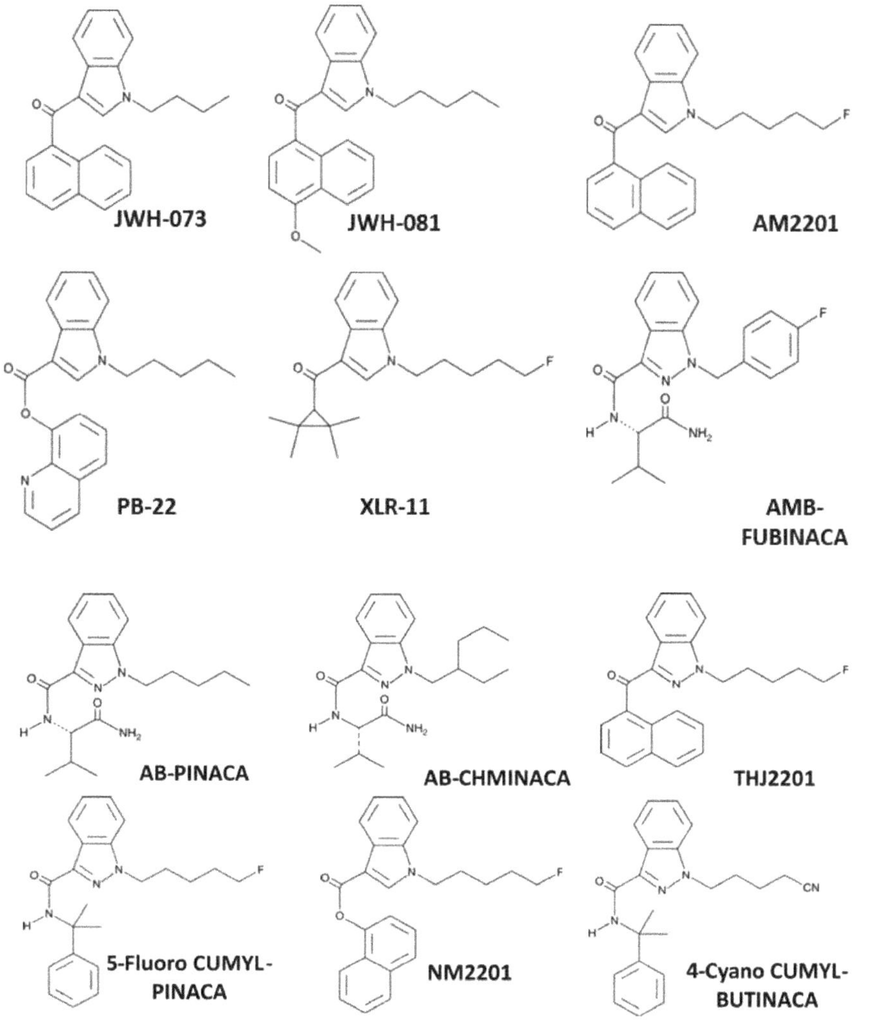

Figure 9. Chemical structures/IUPAC names of selected Schedule I synthetic cannabimimetics.

Over the last few decades, we have witnessed a growing commodification of psychoactive substances, including a diverse range of new chemical entities not controlled under drug laws. During a time of increasing legalization and use of medicinal and recreational cannabis and cannabinoid concentrates, a concurrent drug phenomenon has become largely defined by both the growing number of novel synthetic chemicals being detected from increasingly broad chemical and pharmacological families and the open sale of many of these substances as 'legal highs', 'bath salts', or 'research chemicals' in commercial venues and online shops, as well as by individual street-level drug dealers [103,104]. Over 400 new psychoactive substances were detected in Europe's drug market in 2019, with extremely potent synthetic cannabimimetics, cathinones, arylcyclohexylamines, and opioids being the most prevalent classes of compounds posing significant health and social impact concerns. Reports of cannabis adulterated with new synthetic cannabimimetics, such as MDMB-4en-PINACA, being sold to unsuspecting recreational or medicinal cannabis users highlight the new and potentially growing risks of the inadvertent consumption of these illicit and relatively unknown substances [105]. Thus, the vernacular of designer drugs and new drug substances has been refined and replaced over time with 'new psychoactive substance' (NPS), increasingly being used in the rapidly evolving regulatory framework encompassing the legally contentious concept of use and misuse of psychoactive substances in our society.

The current scheduling of new psychoactive substances in the US includes the specific mention of a variety of compounds as *Schedule I cannabimimetic agents*, "unless specifically exempted or unless listed in another schedule", including "any material, compound, mixture, or preparation which contains any quantity of cannabimimetic agents, or which contains their salts, isomers, and salts of isomers is possible within the specific chemical designation" (Synthetic Drug Abuse Prevention Act of 2012). This act also defines cannabimimetic agents more broadly in terms of elements of their chemical scaffold and their substituents that have been demonstrated to be important for cannabimimetic activity (i.e., pharmacophores)—"The term cannabimimetic agents means any substance that is a cannabinoid receptor type 1 (CB_1 receptor) agonist as demonstrated by binding studies and functional assays within any of the following structural classes:

- 2-(3-hydroxycyclohexyl)phenol with substitution at the 5-position of the phenolic ring by alkyl or alkenyl, whether or not substituted on the cyclohexyl ring to any extent.
- 3-(1-naphthoyl)indole or 3-(1-naphthylmethane)indole by substitution at the nitrogen atom of the indole ring, whether or not further substituted on the indole ring to any extent, whether or not substituted on the naphthoyl or naphthyl ring to any extent.
- 3-(1-naphthoyl)pyrrole by substitution at the nitrogen atom of the pyrrole ring, whether or not further substituted in the pyrrole ring to any extent, whether or not substituted on the naphthoyl ring to any extent.
- 1-(1-naphthylmethylene)indene by substitution of the 3-position of the indene ring, whether or not further substituted in the indene ring to any extent, whether or not substituted on the naphthyl ring to any extent.
- 3-phenylacetylindole or 3-benzoylindole by substitution at the nitrogen atom of the indole ring, whether or not further substituted in the indole ring to any extent, whether or not substituted on the phenyl ring to any extent."

Unfortunately, broad definitions of core structural components may include compounds that have structural similarity to cannabimimetic agents but do not produce cannabimimetic effects. In addition, the inclusion of cannabinoid receptor binding studies and functional assay data as criteria for declaration of a cannabimimetic agent is problematic because these experiments can be complex, must be performed properly by a qualified laboratory with appropriate controls, and the results and conclusions carefully reviewed and confirmed prior to use in a court of law. Finally, the identity of the chemical constituents in the products are often identified, characterized, and banned, but these chemicals may differ dramatically from the chemical exposures that are produced during the use of these compounds, either due to degradation, thermolysis, or rapid metabolic conversion.

When synthetic cannabimimetics are encountered in bulk, the "pure" compounds are commonly in the form of fine crystalline powders but may also be amorphous solids, with colors ranging across white, grey, brown, and yellow hues. The quality of these synthetic chemicals often fails to meet pharmaceutical standards for purity or identification and labeling of all active ingredients, excipients, or impurities exceeding an acceptable standard percentage or estimated daily dose exposure [106–109]. In addition, most of the chemical ingredients are improperly identified on customs declarations, using a variety of inaccurate chemical descriptors or inappropriate descriptions (e.g., herbal incense). The purity of these synthetic preparations varies widely and appears to be poorly controlled. In some instances, seized bulk synthetic cannabimimetic chemicals have been found to be contaminated with a variety of synthetic by-products and intermediates originating from the synthetic procedures employed, and a variety of structural analogs have been shown to degrade at commonly encountered room temperature exposures [110]. The proper handling, storage, separation, and detection of these novel chemicals in complex matrix and elucidation of the exact chemical structure often requires the use of several sophisticated analytical instruments and laboratory techniques and the interpretation of complex datasets that together can provide sufficient integrated molecular information to confirm identity. Moreover, the analytical methods used for legal or forensic purposes must also be validated and shown to provide suitably accurate, specific, and reliable information, which adds to the cost and complexity involved in either targeted or broad-spectrum methods [111,112]. Finally, in vitro and in vivo laboratory studies are increasingly used to provide evidence that novel chemicals that are being encountered on the illicit market are cannabimimetics that bind to and activate CB_1 cannabinoid receptors [113–115].

The evolution of synthetic cannabimimetics has involved modification of both chemical scaffolds and substituents that extend beyond literature precedent or established cannabinoid receptor binding affinity/efficacy studies [116,117] and have tended to produce novel chemicals whose volatility and thermal stability are compromised as compared to JWH-018 [118,119]. Thermolysis and the formation of degradation products of synthetic cannabimimetic chemicals is a function of their chemical structure and high temperature exposure, such as during vaporization or combustion processes employed for inhalation. For example, halogenation of synthetic cannabimimetic analogs has been widely used to evade detection and circumvent law enforcement actions; however, this modification leads to increased thermal lability, specifically, thermolytically induced dehalogenation and desaturation of the alkyl side chain [120,121]. In other instances, synthetic analogs such as UR-144 and XLR-11 containing a sterically strained ring system in lieu of the alkyl sidechain have been shown to rapidly decompose to ring-opened and/or dehalogenated species [110,122–124]. Carboxamide synthetic analogs, including the PICA and FUBINACA analogs associated with fatalities and so-called "zombie outbreaks", have also been shown to undergo rapid thermolytic degradation under elevated temperature exposures that may be relevant to combustion or vaporization and inhalation routes of administration [125]. Even relatively modest changes in chemical structure can have a profound influence on volatility and thermal stability and pharmacokinetics and pharmacodynamics, leading to dramatic differences in chemical exposure due to thermal degradation and transfer of chemicals into the gas vapor phase during heating or combustion and inhalation, or differences in their adsorption, distribution, metabolism, elimination and pharmacodynamic impact over time [126,127].

The variability in the thermal stability of synthetic cannabimimetic analogs appears to span the entire range, from compounds that volatilize intact when heated, with little to no thermal degradation, to compounds that degrade slowly at room temperature and entirely decompose during heating before they can produce a vapor containing the parent compound for inhalation. Thus, when smoking vessels such as pipes or other devices which have been used to combust or vaporize and inhale synthetic cannabimimetic-containing herbal blends are examined for residual chemicals, parent compound(s) may be absent and replaced by degradants and thermolysis products. For example, individuals have been reported to primarily excrete metabolites of the thermal degradants of synthetic cannabimimetics formed during combustion and inhalation of herbal formulations, as opposed to excreting metabolites of the intact drug substance detected on the plant material (i.e., the primary exposure during combustion and inhalation is to the thermal degradant). In this case, the detection of mono-hydroxylated metabolites of UR-144 (LC-MS-MS) and mono-hydroxylated/with hydration metabolites of the UR-144 pyrolysis product (GC-MS) was found to be the most useful method of establishing UR-144 ingestion [128]. Unfortunately, the thermal degradants that are formed during the heating of synthetic cannabimimetics often include compounds with known toxicity. For example, when incrementally heated at 200, 400, 600, and 800 °C, the alkyl indole NNEI decomposes to form a variety of compounds, including naphthylamine (a carcinogen) and pentylindole, whereas the structurally analogous indazole MN-18 appears to volatilize with significantly less thermolysis. However, many of the carboxamide-containing synthetic cannabimimetics also appear to be susceptible to decomposition and liberation of hydrogen cyanide when heated rapidly to 800 °C, which was confirmed and quantified via LC-MS/MS [127]. These results suggest that the liberation of toxic degradants, including hydrogen cyanide released during the heating and inhalation of synthetic indazole carboxamide-type compounds, could have significant health impacts on human users of synthetic cannabimimetic containing herbal formulations [127].

There has been a continued increase in the diversity of both the synthetic cannabimimetic chemicals being manufactured and used and the variety of formulations being encountered in the illicit market, such as in vape pens and tinctures and edible products, in addition to herbal blends and bulk drug substances [105]. These illicit products continue to have no oversight ensuring the accuracy or validity of their label claims and provide little or no guidance on proper storage, indications for use or dose titration, or information on commonly encountered adverse side effects. Thus, the effects that are produced in consumers can vary considerably, and can occasionally be debilitating and lethal, produce dependence and withdrawal, and range dramatically in intensity and duration depending upon dose and route of administration (for example, see [129–133]). Nevertheless, these products remain of considerable interest to individuals who pursue intoxication while enabling their chemical use to remain undetected and clear of legal regulations and criminal consequences (e.g., individuals subjected to periodic urinalysis for employment or military/civil service [134,135]). The evolving supply chain of new chemical scaffolds in designer drugs challenges forensic laboratories and public health resources that rely upon rapid analysis of bulk drug substances, dosage formulations, and drugs and their metabolites in biological fluids to derive an appropriate legal response or treatment strategy. In response, drug-testing laboratories use increasingly sophisticated chemical analysis methods and bioanalytical technologies, which also challenge the clinicians, analytical chemists, and authorities who must properly interpret the complex analytical results and implement appropriate medical or regulatory responses. Even though a chemical prototype may have a long history of use and considerable literature, each new chemical entity is essentially a pharmacological unknown with the inherent potential to produce unanticipated effects in users or their descendants [117]. For example, G-protein promiscuity and signaling bias has been shown to be an important pharmacological property that may differentiate between phytocannabinoids and synthetic cannabimimetics and their relative ability to produce tolerance and dependence and other pharmacological effects [136,137]. Synthetic alkyl indole compounds are able to activate CB_1 and CB_2 cannabinoid receptors, and the selectivity can make a difference in outcomes of G protein signaling (see Figure 10).

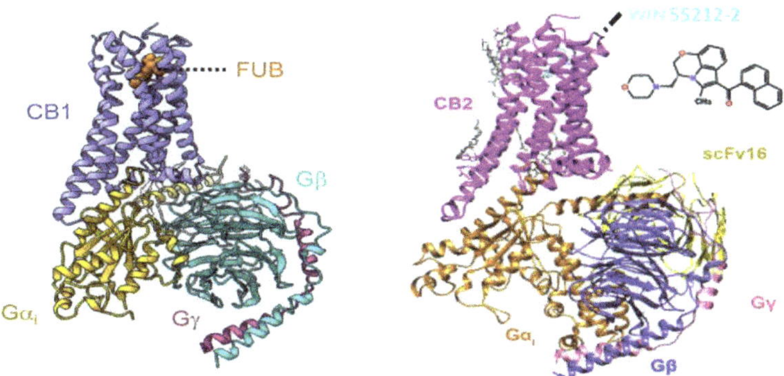

Figure 10. Cryo-electronmicrograph depicting structures of CB1 cannabinoid receptor stimulated by FUBINACA (FUB) and engaging Giα1β1γ2 (**left**). Image reprinted with permission from Cell 176, Krishna Kumar, K. et al., *Structure of a Signaling Cannabinoid Receptor 1 G Protein Complex.*, copyright 2019, with permission from Elsevier [138]; and CB2 cannabinoid receptor stimulated by WIN55212-2 and engaging Giα1β1γ2 (**right**). Image reprinted with permission from Cell 180, Xing, C. et al., *Cryo-EM Structure of the Human Cannabinoid Receptor CB2-G$_i$ Signaling Complex.*, copyright 2020, with permission from Elsevier [139].

Because of the technical difficulty in the detection and characterization of new designer drugs of abuse, estimates of their extent of use and effects produced must be derived using survey data, Poison Control Center data, and many other resources to produce accurate estimates. Thus, there is a significant need for a comprehensive discussion on synthetic cannabimimetic designer drugs that recognizes their potential threat to society, presents the ongoing challenges confronting the various approaches to detection and identification, and informs the development of improved solutions for use in legislation, law enforcement, harm reduction, and clinical treatment.

6. Conclusions: Scientists and Entrepreneurs: Who Takes Social Responsibility?

Researchers in the field of cannabinoid biochemistry, physiology, and pharmacology are conscious of research ethics in developing hypotheses and conducting investigations. Now, we are entering into an era in which consumers are expected to make judgments without having the advantage of education in chemical and biological sciences or training in the scientific method of applying research observations to developing and testing hypotheses, analyzing data, and drawing conclusions. Researchers come to conclusions that are directed at understanding mechanisms and discerning fundamental differences between physiology versus pathophysiology of the disease. Consumers are expected to make conclusions about whether plant products and their derivatives are useful for their health and safe for their use. Entrepreneurs and business developers make conclusions based upon their goal to commercialize cannabinoid plant products and compounds derived therefrom.

The diverse goals for the application of current knowledge are dependent upon stakeholders who are motivated to provide funding support. For scientific researchers, funding support is obtained competitively from governmental sources derived from public tax dollars or foundations directed at curing diseases with public donations. Researchers are accountable to demonstrate that the work will be peer-reviewed and made publically available so that other scientists can build upon the work. Indeed, it is generally expected that the researcher has a history of publishing work before grant proposals are funded.

If funding comes from private sources (i.e., the pharmaceutical industry), researchers are generally expected to keep their work confidential to protect intellectual property. Nearly a dozen pharmaceutical companies contributed to developing and characterizing

AAI compounds described in this review, each espousing the goal to provide consumers with medicines that are effective and safe. Impediments to ultimately marketing a medicine may occur at any of the steps in the drug development process. This is a risk that a legitimate pharmaceutical company is willing to take if it wishes to maintain its reputation for providing safe and effective medicines. Many projects are terminated based upon poor preclinical responses in models, failed clinical outcomes, untoward side effects, or adverse events. It is interesting to note that the preclinical research reviewed herein has not resulted in a marketed medicine.

When the properties of JWH-018 were published, researchers did not anticipate the abuse and misuse of the compound. In hindsight, one can imagine entrepreneurs discussing whether or not the compound could be used to become high. Would it circumvent drug laws that existed at the time? How might it be distributed? There are potential marketing strategies that a less-than-ethical commercial enterprise might consider in its effort to gain profit from the application of available research methods. Entrepreneurs who marketed unscheduled AAI and other cannabimimetic compounds under the guise of "legal marijuana" bear much of the responsibility for the misuse of the compounds. They sold products without testing for safety. They sold an impure product. Even when packaging included a label to the contrary, one can surmise the intention was for users to smoke or ingest the product. While maintaining the "letter of the law" in assuring that compounds they sold were not listed by the US DEA as schedule 1, they bypassed the intent of the law. By indicating that compounds are not illegal, they led consumers to believe that the safety of these products had been tested and that these compounds could be used without harm.

What can scientists do to promote research and avoid public mis- or disinformation? As one government funding goal is to train the next generation of researchers to keep the nation's healthcare capabilities strong, part of the job of scientists is to educate students. However, another part is to educate consumers whose interests are limited to whether a plant product or synthetic compound can treat their maladies and if they can expect "side effects". Scientists need to use accurate wording in scientific communications and avoid terms that are imprecise and lead to generalizations and misunderstandings. An example is "synthetic cannabinoid", which incorrectly includes indole compounds that are neither cannabinoid in structure nor analogs of natural phytocannabinoids. Scientists should communicate to the public at their level of understanding and interest and still take the opportunity to teach consumers about cell biology, physiology, or pathology as it applies to the mechanism of action of medicines. Researchers also need to communicate about the importance of research and the scientific method.

There are no good drugs or bad drugs; rather, there are good uses and bad uses for compounds whether found in nature or synthesized by design. The story of the AAI and analogs described in this review aptly demonstrates this pharmacological principle.

Author Contributions: Conceptualization, original draft preparation, review, and editing were performed by A.C.H., B.F.T. and J.W.H. All authors have read and agreed to the published version of the manuscript.

Funding: This research was funded by the Department of Health and Human Services, National Institutes of Health grant numbers R01-DA042157, R01-DA003672, R01-DA040460, R03-DA031988, National Institute of Justice grant number 2012-R2-CX-K001, and Drug Enforcement Administration contract DJD-14-HQ-P-0713.

Acknowledgments: The authors would like to thank Andrew England for preparing the images of the compounds.

Conflicts of Interest: The authors declare no conflict of interest. The funders had no role in the design of this review article; in the collection, analyses, or interpretation of data discussed herein; the writing of the manuscript, or in the decision to publish the results.

References

1. Howlett, A.C. Inhibition of neuroblastoma adenylate cyclase by cannabinoid and nantradol compounds. *Life Sci.* **1984**, *35*, 1803–1810. [CrossRef]
2. Howlett, A.; Fleming, R.M. Cannabinoid inhibition of adenylate cyclase. Pharmacology of the response in neuroblastoma cell membranes. *Mol. Pharmacol.* **1984**, *26*, 532–538.
3. Milne, G.M., Jr.; Johnson, M.R. Levonantradol: A role for central prostanoid mechanisms? *J. Clin. Pharmacol.* **1981**, *21*, 367s–374s. [CrossRef]
4. Johnson, M.R.; Melvin, L.S.; Althuis, T.H.; Bindra, J.S.; Harbert, C.A.; Milne, G.M.; Weissman, A. Selective and Potent Analgetics Derived from Cannabinoids. *J. Clin. Pharmacol.* **1981**, *21*, 271S–282S. [CrossRef]
5. Jain, A.K.; Ryan, J.R.; McMahon, F.G.; Smith, G. Evaluation of Intramuscular Levonantradol and Placebo in Acute Postoperative Pain. *J. Clin. Pharmacol.* **1981**, *21*, 320S–326S. [CrossRef]
6. Cronin, C.M.; Sallan, S.E.; Gelber, R.; Lucas, V.S.; Laszlo, J. Antiemetic Effect of Intramuscular Levonantradol in Patients Receiving Anticancer Chemotherapy. *J. Clin. Pharmacol.* **1981**, *21*, 43S–50S. [CrossRef]
7. Laszlo, J.; Lucas, V.S.; Hanson, D.C.; Cronin, C.M.; Sallan, S.E. Levonantradol for Chemotherapy-Induced Emesis: Phase I-II Oral Administration. *J. Clin. Pharmacol.* **1981**, *21*, 51S–56S. [CrossRef]
8. Sheidler, V.R.; Ettinger, D.S.; Diasio, R.B.; Enterline, J.P.; Brown, M.D. Double-Blind Multiple-Dose Crossover Study of the Antiemetic Effect of Intramuscular Levonantradol Compared to Prochlorperazine. *J. Clin. Pharmacol.* **1984**, *24*, 155–159. [CrossRef]
9. Stambaugh, J.E.; McAdams, J.; Vreeland, F. Dose Ranging Evaluation of the Antiemetic Efficacy and Toxicity of Intramuscular Levonantradol in Cancer Subjects with Chemotherapy-Induced Emesis. *J. Clin. Pharmacol.* **1984**, *24*, 480–485. [CrossRef]
10. Special Issue: Overview on the Current Status of Therapeutic Opportunities in Cannabinoid Research. *J. Clin. Pharmacol.* **1981**, *21*, 1S–494S.
11. Howlett, A. Cannabinoid inhibition of adenylate cyclase. Biochemistry of the response in neuroblastoma cell membranes. *Mol. Pharmacol.* **1985**, *27*, 429–436.
12. Howlett, A.; Johnson, M.R.; Melvin, L.S.; Milne, G.M. Nonclassical cannabinoid analgetics inhibit adenylate cyclase: Development of a cannabinoid receptor model. *Mol. Pharmacol.* **1988**, *33*, 297–302.
13. Howlett, A.C.; Qualy, J.M.; Khachatrian, L.L. Involvement of Gi in the inhibition of adenylate cyclase by cannabimimetic drugs. *Mol. Pharmacol.* **1986**, *29*, 307–313.
14. Devane, W.A.; Dysarz, F.A.; Johnson, M.R.; Melvin, L.S.; Howlett, A. Determination and characterization of a cannabinoid receptor in rat brain. *Mol. Pharmacol.* **1988**, *34*, 605–613.
15. Howlett, A.; Bidaut-Russell, M.; Devane, W.A.; Melvin, L.S.; Johnson, M.; Herkenham, M. The cannabinoid receptor: Biochemical, anatomical and behavioral characterization. *Trends Neurosci.* **1990**, *13*, 420–423. [CrossRef]
16. Howlett, A.C.; Johnson, M.R.; Melvin, L.S. Classical and Nonclassical Cannabinoids: Mechanism of Action-Brain Binding. *NIDA Res. Monogr.* **1990**, *96*, 100–111.
17. Haubrich, D.R.; Ward, S.J.; Baizman, E.; Bell, M.R.; Bradford, J.; Ferrari, R.; Miller, M.; Perrone, M.; Pierson, A.K.; Saelens, J.K. Pharmacology of pravadoline: A new analgesic agent. *J. Pharmacol. Exp. Ther.* **1990**, *255*, 511–522.
18. Bell, M.R.; D'Ambra, T.E.; Kumar, V.; Eissenstat, M.A.; Herrmann, J.L.; Wetzel, J.R.; Rosi, D.; Philion, R.E.; Daum, S.J. Antinociceptive (aminoalkyl)indoles. *J. Med. Chem.* **1991**, *34*, 1099–1110. [CrossRef]
19. Ward, S.J.; Mastriani, D.; Casiano, F.; Arnold, R. Pravadoline: Profile in isolated tissue preparations. *J. Pharmacol. Exp. Ther.* **1990**, *255*, 1230–1239.
20. Pacheco, M.; Childers, S.R.; Arnold, R.; Casiano, F.; Ward, S.J. Aminoalkylindoles: Actions on specific G-protein-linked receptors. *J. Pharmacol. Exp. Ther.* **1991**, *257*, 170–183.
21. D'Ambra, T.E.; Estep, K.G.; Bell, M.R.; Eissenstat, M.A.; Josef, K.A.; Ward, S.J.; Haycock, D.A.; Baizman, E.R.; Casiano, F.M.; Beglin, N.C.; et al. Conformationally restrained analogues of pravadoline: Nanomolar potent, enantioselective, (aminoalkyl)indole agonists of the cannabinoid receptor. *J. Med. Chem.* **1992**, *35*, 124–135. [CrossRef] [PubMed]
22. Compton, D.R.; Gold, L.H.; Ward, S.J.; Balster, R.L.; Martin, B.R. Aminoalkylindole analogs: Cannabimimetic activity of a class of compounds structurally distinct from delta 9-tetrahydrocannabinol. *J. Pharmacol. Exp. Ther.* **1992**, *263*, 1118–1126. [PubMed]
23. Ward, S.J.; Baizman, E.; Bell, M.; Childers, S.; D'Ambra, T.; Eissenstat, M.; Estep, K.; Haycock, D.; Howlett, A.; Luttinger, D.; et al. Aminoalkylindoles (AAIs): A new route to the cannabinoid receptor? *NIDA Res. Monogr.* **1990**, *105*, 425–426. [PubMed]
24. Kuster, J.E.; Stevenson, J.I.; Ward, S.J.; D'Ambra, T.E.; Haycock, D.A. Aminoalkylindole binding in rat cerebellum: Selective displacement by natural and synthetic cannabinoids. *J. Pharmacol. Exp. Ther.* **1993**, *264*, 1352–1363.
25. Yamada, K.; Rice, K.C.; Flippen-Anderson, J.L.; Eissenstat, M.A.; Ward, S.J.; Johnson, M.R.; Howlett, A.C. (Aminoalkyl)indole Isothiocyanates as Potential Electrophilic Affinity Ligands for the Brain Cannabinoid Receptor. *J. Med. Chem.* **1996**, *39*, 1967–1974. [CrossRef] [PubMed]
26. Jansen, E.M.; Haycock, D.A.; Ward, S.J.; Seybold, V.S. Distribution of cannabinoid receptors in rat brain determined with aminoalkylindoles. *Brain Res.* **1992**, *575*, 93–102. [CrossRef]
27. Zhang, Q.; Ma, P.; Iszard, M.; Cole, R.B.; Wang, W.; Wang, G. In Vitro Metabolism ofR(+)-[2,3-Dihydro-5-methyl-3-[(morpholinyl)methyl]pyrrolo [1,2,3-de]1,4-benzoxazinyl]-(1-naphthalenyl) methanone mesylate, a Cannabinoid Receptor Agonist. *Drug Metab. Dispos.* **2002**, *30*, 1077–1086. [CrossRef]
28. Collins, J.C.; Gwilt, J.R. The Life Cycle of Sterling Drug, Inc. *Bull. Hist. Chem.* **2000**, *25*, 22–27.

29. Huffman, J.W. Cannabimimetic indoles, pyrroles and indenes. *Curr. Med. Chem.* **1999**, *6*, 705–720.
30. Huffman, J.W.; Padgett, L.W. Recent developments in the medicinal chemistry of cannabimimetic indoles, pyrroles and indenes. *Curr. Med. Chem.* **2005**, *12*, 1395–1411. [CrossRef]
31. Shim, J.Y.; Collantes, E.R.; Welsh, W.J.; Howlett, A.C. Unified pharmacophoric model for cannabinoids and aminoalkylindoles derived from molecular superimpositioin of CB1 cannabinoid receptor agonists CP55244 and WIN55212-2. In *Rational Drug Design: Novel Methodology and Practical Applications*; Parrill, A.L., Ed.; American Chemical Society: Washington, DC, USA, 1999; pp. 165–184.
32. Shim, J.-Y.; Collantes, E.R.; Welsh, W.J.; Subramaniam, B.; Howlett, A.C.; Eissenstat, M.A.; Ward, S.J. Three-Dimensional Quantitative Structure–Activity Relationship Study of the Cannabimimetic (Aminoalkyl)indoles Using Comparative Molecular Field Analysis. *J. Med. Chem.* **1998**, *41*, 4521–4532. [CrossRef]
33. Eissenstat, M.A.; Bell, M.R.; D'Ambra, T.E.; Alexander, E.J.; Daum, S.J.; Ackerman, J.H.; Gruett, M.D.; Kumar, V.; Estep, K.G. Aminoalkylindoles: Structure-Activity Relationships of Novel Cannabinoid Mimetics. *J. Med. Chem.* **1995**, *38*, 3094–3105. [CrossRef]
34. D'Ambra, T.E.; Eissenstat, M.A.; Abt, J.; Ackerman, J.H.; Bacon, E.R.; Bell, M.R.; Carabateas, P.M.; Josef, K.A.; Kumar, V.; Weaver, J.D.; et al. C-Attached aminoalkylindoles: Potent cannabinoid mimetics. *Bioorg. Med. Chem. Lett.* **1996**, *6*, 17–22. [CrossRef]
35. Shim, J.-Y.; Welsh, W.J.; Cartier, E.; Edwards, J.L.; Howlett, A. Molecular Interaction of the Antagonist N-(Piperidin-1-yl)-5-(4-chlorophenyl)-1-(2,4-dichlorophenyl)-4-methyl-1H-pyrazole-3-carboxamide with the CB1 Cannabinoid Receptor. *J. Med. Chem.* **2002**, *45*, 1447–1459. [CrossRef]
36. Xiet, X.-Q.; Eissenstat, M.; Makriyannis, A. Common cannabimimetic pharmacophoric requirements between aminoalkyl indoles and classical cannabinoids. *Life Sci.* **1995**, *56*, 1963–1970. [CrossRef]
37. Xie, X.-Q.; Han, X.-W.; Chen, J.-Z.; Eissenstat, M.; Makriyannis, A. High-resolution NMR and computer modeling studies of the cannabimimetic aminoalkylindole prototype WIN-55212-2. *J. Med. Chem.* **1999**, *42*, 4021–4027. [CrossRef]
38. Wiley, J.; Compton, D.R.; Dai, D.; Lainton, J.A.; Phillips, M.; Huffman, J.W.; Martin, B.R. Structure-activity relationships of indole- and pyrrole-derived cannabinoids. *J. Pharmacol. Exp. Ther.* **1998**, *285*, 995–1004.
39. Huffman, J.W.; Dai, D.; Martin, B.R.; Compton, D.R. Design, Synthesis and Pharmacology of Cannabimimetic Indoles. *Bioorg. Med. Chem. Lett.* **1994**, *4*, 563–566. [CrossRef]
40. Huffman, J.W.; Lu, J.; Dai, D.; Kitaygorodskiy, A.; Wiley, J.L.; Martin, B.R. Synthesis and pharmacology of a hybrid cannabinoid. *Bioorg. Med. Chem.* **2000**, *8*, 439–447. [CrossRef]
41. Reggio, P.H.; Basu-Dutt, S.; Barnett-Norris, J.; Castro, M.T.; Hurst, W.P.; Seltzman, H.H.; Roche, M.J.; Gilliam, A.F.; Thomas, B.F.; Stevenson, L.A.; et al. The Bioactive Conformation of Aminoalkylindoles at the Cannabinoid CB1 and CB2 Receptors: Insights Gained from (E)- and (Z)-Naphthylidene Indenes. *J. Med. Chem.* **1998**, *41*, 5177–5187. [CrossRef]
42. Huffman, J.W.; Mabon, R.; Wu, M.-J.; Lu, J.; Hart, R.; Hurst, D.P.; Reggio, P.H.; Wiley, J.L.; Martin, B.R. 3-Indolyl-1-naphthylmethanes: New cannabimimetic indoles provide evidence for aromatic stacking interactions with the CB1 cannabinoid receptor. *Bioorg. Med. Chem.* **2003**, *11*, 539–549. [CrossRef]
43. Song, Z.H.; Bonner, T.I. A lysine residue of the cannabinoid receptor is critical for receptor recognition by several agonists but not WIN55212-2. *Mol. Pharmacol.* **1996**, *49*, 891–896. [PubMed]
44. Tao, Q.; Abood, M. Mutation of a highly conserved aspartate residue in the second transmembrane domain of the cannabinoid receptors, CB1 and CB2, disrupts G-protein coupling. *J. Pharmacol. Exp. Ther.* **1998**, *285*, 651–658. [PubMed]
45. McAllister, S.D.; Tao, Q.; Barnett-Norris, J.; Buehner, K.; Hurst, D.P.; Guarnieri, F.; Reggio, P.H.; Harmon, K.W.N.; Cabral, G.A.; Abood, M.E. A critical role for a tyrosine residue in the cannabinoid receptors for ligand recognition. *Biochem. Pharmacol.* **2002**, *63*, 2121–2136. [CrossRef]
46. Song, Z.H.; Slowey, C.A.; Hurst, D.P.; Reggio, P.H. The difference between the CB(1) and CB(2) cannabinoid receptors at position 5.46 is crucial for the selectivity of WIN55212-2 for CB(2). *Mol. Pharmacol.* **1999**, *56*, 834–840.
47. Shim, J.-Y.; Howlett, A.C. WIN55212-2 Docking to the CB1 Cannabinoid Receptor and Multiple Pathways for Conformational Induction. *J. Chem. Inf. Model.* **2006**, *46*, 1286–1300. [CrossRef] [PubMed]
48. Shim, J.-Y.; Howlett, A.C. Steric Trigger as a Mechanism for CB1 Cannabinoid Receptor Activation. *J. Chem. Inf. Comput. Sci.* **2004**, *44*, 1466–1476. [CrossRef]
49. Shim, J.-Y.; Welsh, W.J.; Howlett, A.C. Homology model of the CB1 cannabinoid receptor: Sites critical for nonclassical cannabinoid agonist interaction. *Biopolymers* **2003**, *71*, 169–189. [CrossRef]
50. Guindon, J.; Hohmann, A.G. Cannabinoid CB2 receptors: A therapeutic target for the treatment of inflammatory and neuropathic pain. *Br. J. Pharmacol.* **2008**, *153*, 319–334. [CrossRef]
51. Beltramo, M. Cannabinoid Type 2 Receptor as a Target for Chronic-Pain. *Mini-Rev. Med. Chem.* **2009**, *9*, 11–25. [CrossRef]
52. Atwood, B.K.; Straiker, A.; Mackie, K. CB₂: Therapeutic target-in-waiting. *Prog. Neuropsychopharmacol. Biol. Psychiatry* **2012**, *38*, 16–20. [CrossRef] [PubMed]
53. Huffman, J.W. CB2 receptor ligands. *Mini-Rev. Med. Chem.* **2005**, *5*, 641–649. [CrossRef] [PubMed]
54. Huffman, J.W. The Search for Selective Ligands for the CB2 Receptor. *Curr. Pharm. Des.* **2000**, *6*, 1323–1337. [CrossRef] [PubMed]
55. Huffman, J.W.; Marriott, K.-S.C. Recent Advances in the Development of Selective Ligands for the Cannabinoid CB2 Receptor. *Curr. Top. Med. Chem.* **2008**, *8*, 187–204. [CrossRef] [PubMed]

56. Felder, C.C.; Joyce, K.E.; Briley, E.M.; Mansouri, J.; Mackie, K.; Blond, O.; Lai, Y.; Ma, A.L.; Mitchell, R.L. Comparison of the pharmacology and signal transduction of the human cannabinoid CB1 and CB2 receptors. *Mol. Pharmacol.* **1995**, *48*, 443–450. [PubMed]

57. Showalter, V.M.; Compton, D.R.; Martin, B.R.; Abood, M. Evaluation of binding in a transfected cell line expressing a peripheral cannabinoid receptor (CB2): Identification of cannabinoid receptor subtype selective ligands. *J. Pharmacol. Exp. Ther.* **1996**, *278*, 989–999.

58. Aung, M.M.; Griffin, G.; Huffman, J.W.; Wu, M.-J.; Keel, C.; Yang, B.; Showalter, V.M.; Abood, M.; Martin, B.R. Influence of the N-1 alkyl chain length of cannabimimetic indoles upon CB1 and CB2 receptor binding. *Drug Alcohol Depend.* **2000**, *60*, 133–140. [CrossRef]

59. Huffman, J.W.; Zengin, G.; Wu, M.J.; Lu, J.; Hynd, G.; Bushell, K.; Thompson, A.L.S.; Bushell, S.; Tartal, C.; Hurst, D.P.; et al. Structure-activity relationships for 1-alkyl-3-(1-naphthoyl)indoles at the cannabinoid CB(1) and CB(2) receptors: Steric and electronic effects of naphthoyl substituents. New highly selective CB(2) receptor agonists. *Bioorg. Med. Chem.* **2005**, *13*, 89–112. [CrossRef]

60. Wiley, J.L.; Smith, V.J.; Chen, J.; Martin, B.R.; Huffman, J.W. Synthesis and pharmacology of 1-alkyl-3-(1-naphthoyl)indoles: Steric and electronic effects of 4- and 8-halogenated naphthoyl substituents. *Bioorg. Med. Chem.* **2012**, *20*, 2067–2081. [CrossRef]

61. Gallant, M.; Dufresne, C.; Gareau, Y.; Guay, D.; Leblanc, Y.; Prasit, P.; Rochette, C.; Sawyer, N.; Slipetz, D.M.; Tremblay, N.; et al. New class of potent ligands for the human peripheral cannabinoid receptor. *Bioorg. Med. Chem. Lett.* **1996**, *6*, 2263–2268. [CrossRef]

62. Valenzano, K.J.; Tafesse, L.; Lee, G.; Harrison, J.E.; Boulet, J.M.; Gottshall, S.L.; Mark, L.; Pearson, M.S.; Miller, W.; Shan, S.; et al. Pharmacological and pharmacokinetic characterization of the cannabinoid receptor 2 agonist, GW405833, utilizing rodent models of acute and chronic pain, anxiety, ataxia and catalepsy. *Neuropharmacology* **2005**, *48*, 658–672. [CrossRef] [PubMed]

63. Clayton, N.; Marshall, F.H.; Bountra, C.; O'Shaughnessy, C.T. CB1 and CB2 cannabinoid receptors are implicated in inflammatory pain. *Pain* **2002**, *96*, 253–260. [CrossRef]

64. Beltramo, M.; Bernardini, N.; Bertorelli, R.; Campanella, M.; Nicolussi, E.; Fredduzzi, S.; Reggiani, A. CB2 receptor-mediated antihyperalgesia: Possible direct involvement of neural mechanisms. *Eur. J. Neurosci.* **2006**, *23*, 1530–1538. [CrossRef] [PubMed]

65. Ibrahim, M.M.; Deng, H.; Zvonok, A.; Cockayne, D.A.; Kwan, J.; Mata, H.P.; Vanderah, T.W.; Lai, J.; Porreca, F.; Makriyannis, A.; et al. Activation of CB2 cannabinoid receptors by AM1241 inhibits experimental neuropathic pain: Pain inhibition by receptors not present in the CNS. *Proc. Natl. Acad. Sci. USA* **2003**, *100*, 10529–10533. [CrossRef]

66. Bingham, B.; Jones, P.G.; Uveges, A.J.; Kotnis, S.; Lu, P.; Smith, V.A.; Sun, S.-C.; Resnick, L.; Chlenov, M.; He, Y.; et al. Species-specific in vitro pharmacological effects of the cannabinoid receptor 2 (CB2) selective ligand AM1241 and its resolved enantiomers. *Br. J. Pharmacol.* **2007**, *151*, 1061–1070. [CrossRef] [PubMed]

67. Mancini, I.; Brusa, R.; Quadrato, G.; Foglia, C.; Scandroglio, P.; Silverman, L.; Tulshian, D.; Reggiani, A.; Beltramo, M. Constitutive activity of cannabinoid-2 (CB2) receptors plays an essential role in the protean agonism of (+)AM1241 and L768242. *Br. J. Pharmacol.* **2009**, *158*, 382–391. [CrossRef]

68. Quartilho, A.; Mata, H.P.; Ibrahim, M.M.; Vanderah, T.W.; Porreca, F.; Makriyannis, A.; Malan, T.P. Inhibition of Inflammatory Hyperalgesia by Activation of Peripheral CB2Cannabinoid Receptors. *Anesthesiology* **2003**, *99*, 955–960. [CrossRef]

69. Hynes, J.; Leftheris, K.; Wu, H.; Pandit, C.; Chen, P.; Norris, D.J.; Chen, B.-C.; Zhao, R.; Kiener, P.A.; Chen, X.; et al. C-3 Amido-Indole cannabinoid receptor modulators. *Bioorg. Med. Chem. Lett.* **2002**, *12*, 2399–2402. [CrossRef]

70. Dart, M.; Frost, J.; Tietje, K.; Daza, A.; Grayson, G.; Fan, Y.; Mukherjee, S.; Garrison, T.R.; Yao, B.; Meyer, M. 1-Alkyl-3-ketoindoles: Identification and in vitro characterization of a series of potent cannabinoid ligands. In *Symposium on the Cannabinoids*; International Cannabinoid Research Society: Burlington, VT, USA, 2006.

71. Pertwee, R.; Griffin, G.; Fernando, S.; Li, X.; Hill, A.; Makriyannis, A. AM630, a competitive cannabinoid receptor antagonist. *Life Sci.* **1995**, *56*, 1949–1955. [CrossRef]

72. Pertwee, R.G.; Fernando, S.R.; Nash, J.E.; Coutts, A.A. Further evidence for the presence of cannabinoid CB1 receptors in guinea-pig small intestine. *Br. J. Pharmacol.* **1996**, *118*, 2199–2205. [CrossRef] [PubMed]

73. Hosohata, Y.; Quock, R.M.; Hosohata, K.; Makriyannis, A.; Consroe, P.; Roeske, W.R.; Yamamura, H.I. AM630 antagonism of cannabinoid-stimulated [35S] GTP gamma S binding in the mouse brain. *Eur. J. Pharmacol.* **1997**, *321*, R1–R3. [CrossRef]

74. Hosohata, K.; Quock, R.M.; Hosohata, Y.; Burkey, T.H.; Makriyannis, A.; Consroe, P.; Roeske, W.R.; Yamamura, H.I. AM630 is a competitive cannabinoid receptor antagonist in the guinea pig brain. *Life Sci.* **1997**, *61*, PL115–PL118. [CrossRef]

75. Landsman, R.S.; Makriyannis, A.; Deng, H.; Consroe, P.; Roeske, W.R.; Yamamura, H.I. AM630 is an inverse agonist at the human cannabinoid CB1 receptor. *Life Sci.* **1998**, *62*, PL109–PL113. [CrossRef]

76. Ross, R.A.; Brockie, H.C.; Stevenson, L.A.; Murphy, V.L.; Templeton, F.; Makriyannis, A.; Pertwee, R.G. Agonist-inverse agonist characterization at CB1 and CB2 cannabinoid receptors of L759633, L759656 and AM630. *Br. J. Pharmacol.* **1999**, *126*, 665–672. [CrossRef] [PubMed]

77. Dhopeshwarkar, A.; Mackie, K. Functional Selectivity of CB2 Cannabinoid Receptor Ligands at a Canonical and Noncanonical Pathway. *J. Pharmacol. Exp. Ther.* **2016**, *358*, 342–351. [CrossRef] [PubMed]

78. Chin, C.N.; Murphy, J.W.; Huffman, J.W.; Kendall, D.A. The third transmembrane helix of the cannabinoid receptor plays a role in the selectivity of aminoalkylindoles for CB2, peripheral cannabinoid receptor. *J. Pharmacol. Exp. Ther.* **1999**, *291*, 837–844. [PubMed]

79. Tao, Q.; McAllister, S.D.; Andreassi, J.; Nowell, K.W.; Cabral, G.A.; Hurst, D.P.; Bachtel, K.; Ekman, M.C.; Reggio, P.H.; Abood, M. Role of a conserved lysine residue in the peripheral cannabinoid receptor (CB2): Evidence for subtype specificity. *Mol. Pharmacol.* **1999**, *55*, 605–613. [PubMed]
80. Rhee, M.-H.; Nevo, I.; Levy, R.; Vogel, Z. Role of the highly conserved Asp-Arg-Tyr motif in signal transduction of the CB2 cannabinoid receptor. *FEBS Lett.* **2000**, *466*, 300–304. [CrossRef]
81. Feng, W.; Song, Z. Effects of D3.49A, R3.50A, and A6.34E mutations on ligand binding and activation of the cannabinoid-2 (CB2) receptor. *Biochem. Pharmacol.* **2003**, *65*, 1077–1085. [CrossRef]
82. Rhee, M.H.; Nevo, I.; Bayewitch, M.L.; Zagoory, O.; Vogel, Z. Functional role of tryptophan residues in the fourth transmembrane domain of the CB(2) cannabinoid receptor. *J. Neurochem.* **2000**, *75*, 2485–2491. [CrossRef]
83. Atwood, B.K.; Wager-Miller, J.; Haskins, C.; Straiker, A.; Mackie, K. Functional Selectivity in CB2 Cannabinoid Receptor Signaling and Regulation: Implications for the Therapeutic Potential of CB2 Ligands. *Mol. Pharmacol.* **2012**, *81*, 250–263. [CrossRef]
84. Stark, S.; Pacheco, M.A.; Childers, S.R. Binding of aminoalkylindoles to noncannabinoid binding sites in NG108-15 cells. *Cell. Mol. Neurobiol.* **1997**, *17*, 483–493. [CrossRef]
85. Ho, B.Y.; Zhao, J. Determination of the cannabinoid receptors in mouse x rat hybridoma NG108-15 cells and rat GH4C1 cells. *Neurosci. Lett.* **1996**, *212*, 123–126. [CrossRef]
86. Devane, W.A.; Spain, J.W.; Coscia, C.J.; Howlett, A.C. An Assessment of the Role of Opioid Receptors in the Response to Cannabimimetic Drugs. *J. Neurochem.* **1986**, *46*, 1929–1935. [CrossRef]
87. Breivogel, C.S.; Griffin, G.; Di Marzo, V.; Martin, B.R. Evidence for a New G Protein-Coupled Cannabinoid Receptor in Mouse Brain. *Mol. Pharmacol.* **2001**, *60*, 155–163. [CrossRef] [PubMed]
88. Monory, K.; Tzavara, E.T.; Lexime, J.; Ledent, C.; Parmentier, M.; Borsodi, A.; Hanoune, J. Novel, Not Adenylyl Cyclase-Coupled Cannabinoid Binding Site in Cerebellum of Mice. *Biochem. Biophys. Res. Commun.* **2002**, *292*, 231–235. [CrossRef] [PubMed]
89. Hájos, N.; Ledent, C.; Freund, T. Novel cannabinoid-sensitive receptor mediates inhibition of glutamatergic synaptic transmission in the hippocampus. *Neuroscience* **2001**, *106*, 1–4. [CrossRef]
90. Németh, B.; Ledent, C.; Freund, T.F.; Hájos, N. CB1 receptor-dependent and -independent inhibition of excitatory postsynaptic currents in the hippocampus by WIN 55212-2. *Neuropharmacology* **2008**, *54*, 51–57. [CrossRef]
91. Accorsi-Mendonça, D.; Almado, C.E.; Dagostin, A.L.; Machado, B.H.; Leão, R.M. Inhibition of spontaneous neurotransmission in the nucleus of solitary tract of the rat by the cannabinoid agonist WIN 55212-2 is not via CB1 or CB2 receptors. *Brain Res.* **2008**, *1200*, 1–9. [CrossRef]
92. Stella, N. Cannabinoid and cannabinoid-like receptors in microglia, astrocytes, and astrocytomas. *Glia* **2010**, *58*, 1017–1030. [CrossRef]
93. Fung, S.; Cherry, A.E.; Xu, C.; Stella, N. Alkylindole-sensitive receptors modulate microglial cell migration and proliferation. *Glia* **2015**, *63*, 1797–1808. [CrossRef]
94. Fung, S.; Xu, C.; Hamel, E.; Wager-Miller, J.B.; Woodruff, G.; Miller, A.; Sanford, C.; Mackie, K.; Stella, N. Novel indole-based compounds that differentiate alkylindole-sensitive receptors from cannabinoid receptors and microtubules: Characterization of their activity on glioma cell migration. *Pharmacol. Res.* **2017**, *115*, 233–241. [CrossRef] [PubMed]
95. Cherry, A.E.; Haas, B.R.; Naydenov, A.V.; Fung, S.; Xu, C.; Swinney, K.; Wagenbach, M.; Freeling, J.; Canton, D.A.; Coy, J.; et al. ST-11: A New Brain-Penetrant Microtubule-Destabilizing Agent with Therapeutic Potential for Glioblastoma Multiforme. *Mol. Cancer Ther.* **2016**, *15*, 2018–2029. [CrossRef] [PubMed]
96. Morales, P.; Reggio, P.H. An Update on Non-CB(1), Non-CB(2) Cannabinoid Related G-Protein-Coupled Receptors. *Cannabis Cannabinoid Res.* **2017**, *2*, 265–273. [CrossRef] [PubMed]
97. Adams, R.; Aycock, B.F.; Loewe, S. Tetrahydrocannabinol homologs. XVII. *J. Am. Chem. Soc.* **1948**, *70*, 662–664. [CrossRef]
98. Adams, R.; Loewe, S.; Smith, C.M.; McPhee, W.D. Tetrahydrocannabinol homologs with marihuana activity. XIII. *J. Am. Chem. Soc.* **1942**, *64*, 694–698. [CrossRef]
99. Adams, R.; MacKenzie, S.; Loewe, S. Tetrahydrocannabinol Homologs with Doubly Branched Alkyl Groups in the 3-Position. XVIII1. *J. Am. Chem. Soc.* **1948**, *70*, 664–668. [CrossRef] [PubMed]
100. Appendino, G. The early history of cannabinoid research. *Rend. Fis. Acc. Lincei* **2020**, *31*, 919–929. [CrossRef]
101. Williams, E.G.; Himmelsbach, C.K.; Wikler, A.; Ruble, D.C.; Lloyd, B.J. Studies on Marihuana and Pyrahexyl Compound. *Public Health Rep.* **1946**, *61*, 1059. [CrossRef] [PubMed]
102. Ketchum, J.S. *Chemical Warfare Secrets Almost Forgotten: A Personal Story of Medical Testing of Army Volunteers with Incapacitating Chemical Agents during the Cold War (1955–1975)*; Chembooks: Santa Rosa, CA, USA, 2006.
103. Carroll, F.I.; Lewin, A.H.; Mascarella, S.W.; Seltzman, H.H.; Reddy, P.A. Designer drugs: A medicinal chemistry perspective (II). *Ann. N. Y. Acad. Sci.* **2021**, *1489*, 48–77. [CrossRef] [PubMed]
104. Dolliver, D.S.; Kuhns, J.B. The Presence of New Psychoactive Substances in a Tor Network Marketplace Environment. *J. Psychoact. Drugs* **2016**, *48*, 321–329. [CrossRef] [PubMed]
105. European Monitoring Centre for Drugs and Drug Addiction. *European Drug Report 2021: Trends and Developments 2021*; Publications Office of the European Union: Luxembourg, 2021.
106. Cox, A.O.; Daw, R.C.; Mason, M.D.; Grabenauer, M.; Pande, P.G.; Davis, K.H.; Wiley, J.L.; Stout, P.R.; Thomas, B.F.; Huffman, J.W. Use of SPME-HS-GC-MS for the Analysis of Herbal Products Containing Synthetic Cannabinoids. *J. Anal. Toxicol.* **2012**, *36*, 293–302. [CrossRef] [PubMed]

107. Ginsburg, B.C.; McMahon, L.R.; Sanchez, J.J.; Javors, M.A. Purity of Synthetic Cannabinoids Sold Online for Recreational Use. *J. Anal. Toxicol.* **2012**, *36*, 66–68. [CrossRef]

108. Münster-Müller, S.; Matzenbach, I.; Knepper, T.; Zimmermann, R.; Pütz, M. Profiling of synthesis-related impurities of the synthetic cannabinoid Cumyl-5F-PINACA in seized samples of e-liquids via multivariate analysis of UHPLC-MS(n) data. *Drug Test Anal.* **2020**, *12*, 119–126. [CrossRef]

109. Moore, K.N.; Garvin, D.; Thomas, B.F.; Grabenauer, M. Identification of Eight Synthetic Cannabinoids, Including 5F-AKB48 in Seized Herbal Products Using DART-TOF-MS and LC-QTOF-MS as Nontargeted Screening Methods. *J. Forensic Sci.* **2017**, *62*, 1151–1158. [CrossRef]

110. Kavanagh, P.; Grigoryev, A.; Savchuk, S.; Mikhura, I.; Formanovsky, A. UR-144 in products sold via the Internet: Identification of related compounds and characterization of pyrolysis products. *Drug Test Anal.* **2013**, *5*, 683–692. [CrossRef]

111. Grabenauer, M.; Krol, W.L.; Wiley, J.L.; Thomas, B.F. Analysis of Synthetic Cannabinoids Using High-Resolution Mass Spectrometry and Mass Defect Filtering: Implications for Nontargeted Screening of Designer Drugs. *Anal. Chem.* **2012**, *84*, 5574–5581. [CrossRef] [PubMed]

112. Thomas, B.F.; Pollard, G.T.; Grabenauer, M. Analytical surveillance of emerging drugs of abuse and drug formulations. *Life Sci.* **2013**, *92*, 512–519. [CrossRef]

113. Wiley, J.L.; Lefever, T.W.; Marusich, J.A.; Grabenauer, M.; Moore, K.N.; Huffman, J.W.; Thomas, B.F. Evaluation of first generation synthetic cannabinoids on binding at non-cannabinoid receptors and in a battery of in vivo assays in mice. *Neuropharmacology* **2016**, *110*, 143–153. [CrossRef]

114. Wiley, J.L.; Marusich, J.A.; Lefever, T.W.; Grabenauer, M.; Moore, K.N.; Thomas, B. Cannabinoids in disguise: Δ9-Tetrahydrocannabinol-like effects of tetramethylcyclopropyl ketone indoles. *Neuropharmacology* **2013**, *75*, 145–154. [CrossRef]

115. Wiley, J.L.; Marusich, J.A.; Thomas, B.F. Combination Chemistry: Structure–Activity Relationships of Novel Psychoactive Cannabinoids. *Curr. Top. Behav. Neurosci.* **2017**, *32*, 231–248. [CrossRef]

116. Wiley, J.L.; Marusich, J.A.; Huffman, J.W. Moving around the molecule: Relationship between chemical structure and in vivo activity of synthetic cannabinoids. *Life Sci.* **2014**, *97*, 55–63. [CrossRef]

117. Ford, B.M.; Tai, S.; Fantegrossi, W.E.; Prather, P.L. Synthetic Pot: Not Your Grandfather's Marijuana. *Trends Pharmacol. Sci.* **2017**, *38*, 257–276. [CrossRef] [PubMed]

118. Banister, S.D.; Connor, M. The Chemistry and Pharmacology of Synthetic Cannabinoid Receptor Agonists as New Psychoactive Substances: Origins. *Handb. Exp. Pharmacol.* **2018**, *252*, 165–190. [CrossRef] [PubMed]

119. Banister, S.D.; Connor, M. The Chemistry and Pharmacology of Synthetic Cannabinoid Receptor Agonist New Psychoactive Substances: Evolution. *Handb. Exp. Pharmacol.* **2018**, *252*, 191–226. [CrossRef] [PubMed]

120. Thomas, B.F.; Daw, R.C.; Pande, P.G.; Cox, A.O.; Kovach, A.L.; Davis, K.H., Jr.; Grabenauer, M. Analysis of Smoke Condensate From Combustion of Synthetic Cannabinoids in Herbal Products. In Proceedings of the 23rd Annual International Cannabinoid Research Society Symposium on the Cannabinoids, Vancouver, BC, Canada, 21–26 June 2013; pp. 4–24.

121. Daw, R.; Grabenauer, M.; Pande, P.G.; Cox, A.; Kovach, A.; Davis, K.; Wiley, J.; Stout, P.; Thomas, B. Pyrolysis studies of synthetic cannabinoids in herbal products. *Drug Alcohol Depend.* **2014**, *140*, e44. [CrossRef]

122. Tsujikawa, K.; Yamamuro, T.; Kuwayama, K.; Kanamori, T.; Iwata, Y.T.; Inoue, H. Thermal degradation of a new synthetic cannabinoid QUPIC during analysis by gas chromatography–mass spectrometry. *Forensic Toxicol.* **2014**, *32*, 201–207. [CrossRef]

123. Thomas, B.F.; Lefever, T.W.; Cortes, R.A.; Grabenauer, M.; Kovach, A.; Cox, A.O.; Patel, P.R.; Pollard, G.T.; Marusich, J.A.; Kevin, R.C.; et al. Thermolytic Degradation of Synthetic Cannabinoids: Chemical Exposures and Pharmacological Consequences. *J. Pharmacol. Exp. Ther.* **2017**, *361*, 162–171. [CrossRef]

124. Thomas, B.F.; Wiley, J.L.; Endres, G.W. Synthetic cannabinoids are recurring chemical threats. *Cayman Curr.* **2015**, *26*, 1–3.

125. Gamage, T.F.; Farquhar, C.E.; Lefever, T.W.; Marusich, J.A.; Kevin, R.C.; Mcgregor, I.; Wiley, J.; Thomas, B.F. Molecular and Behavioral Pharmacological Characterization of Abused Synthetic Cannabinoids MMB- and MDMB-FUBINACA, MN-18, NNEI, CUMYL-PICA, and 5-Fluoro-CUMYL-PICA. *J. Pharmacol. Exp. Ther.* **2018**, *365*, 437–446. [CrossRef]

126. Patton, A.L.; Seely, K.A.; Yarbrough, A.L.; Fantegrossi, W.; James, L.P.; McCain, K.R.; Fujiwara, R.; Prather, P.L.; Moran, J.H.; Radominska-Pandya, A. Altered metabolism of synthetic cannabinoid JWH-018 by human cytochrome P450 2C9 and variants. *Biochem. Biophys. Res. Commun.* **2018**, *498*, 597–602. [CrossRef]

127. Kevin, R.C.; Kovach, A.; Lefever, T.W.; Gamage, T.F.; Wiley, J.L.; McGregor, I.S.; Thomas, B.F. Toxic by design? Formation of thermal degradants and cyanide from carboxamide-type synthetic cannabinoids CUMYL-PICA, 5F-CUMYL-PICA, AMB-FUBINACA, MDMB-FUBINACA, NNEI, and MN-18 during exposure to high temperatures. *Forensic Toxicol.* **2019**, *37*, 17–26. [CrossRef] [PubMed]

128. Grigoryev, A.; Kavanagh, P.; Melnik, A.; Savchuk, S.; Simonov, A. Gas and Liquid Chromatography-Mass Spectrometry Detection of the Urinary Metabolites of UR-144 and Its Major Pyrolysis Product. *J. Anal. Toxicol.* **2013**, *37*, 265–276. [CrossRef] [PubMed]

129. Cooper, Z.D. Adverse Effects of Synthetic Cannabinoids: Management of Acute Toxicity and Withdrawal. *Curr. Psychiatry Rep.* **2016**, *18*, 52. [CrossRef]

130. Pacher, P.; Steffens, S.; Haskó, G.; Schindler, T.H.; Kunos, G. Cardiovascular effects of marijuana and synthetic cannabinoids: The good, the bad, and the ugly. *Nat. Rev. Cardiol.* **2018**, *15*, 151–166. [CrossRef] [PubMed]

131. Grigg, J.; Manning, V.; Arunogiri, S.; Lubman, D.I. Synthetic cannabinoid use disorder: An update for general psychiatrists. *Australas. Psychiatry* **2019**, *27*, 279–283. [CrossRef]

132. Morrow, P.L.; Stables, S.; Kesha, K.; Tse, R.; Kappatos, D.; Pandey, R.; Russell, S.; Linsell, O.; McCarthy, M.J.; Spark, A.; et al. An outbreak of deaths associated with AMB-FUBINACA in Auckland NZ. *EClinicalMedicine* **2020**, *25*, 100460. [CrossRef]

133. Tiemensma, M.; Rutherford, J.D.; Scott, T.; Karch, S. Emergence of Cumyl-PEGACLONE-related fatalities in the Northern Territory of Australia. *Forensic Sci. Med. Pathol.* **2021**, *17*, 3–9. [CrossRef]

134. Ralphs, R.; Williams, L.; Askew, R.; Norton, A. Adding Spice to the Porridge: The development of a synthetic cannabinoid market in an English prison. *CrimRxiv* **2016**. [CrossRef]

135. Wohlfarth, A.; Scheidweiler, K.B.; Castaneto, M.; Gandhi, A.S.; Desrosiers, N.A.; Klette, K.L.; Martin, T.M.; Huestis, M.A. Urinary prevalence, metabolite detection rates, temporal patterns and evaluation of suitable LC-MS/MS targets to document synthetic cannabinoid intake in US military urine specimens. *Clin. Chem. Lab. Med.* **2015**, *53*, 423–434. [CrossRef]

136. Finlay, D.B.; Manning, J.J.; Ibsen, M.S.; Macdonald, C.E.; Patel, M.; Javitch, J.A.; Banister, S.D.; Glass, M. Do Toxic Synthetic Cannabinoid Receptor Agonists Have Signature in Vitro Activity Profiles? A Case Study of AMB-FUBINACA. *ACS Chem. Neurosci.* **2019**, *10*, 4350–4360. [CrossRef] [PubMed]

137. Sachdev, S.; Banister, S.; Santiago, M.J.; Bladen, C.; Kassiou, M.; Connor, M. Differential activation of G protein-mediated signaling by synthetic cannabinoid receptor agonists. *Pharmacol. Res. Perspect.* **2020**, *8*, e00566. [CrossRef] [PubMed]

138. Kumar, K.K.; Shalev-Benami, M.; Robertson, M.J.; Hu, H.; Banister, S.; Hollingsworth, S.A.; Latorraca, N.R.; Kato, H.; Hilger, D.; Maeda, S.; et al. Structure of a Signaling Cannabinoid Receptor 1-G Protein Complex. *Cell* **2019**, *176*, 448–458.e12. [CrossRef] [PubMed]

139. Xing, C.; Zhuang, Y.; Xu, T.-H.; Feng, Z.; Zhou, X.E.; Chen, M.; Wang, L.; Meng, X.; Xue, Y.; Wang, J.; et al. Cryo-EM Structure of the Human Cannabinoid Receptor CB2-Gi Signaling Complex. *Cell* **2020**, *180*, 645–654.e13. [CrossRef]

molecules

Article

Effects of a Peripherally Restricted Hybrid Inhibitor of CB1 Receptors and iNOS on Alcohol Drinking Behavior and Alcohol-Induced Endotoxemia

Luis Santos-Molina [1], Alexa Herrerias [1], Charles N. Zawatsky [2], Ozge Gunduz-Cinar [3], Resat Cinar [2], Malliga R. Iyer [4], Casey M. Wood [4], Yuhong Lin [5], Bin Gao [5], George Kunos [1] and Grzegorz Godlewski [1,*]

[1] Laboratory of Physiologic Studies, National Institute on Alcohol Abuse and Alcoholism, National Institutes of Health, Bethesda, MD 20892, USA; luis.f.santos29@gmail.com (L.S.-M.); alexa.herrerias@nih.gov (A.H.); george.kunos@nih.gov (G.K.)

[2] Section on Fibrotic Disorders, National Institute on Alcohol Abuse and Alcoholism, National Institutes of Health, Bethesda, MD 20892, USA; nickzawatsky@gmail.com (C.N.Z.); resat.cinar@nih.gov (R.C.)

[3] Laboratory of Behavioral and Genomic Neuroscience, National Institute on Alcohol Abuse and Alcoholism, National Institutes of Health, Bethesda, MD 20892, USA; ozge.gunduzcinar@nih.gov

[4] Section on Medicinal Chemistry, National Institute on Alcohol Abuse and Alcoholism, National Institutes of Health, Bethesda, MD 20892, USA; malliga.iyer@nih.gov (M.R.I.); casey.wood@nih.gov (C.M.W.)

[5] Laboratory of Liver Diseases, National Institute on Alcohol Abuse and Alcoholism, National Institutes of Health, Bethesda, MD 20892, USA; yulin@mail.nih.gov (Y.L.); bgao@mail.nih.gov (B.G.)

* Correspondence: greg.godlewski@nih.gov; Tel.: +1-301-443-4089

Citation: Santos-Molina, L.; Herrerias, A.; Zawatsky, C.N.; Gunduz-Cinar, O.; Cinar, R.; Iyer, M.R.; Wood, C.M.; Lin, Y.; Gao, B.; Kunos, G.; et al. Effects of a Peripherally Restricted Hybrid Inhibitor of CB1 Receptors and iNOS on Alcohol Drinking Behavior and Alcohol-Induced Endotoxemia. *Molecules* **2021**, *26*, 5089. https://doi.org/10.3390/molecules26165089

Academic Editor: Mauro Maccarrone

Received: 27 July 2021
Accepted: 19 August 2021
Published: 22 August 2021

Publisher's Note: MDPI stays neutral with regard to jurisdictional claims in published maps and institutional affiliations.

Abstract: Alcohol consumption is associated with gut dysbiosis, increased intestinal permeability, endotoxemia, and a cascade that leads to persistent systemic inflammation, alcoholic liver disease, and other ailments. Craving for alcohol and its consequences depends, among other things, on the endocannabinoid system. We have analyzed the relative role of central vs. peripheral cannabinoid CB1 receptors (CB1R) using a "two-bottle" as well as a "drinking in the dark" paradigm in mice. The globally acting CB1R antagonist rimonabant and the non-brain penetrant CB1R antagonist JD5037 inhibited voluntary alcohol intake upon systemic but not upon intracerebroventricular administration in doses that elicited anxiogenic-like behavior and blocked CB1R-induced hypothermia and catalepsy. The peripherally restricted hybrid CB1R antagonist/iNOS inhibitor *S*-MRI-1867 was also effective in reducing alcohol consumption after oral gavage, while its *R* enantiomer (CB1R inactive/iNOS inhibitor) was not. The two MRI-1867 enantiomers were equally effective in inhibiting an alcohol-induced increase in portal blood endotoxin concentration that was caused by increased gut permeability. We conclude that (i) activation of peripheral CB1R plays a dominant role in promoting alcohol intake and (ii) the iNOS inhibitory function of MRI-1867 helps in mitigating the alcohol-induced increase in endotoxemia.

Keywords: cannabinoid; MRI-1867; hybrid ligand; CB1 receptor antagonist; iNOS inhibitor; rimonabant; intracerebroventricular administration; alcohol craving; two-bottle paradigm; drinking in the dark

1. Introduction

Chronic alcohol consumption poses a serious public health problem in the United States and worldwide. An estimated 8.6% Americans remain addicted to alcohol or drugs and there are 15 million new cases of alcohol use disorder (AUD) each year in the US alone, representing an economic burden of nearly a quarter of a trillion dollars [1,2]. The frequency of drinking has been accelerated during the COVID-19 pandemic [3].

Alcohol dependence has traditionally been viewed as a brain disorder caused by neuroadaptations of the reward circuits to alcohol [4]. Despite efforts to develop effective medications, pharmacotherapy to rebalance central neurotransmission has done little to

53

improve drinking outcomes [5,6]. More recent evidence links alcoholism to peripherally born endotoxemia and gut-derived inflammation [7]. One of the key components of the inflammatory cascade is gut microflora lipopolysaccharide (LPS, also termed endotoxin). Alcohol drinking has been shown to induce gut dysbiosis and bacterial overgrowth, impair intestinal permeability, and increase the translocation of bacterial products from the gut into the systemic circulation in rodents [8–10], heavy drinkers [11,12], and healthy individuals [13]. Once in circulation, LPS can bind to Toll-like receptors in the liver and innate immune cells to alter the cytokine milieu in favor of inflammatory species, e.g., TNF-α, interleukin (IL)-1β, and IL-6 [14]. These cytokines may be transported to the brain and other remote tissues causing systemic inflammation and tissue damage [15,16]. The reduction in the intestinal bacterial load or alcohol withdrawal has proven effective in attenuating the severity of inflammation and alcohol dependence [17,18]. Ethanol also up-regulates inducible nitric oxide synthase (iNOS), which leads to the dysfunction of intestinal tight junctions [19], gut leakiness, endotoxemia, and liver injury [20–22], effects not seen in iNOS-deficient mice [23].

Recent decades have seen growing interest in exploring the endocannabinoid system (ECS) as a target in the treatment of addiction and AUD, driven in part by the synergistic rewarding properties of alcohol and Δ^9-tetrahydrocannabinol (THC)—the psychoactive component of marijuana that binds to the same cannabinoid receptors as do the endogenous ligands anandamide and 2-arachidonoyl glycerol [24–26]. Early preclinical data focused mainly on the therapeutic potential of the prototypical cannabinoid-1 receptor (CB1R) antagonists that cross the blood–brain barrier. Accordingly, blockade of CB1R has been shown to reduce alcohol consumption [27], its rewarding properties [28,29], and to diminish inflammation in the central nervous system (CNS) [30] and in the gut [31]. However, clinical trials to counteract metabolic obesity with SR14716A (rimonabant), the prototype brain penetrant CB1R receptor antagonist/inverse agonist, showed that it produced serious neuropsychiatric adverse events [32], which halted the therapeutic development of this class of compounds.

We have recently reported that the brain non-penetrant CB1R antagonist, JD5037, representing the second generation of CB1R antagonists, significantly suppresses alcohol preference and proposed that endocannabinoids engage CB1R in ghrelin-producing cells of the stomach to promote alcohol drinking in a manner sensitive to blockade by JD5037 [33]. The ability to influence drinking behavior by CB1R outside the CNS has renewed interest in the therapeutic potential of CB1R antagonists. It also raised questions about the role of central versus peripheral CB1R in controlling voluntary ethanol intake, and whether CB1R antagonists could also be beneficial in mitigating other AUD features, e.g., endotoxemia or gut permeability. To address these aspects, we compared here the effects of rimonabant, JD5037, and the two stereoisomers of a newly developed peripherally active hybrid ligand: *S*-MRI-1867 (CB1R antagonist/iNOS inhibitor) and its enantiomer *R*-MRI-1867 (CB1R inactive/iNOS inhibitor) [34,35] in murine models of alcohol drinking. We found that alcohol intake was significantly inhibited by all CB1R antagonists upon systemic, but not upon intracerebroventricular (i.c.v.), administration, and was unaffected by iNOS inhibition, whereas the two MRI-1867 enantiomers were equally effective at inhibiting alcohol-induced increase in blood endotoxin concentration.

2. Results

2.1. Central Administration of Rimonabant Inhibits CP 55,940-Induced Catalepsy and Hypothermia in Mice

Acute administration of CB1R agonists, such as CP55,940, induces four behavioral phenotypes including hypothermia, hypoalgesia, catalepsy, and hypomotility, with the latter two being exclusively mediated by central CB1R. Thus, these responses are reversible by oral administration of rimonabant and largely insensitive to non-brain penetrant CB1R antagonists [36,37]. To further document the role of central CB1R in these effects, we tested the ability of i.c.v.-administered rimonabant in antagonizing CP55,940-induced catalepsy and hypothermia.

Intraperitoneal (i.p.) injection of CP55,940 (0.1–3 mg/kg) produced a dose-dependent decrease in body temperature (Figure 1a) and cataleptic behavior (Figure 1b) in wild-type mice. These responses were selectively mediated by CB1R, as CB1R-deficient animals remained completely insensitive to treatment apart from the highest CP55,940 dose (3 mg/kg; Figure 1a,b). The i.c.v. infusion of 2 µg of rimonabant effectively blocked the hypothermic (Figure 1c) and cataleptic (Figure 1d) effects of two different doses of CP55,940.

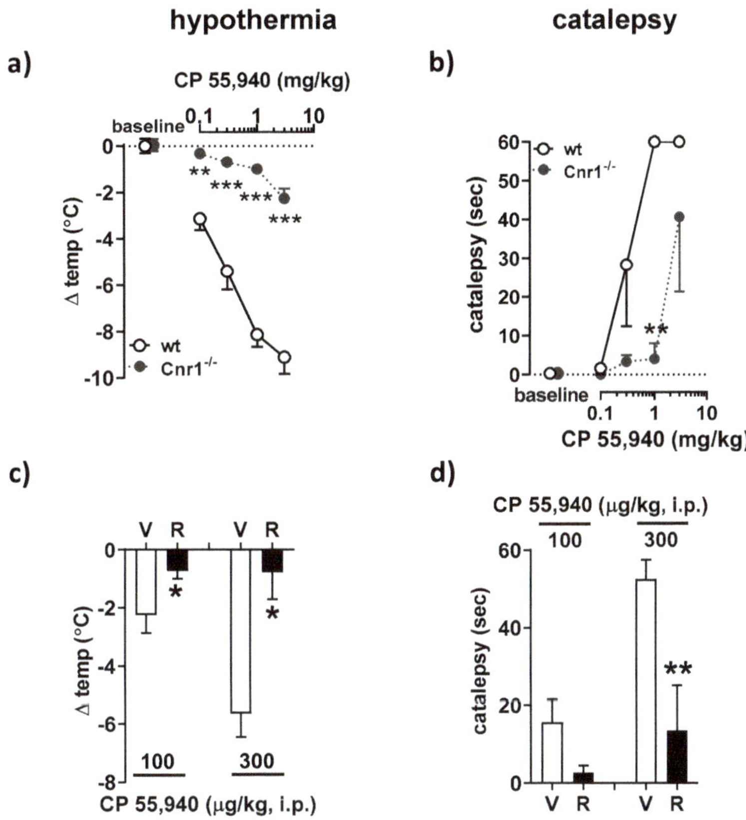

Figure 1. Effect of central administration of rimonabant on CP 55,940-induced catalepsy and hypothermia in mice. (**a**,**b**) Pharmacologically naïve $Cnr1^{-/-}$ (ko) mice and wild-type littermates (wt) were injected with CP 55,940 (0.1, 0.3, 1 and 3 mg/kg; i.p.) at 30 min intervals. Body temperature (**a**) and cataleptic behavior (**b**) were evaluated before administering the next dose of CP 55,940. The respective baseline body temperature in wt and ko groups prior to CP 55,940 injection were 37.1 ± 0.3 °C ($n = 3$) and 37.2 ± 0.3 °C ($n = 3$). Mice held the bar for 0.3 ± 0.3 s and 0.3 ± 0.3 s in wt and ko groups, respectively. (**c**,**d**) Conscious freely moving C57BL/6J mice ($n = 5$ animals per treatment group) were infused intracerebroventricularly (i.c.v.) with rimonabant (2 µg, R) or its solvent (V), followed 30 min later by intraperitoneal (i.p.) injection of CP 55,940 (0.1 or 0.3 mg/kg). Another 30 min passed before the hypothermic (**c**) and cataleptic (**d**) responses were measured. The respective baseline body temperatures before CP 55,940 injection to V- and R-infused groups were 37.6 ± 0.1 °C and 37.5 ± 0.1 °C ($n = 10$). Mice held the bar for 0.3 ± 0.3 s and 0.1 ± 0.1 s in V- and R-treated groups, respectively. Results are means $\pm$ s.e.m. ** $p < 0.01$, *** $p < 0.01$ compared to $Cnr1^{-/-}$ mice (**a**,**b**); * $p < 0.05$, ** $p < 0.01$ compared to vehicle (**c**,**d**).

2.2. Central Administration of JD5037 Increases Anxiety-Like Behavior in Mice in the Elevated plus Maze Test

The elevated plus maze tests the natural spontaneous exploratory behavior of rodents in novel environments. This trait can be impaired by genetic deletion of CB1R or its pharmacological inhibition by rimonabant.

To test if the central administration of JD5037 triggers an anxiety-like behavior in mice, 1 µg JD5037 or its solvent was delivered directly into the 3rd ventricle. Figure 2 shows the heatmaps of overall activity in drug- and vehicle-treated groups of animals that were tested in the elevated plus maze (Figure 2a) and activities in individual maze compartments (Figure 2b–g). Thus, during the five-minute run in the maze, control mice traveled an average distance of 1189.0 ± 111.6 cm (Figure 2b, $n = 6$) and moved in the arena at the speed of 4.0 ± 0.4 cm/s (Figure 2c, $n = 6$). In contrast, JD5037-treated animals were significantly less active, covering the distance of 740.5 ± 130.6 cm (Figure 2b, $p < 0.05$, $n = 6$) at the velocity of 2.4 ± 0.4 cm/s (Figure 2c, $p < 0.05$, $n = 6$). Drug-treated animals also stayed much longer in the closed arms (Figure 2d) and moved less frequently between the two closed arms of the maze (Figure 2e) than their vehicle-treated counterparts. They refrained from exploring open arms, which is reflected by the failure to enter open arms (Figure 2f) and explore them (Figure 2g), a clear indication of increased anxiety.

a)

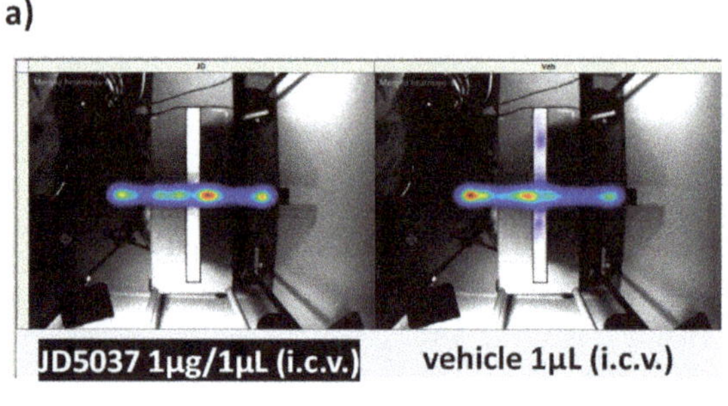

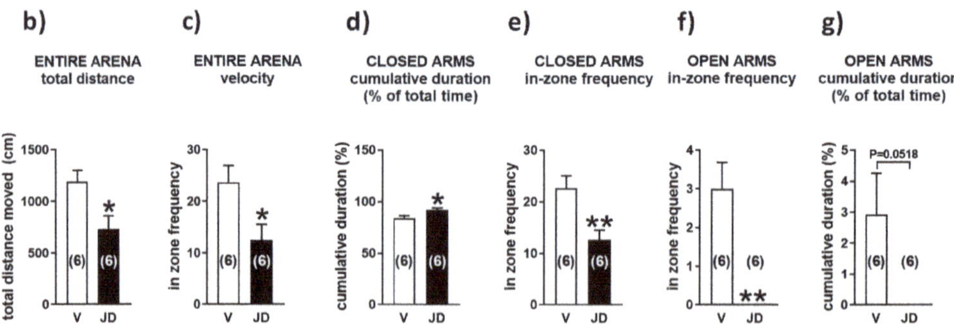

Figure 2. Intracerebroventricular microinfusion of JD5037 increases anxiety-related behavior in the elevated plus maze test. Heat maps (**a**) and summary of mouse activities in individual compartments (**b–g**) of the elevated plus maze. Animals received i.c.v. infusion of JD5037 (1 µg; JD) or its vehicle (3% DMSO, 8% Tween 80, 30% PEG-400, 59% saline; V). They were tested in the elevated plus maze 1 h later. The computerized EthoVision video tracking system was used for data collection and analysis. Bars are mean $\pm$ s.e.m. * $p < 0.05$, ** $p < 0.01$, compared with vehicle.

2.3. Rimonabant and JD5037 Inhibit Voluntary Ethanol Intake Only via the Peripheral Administration Route

To assess the relative role of central versus peripheral CB1R in the control of voluntary ethanol intake, animals receiving CB1R antagonists were tested in a restricted-access drinking-in-the-dark paradigm. Mice exposed to 20% alcohol for a short period at night tend to drink to inebriation, reflected by high blood levels of ethanol [33]. The i.c.v. infusion of rimonabant in a dose that inhibited hypothermia and catalepsy did not alter alcohol drinking (Figure 3a). In contrast, alcohol drinking was markedly reduced when animals received rimonabant by oral gavage. This effect was CB1R-dependent as it did not occur in CB1R KO mice. This is also reflected by the changes in blood alcohol concentration (BAC) in wt and CB1R KO mice (Figure 3b).

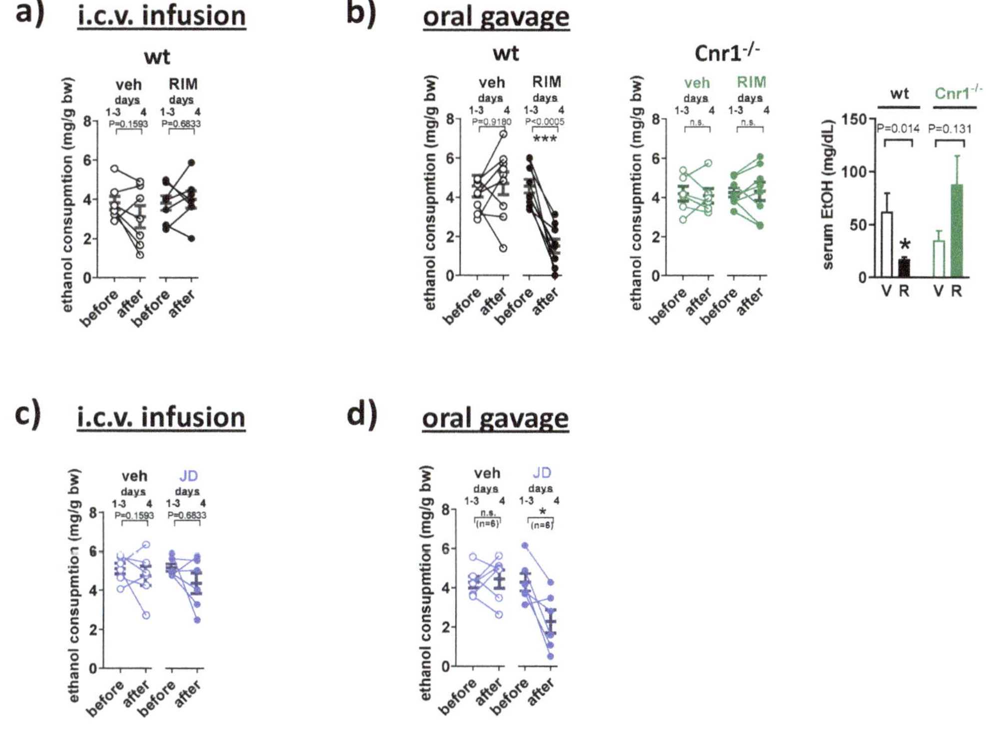

Figure 3. Relative involvement of central vs. peripheral CB1R blockade in the inhibition of voluntary ethanol intake by CB1 antagonists. Mice had access to 20% ethanol for 4 h each day. On day 4, one hour before the dark period, mice were infused intracerebroventricularly (i.c.v.) with (**a**) rimonabant (2 μg, RIM, R), (**b**) JD5037 (1 μg, JD) or their solvents (veh, V) and drinking session was repeated one more time. Another cohort of wild-type mice (wt) and/or their CB1 receptor-deficient (Cnr$^{-/-}$) counterparts received (**c**) rimonabant (10 mg/kg), (**d**) JD5037 (3 mg/kg), or vehicle by oral gavage. Drinking behavior in individual animals is expressed as points before (average of days 1–3) and after treatment (day 4). The corresponding serum ethanol values are shown as mean ± s.e.m. * $p < 0.05$; *** $p < 0.001$, compared with before treatment (Student's *t*-test for paired samples), [#] $p < 0.05$ compared with the vehicle (Student's *t*-test for unpaired samples), n.s. not significant.

Likewise, i.c.v. administration of JD5037 in a dose that caused anxiety turned out to be ineffective in reducing alcohol drinking (Figure 3c). The drug effectively reduced alcohol drinking only when given by oral gavage (Figure 3d). This observation is consistent with our earlier finding, which also showed that the effect of JD5037 was CB1R-dependent as it did not occur in CB1R-deficient mice [33].

2.4. MRI-1867 Regulates Alcohol Consumption in Mice through the Inhibition of Peripheral CB1R but Not iNOS

The effect of MRI-1867 on alcohol intake in mice was isomer-specific in two experimental models. Thus, in the drinking-in-the-dark paradigm, oral administration of the S-MRI-1867 (CB1R antagonist/iNOS inhibitor) inhibited alcohol consumption in a dose-dependent fashion, whereas alcohol intake was unaffected by similar treatment with R-MRI-1867 (CB1R inactive/iNOS inhibitor). The trend is also reflected by the changes in serum acetaldehyde and alcohol levels (Figure 4a), indicating that the drug does not affect the rate of alcohol metabolism.

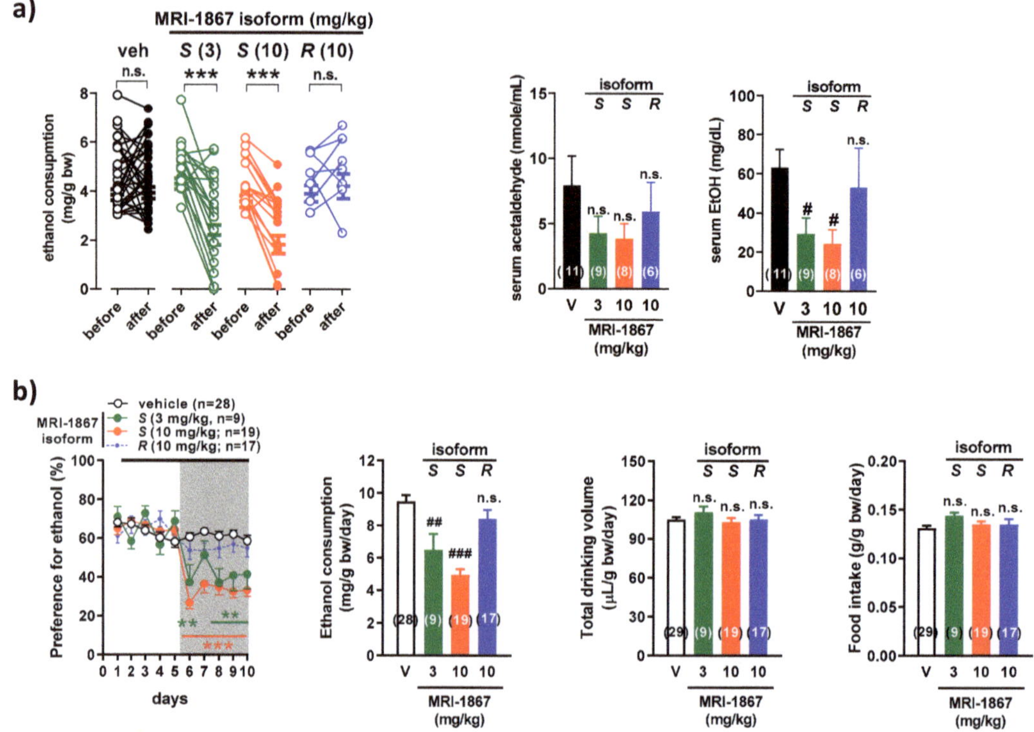

Figure 4. MRI-1867 decreases alcohol consumption after oral administration in two drinking models. (**a**) C57BL/6J mice had access to 20% ethanol for 4 h daily. On day 4, one hour before the dark period, mice received *S*-1867, (3, 10 mg/kg; S), *R*-MRI-1867 (10 mg/kg; R), or vehicle (V) by oral gavage and drinking session was repeated. Serum level of acetaldehyde and alcohol from blood obtained at the end of the drinking session. (**b**) Mice had free access to a 15% ethanol solution and water, using a two-bottle free-choice paradigm. From days 6 to 10, mice received daily *S*-MRI-1867 (3, 10 mg/kg; S), *R*-MRI-1867 (10 mg/kg; R), or vehicle (V) by oral gavage. Drinking behavior in individual animals (**a**) is expressed as points before (average of days 1–3) and after treatment (day 4). Other points and bars (**a**,**b**) are mean ± s.e.m. of daily to 5-day drinking behavior, respectively. ** $p < 0.01$, *** $p < 0.001$, compared with before treatment (Student's *t*-test for paired samples) (**a**) or with vehicle (two-way ANOVA followed by Tukey's multiple comparisons test) (**b**); # $p < 0.05$, ## $p < 0.01$, ### $p < 0.001$ compared to vehicle (one-way ANOVA followed by Dunnett's post hoc test), n.s. not significant.

The same pattern is evident when using the 2-bottle choice test. Consistent with our earlier study [33], male C57BL/6J mice with continuous access to water and 15% ethanol solution displayed high preference for alcohol (64.2 ± 1.2%), resulting in an average daily intake of 10.3 ± 0.2 mg ethanol/g body weight (*n* = 74). The high alcohol preference and intake remained unaffected by daily gavage with vehicle or R-MRI-1867, whereas S-MRI-1867 was effective in reducing alcohol preference and intake, without affecting total liquid consumption or food intake (Figure 4b).

2.5. Inhibition of iNOS by MRI-1867 Enantiomers Decreases Serum Endotoxin Level in Acutely Alcohol-Intoxicated Mice

Acute challenge of mice with ethanol is known to increase gut permeability [9]. We used changes in the amount of endotoxin measured in the portal vein that delivers blood to the liver as an indicator of gut permeability. As expected, endotoxin level was significantly higher in animals exposed to alcohol. Pretreatment of mice with either MRI-1867 enantiomer significantly reduced endotoxin level in the blood (Figure 5).

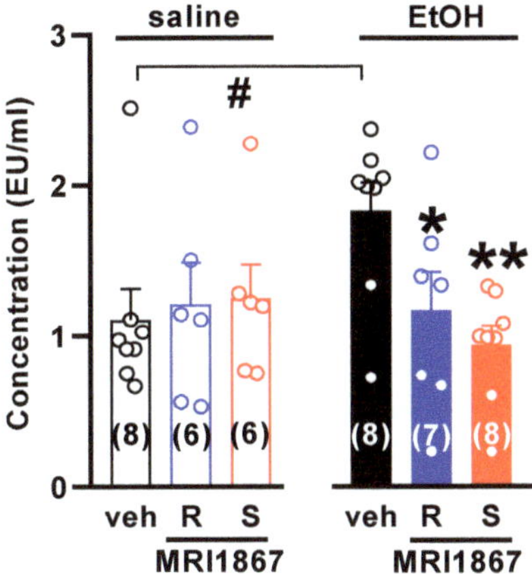

Figure 5. The iNOS inhibitor *R*-MRI-1867 decreases serum endotoxin level in acutely alcohol-intoxicated mice. Mice received *S*-MRI-1867 10 mg/kg (S) or *R*-MRI-1867 10 mg/kg (R) or vehicle (veh) by oral gavage (time 0), followed by intragastric administration of 6 g/kg ethanol (30% *w/v*; EtOH) or saline at 30 min. Endotoxin was measured in the serum obtained from the portal vein 1 h after the acute alcohol challenge. Results are expressed as mean ± s.e.m. * $p < 0.05$, ** $p < 0.01$, compared with vehicle in alcohol-treated group (one-way ANOVA followed by Dunnett's post hoc test), [#] $p < 0.05$, compared with saline-treated mice.

Two hours after oral administration of 10 mg/kg *S*-MRI-1867, drug concentration measured in different segments of the gastrointestinal tract was on average ~36 μmol/g wet tissue weight (Table 1, *n* = 3).

Table 1. Concentration of *S*-MRI-1867 in different segments of the gastrointestinal tract.

Tissue	Concentration (μmol/g Wet Tissue Weight)
stomach	30.57 ± 8.05
duodenum	36.88 ± 11.63
jejunum	44.14 ± 9.18
ileum	52.78 ± 11.50
colon	17.62 ± 5.66

3. Discussion

Evidence has accumulated over the years to implicate endocannabinoids acting via CB1R in the control of alcohol seeking behavior. Increasing endocannabinoid 'tone' by genetic deletion or pharmacologic inhibition of enzymes involved in endocannabinoid degradation was reported to increase alcohol intake and preference in rodent drinking paradigms [38]. Conversely, alcohol preference and intake are suppressed by genetic knockout or pharmacologic blockade of CB1R. Specifically, the globally acting CB1R antagonist/inverse agonist rimonabant reduced alcohol preference and intake not only upon systemic administration [27], but also when it was microinjected into limbic structures believed to regulate alcohol drinking behavior, such as the ventral tegmental area (VTA), nucleus accumbens, or prefrontal cortex [39–41], providing strong support for the role of these central structures in the control of addictive drinking. Therefore, the recent report that a peripherally restricted CB1R antagonist was as effective as rimonabant in inhibiting alcohol preference and intake in mice was unexpected, as it suggested that alcohol drinking behavior can be disrupted by blocking CB1R at a peripheral site(s) [33]. Further evidence indicated the existence of a gut–brain axis, whereby endocannabinoids acting via CB1R on ghrelin-producing cells in the stomach promote the posttranslational activation of ghrelin and its signaling to the brain via ghrelin receptors on vagal afferent terminals in the stomach [33]. However, the relative contribution of peripheral vs. central CB1R in the control of alcohol-seeking behavior has remained unclear.

In the current study, we have explored the relative role of central vs. peripheral CB1R in promoting alcohol drinking in mice using the two-bottle free choice as well as the drinking-in-the-dark paradigm in mice. We found that the brain penetrant CB1R antagonist rimonabant and its non-brain penetrant counterpart JD5037 significantly inhibited voluntary alcohol intake upon systemic administration. The response was CB1R-dependent as it was absent in CB1R KO mice. It was also reflected by the reduction in inebriating blood alcohol levels in wild-type, but not in CB1R KO mice. However, both drugs lost their efficacy to modulate alcohol drinking when administered intracerebroventricularly, even though they elicited an anxiogenic-like response and blocked CB1-induced hypothermia and catalepsy, indicating that they were able to engage CB1R in the CNS. Thus, our observations do not support a significant role of the central CB1R in the control of alcohol drinking in mice. This conclusion is also compatible with the earlier finding that systemically administered rimonabant lost its ability to inhibit alcohol preference and intake in mice with vagal afferent denervation [33]. However, the input of central CB1R in mediating other symptoms of AUD, e.g., alcohol tolerance, cannot be entirely excluded as it may rest on the experiment model. It has been shown that mice chronically exposed to alcohol display considerably lower sensitivity to cannabinoid-induced hypomotility, hypothermia, and antinociception because of lower CB1R density in the hypothalamus, VTA, and other brain areas [42].

A second objective of this study was to further test the role of peripheral CB1R in the control of alcohol drinking behavior and explore potential additional mechanisms. As detailed in the introduction, chronic alcohol consumption has been associated with low-grade inflammation, including intestinal inflammation, and the resulting increase in gut permeability has been causally linked to increased expression and activity of iNOS in the intestinal mucosa [7,18,23]. Endocannabinoids have also been shown to increase gut permeability via CB1R activation, an effect reversible by CB1R antagonists [31]. The alcohol-induced dysfunction of the intestinal barrier results in the translocation of bacterial endotoxin (LPS) and gut-derived microbial products into the circulation where their presence correlated with increased expression of inflammatory cytokines in peripheral blood mononuclear cells [43]. These changes have been implicated in alcoholic steatohepatitis and were also found to correlate with alcohol craving and consumption by alcohol-dependent individuals [44]. The recent development of a peripherally restricted hybrid inhibitor of CB1R and iNOS, *S*-MRI-1867 has enabled us to assess the relative contribution of these two molecules to alcohol drinking behavior and alcohol-induced intestinal barrier dysfunction in mouse models. The incorporation of acetamidine, a known inhibitor of iNOS, into the side chain of the same ibipinabant scaffold used to generate JD5037, imparted iNOS inhibitory activity to both the *S*- and *R*-enantiomers of MRI-1867, whereas the nanomolar CB1R inhibitory potency uniquely resides in the *S*-enantiomer [34,45]. Similar to rimonabant and the single-target peripheral CB1R antagonist JD5037, orally administered *S*-MRI-1867 potently inhibited alcohol drinking in both drinking models used whereas *R*-MRI-1867 was without such effect, indicating the exclusive role of peripheral CB1R in this effect. In contrast, the alcohol-induced increase in plasma levels of LPS was similarly inhibited by *S*- and *R*-MRI-1867, which strongly suggests the dominant role of iNOS rather than CB1R inhibition in this effect. This is further supported by the fact that the drug concentration measured across the gastrointestinal tract was well above the IC50 value for MRI-1867 enantiomers ($\geq$10 μM) that inhibits iNOS activity in in vitro assays and in mouse tissue homogenates [35,45]. Since both drug enantiomers were used at their maximally effective doses and both were active iNOS inhibitors, we could not assess the relative contribution of iNOS and CB1R components of MRI-1867 to endotoxemia and intestinal permeability, which would require the drug to be administered at submaximal doses. The role of CB1R should be considered in the light of the fact that endocannabinoids increase intestinal permeability in Caco-2 cells [46,47] as well as in obese mice [48], where LPS acts as a master switch to control adipose tissue metabolism, sensitive to blockade by rimonabant. Therefore, a possible crosstalk between iNOS and CB1R in the control intestinal barrier integrity remains to be explored.

In conclusion, our observations support the predominant role of peripheral CB1R in the control of alcohol drinking behavior. Furthermore, our findings using a novel, peripherally restricted, hybrid inhibitor of CB1R and iNOS indicate that engaging these two distinct targets, with respective roles in the drive to drink and alcohol-induced organ toxicity, by a single chemical entity could represent an attractive therapeutic approach to simultaneously mitigate the urge to consume alcohol and some of its harmful peripheral effects. So far, simultaneous inhibition of CB1R and iNOS has been a promising therapeutic strategy for the treatment of pulmonary fibrosis [45], Hermansky-Pudlak syndrome, pulmonary fibrosis [49] liver fibrosis [34], obesity-related dyslipidemia [35], and chronic kidney disease [50], and the hybrid inhibitor featured in all these studies is in early clinical development. The current study emphasizes its potential therapeutic use in AUD and alcohol-induced organ injury [51].

4. Materials and Methods

4.1. Animals

All animal procedures were approved by the Institutional Animal Care and Use Committee of NIAAA, NIH (Animal Experimentation permit number LPS-GK-1), and the experiments were carried out in accordance with its guidelines. C57BL/6J mice were purchased from The Jackson Laboratory (USA). Cnr1$^{-/-}$ were generated as described [52] and were propagated by heterozygote breeding, using corresponding wild-type littermates as controls. The strain had been backcrossed at least 10 times to maintain the C57BL/6J background. Animals were housed 4 per cage on a 12-/12 h light/dark cycle, had free access to food (rodent sterilizable diet; Harlan Teklad, USA) and water, and were experimentally naïve before testing. Mice were housed individually for the two-bottle alcohol preference and drinking-in-the-dark tests. They were allowed at least 5–7 days to habituate to the experimental conditions and handling prior to testing.

4.2. Cannulation and Intracerebroventricular Microinfusion of CB1R Antagonists

For experiments requiring intracerebroventricular (i.c.v.) drug infusion, pre-canulated (third ventricle) C57BL/6J mice were received from The Jackson Laboratory. Animals had internal guide cannula (2.5 mm long; Plastics One, USA) mounted in the 3rd ventricle (standard coordinates—ML: +1.0, RC: -0.4, DV: 2.0 mm) and protected by a dummy cap until the experiment. For more information on the cannulation procedure, animal care, and use, see the link: https://www.jax.org/-/media/jaxweb/files/jax-mice-and-services/brain-cannulation-information-care-use.pdf?la=en&hash=FD75F73AB0CD7A47808C78D0FC405AB3AF123F3B (accessed on 14 June 2021).

Conscious freely moving mice were infused i.c.v. with rimonabant (2 μg), JD5037 (1 μg), or their solvents. Drugs were applied in a volume of 1 μL over the course of 2 min via 33 g internal injector (P1 Technologies, Roanoke, VA, USA) connected with the 2 μL precision glass Hamilton syringe (USA) by a PE-20 tubing (Fisher Scientific, Hampton, NH, USA). The infusion rate and volume were controlled through the use of the syringe pump (model PHD 22/2000, Harvard Apparatus, Cambridge, MA, USA). Following 2 min infusion, injectors were left in place for additional 3 min to allow a passive diffusion of drugs into the tissue.

4.3. Catalepsy and Hypothermia

Catalepsy was assessed using the bar test described [36] with modifications. Mice were removed from home cages and their forepaws were placed on a horizontal stainless-steel rod, 0.5 cm in diameter, positioned 3.5 cm above the bench surface. Cataleptic behavior was defined as the time the animals remained motionless holding on to the bar. Vehicle-treated mice routinely went off the bar within ~2 s. The arbitrary cutoff time for cataleptic mice was 60 s. Hypothermia was then evaluated by measuring core body temperature with a rectal probe (Ellab Inc., Denver, CO, USA).

To determine a working range of CP 55,940 doses that induce catalepsy and hypothermia through CB1R, a dose–response curve for CP 55,940 was constructed first. Pharmacologically naïve Cnr1$^{-/-}$ and wild-type littermates were injected with increasing doses of CP 55,940 (0.1, 0.3, 1, and 3 mg/kg; i.p.) at 30 min intervals. Hypothermia and catalepsy tests were performed every 30 min before ordering the next dose of CP 55,940. Subsequently, conscious freely moving C57BL/6J mice were infused i.c.v. with rimonabant (2 μg), S-MRI1867 (3 μg), or vehicle followed by an injection of a single dose of CP 55,940 (0.1 or 0.3 mg/kg, i.p.) 30 min later. The cataleptic behavior and body temperature were evaluated before and 30 min after CP 55,940 injection.

4.4. Elevated plus Maze Test

Anxiety-related behavior was assessed using the elevated plus maze (EPM) test as described [53], with modifications. C57BL/6J mice were allowed to acclimate in their home cages for 2 h prior to the procedure. Animals received i.c.v infusion of JD5037 1 μg or vehicle and were tested in the EPM one hour later.

Mice were given 5 min to explore an elevated platform (72 cm above the floor) consisting of two opposing open and closed arms (each 30 × 5 cm) crossing each other. Illumination in the open and closed arms was 20 and 90 Lux, respectively. Mice were placed individually in the center of the platform (5 × 5 cm) facing an open arm. The behavior of each mouse was monitored by a computer-assisted video tracking system EthoVision XT (Noldus, Leesburg, VA, USA). The tested categories include total distance (cm) and average velocity (cm/s) of the run in the entire maze, cumulative duration (% of total time) spent in closed and open arms, number of entries into each closed and open arm of the maze (in-zone frequency). Arm entries were defined as crossing of the center point (located at approximately two thirds of the mouse body) into the arm. Number of entries, time spent in the open or closed arms, and distance travelled were measured by an automated HindSight software system (Hindsight, version 1.4, Hindsight Software Solutions Inc., Frisco, TX, USA).

4.5. Two-Bottle Alcohol Preference Test

The procedure was performed as described [27], with modifications. Animals were individually housed and acclimated to the paradigm for 5–7 days by having access to two identical water bottles and handled daily to minimize the stress associated with drug testing. Animals were first subjected to a gradual increase in ethanol concentration in a drinking bottle (3%, 6%, 9%, 12%, 15%), while the other bottle contained water. The position of the bottles was changed every day, and alcohol and water bottles were replaced every 4 days. Once alcohol concentration reached 15%, animals remained on the paradigm for 10 days. Starting on day 2, mice received vehicle by oral gavage for 4 days, followed by a daily treatment with the global or peripheral CB1 antagonist (*S*-MRI-1867 1, 3, 10 mg/kg; *R*-MRI-1867 10 mg/kg,) or vehicle (control group) for another 5 days, one hour before the dark period. All animals were sacrificed for blood and tissue collection 12–16 h after treatment.

4.6. Drinking in the Dark

The procedure was performed as described [53], with modifications. Mice were individually housed and acclimated to the room for 5–7 days before testing and randomly assigned to treatment groups. Starting 3 h into the dark cycle, water bottles were replaced with 20% ethanol in 25 × 100 mm glass tubes fitted with metal sippers, Access to the ethanol solution was limited to 4 h every day. One hour before the dark period on day 4, animals received a single dose of CB1R antagonist by oral gavage (*S*-MRI-1867 1, 3, 10 mg/kg; *R*-MRI-1867 10 mg/kg, rimonabant 10 mg/kg; JD5037 3 mg/kg or vehicle) or i.c.v. (*S*-MRI-1867 2 μg; rimonabant 2 μg; JD5037 1 μg or vehicle), and the alcohol session was repeated. Immediately after the drinking session, mice were deeply anesthetized with isoflurane. Blood was collected by traumatic avulsion of the orbital globe and kept for 10 min at room temperature in Eppendorf tubes followed by centrifugation at $3000 \times g$ for 10 min at 4 °C. Serum was collected and kept frozen in sealed vials at −80 °C until analysis.

4.7. Acute Ethanol Intoxication

The acute intoxication was performed on 8-week-old C57BL/6J mice as developed by [54], with modifications. This model was designed to achieve blood alcohol levels that would produce physiological effects comparable to human binge drinking. The animals were fasted overnight. On the day of the experiment, they were transferred to the procedure room and allowed to acclimate to the new environment for 2 h while remaining in their home cages. Animals were divided into six treatment groups. Each mouse received *S*-MRI-1867 10 mg/kg or *R*-MRI-1867 10 mg/kg or vehicle by oral gavage, followed 6 g/kg ethanol (30% *w/v*) or saline by the same administration route 30 min later. Mice were anaesthetized by isoflurane 90 min later and sacrificed for blood withdrawal and tissue collection. To determine endotoxin levels, blood was drawn aseptically from the portal vein without anticoagulant and clotted for 30 min at room temperature. Samples were then centrifuged at $2000 \times g$ for 15 min at room temperature. Serum was pipetted off aseptically into new sterile vials and kept frozen at −80 °C until analysis. For the collection of tissue specimens, the entire alimentary tract was removed and cleaned from the mesentery from mice treated with S-MRI-1867 and saline. Stomach and ~5 cm segments of duodenum, jejunum, ileum, and colon were dissected out, flushed three times with 20 mL ice-cold PBS, and kept in −80 °C freezer till analysis.

4.8. Blood Alcohol and Acetaldehyde Assays

Alcohol concentration was measured in serum using a sample analyzer (Model GM7 Micro-Stat, Analox Instruments Ltd., Amblecote, United Kingdom) or an alcohol dehydrogenase kit (procedure 332-UV; Sigma-Aldrich, St. Louis, MO USA) according to the manufacturer's instructions.

The acetaldehyde measurement by gas chromatography-mass spectrometry (GC/MS) was conducted as described by Jin et al., [55] with modifications. In brief, 50 μL of serum or about 50 mg of liver tissue were mixed with 5 μM of 2H6–EtOH (internal standard for ethanol) and 0.04 μM of $^2H_4$-acetaldehyde (internal standard for acetaldehyde) prior to adding 200 μL of 0.6 N perchloric acid into each sample. Serum samples were centrifuged at $1780 \times g \times 15$ min at 2 °C after vortexed for 30 s. Liver samples were homogenized and then centrifuged at $13,200 \times g \times 15$ min at 2 °C. The supernatant of each sample was quantitatively transferred into a 20 mL headspace vial and capped immediately. Headspace vials were then loaded onto the 111-vial tray of a Headspace Sampler coupled to GC/MS (Agilent Technologies, Santa Clara, CA, USA). The concentrations of acetaldehyde in serum were calculated by comparing the integrated areas of ethanol and acetaldehyde peaks on the gas chromatograms with those of known concentrations of internal standards added in each sample.

4.9. Serum Endotoxin Measurement

Serum samples were thawed and diluted 1:5 with sterile water. Samples were then heat shocked at 75 °C in Eppendorf ThermoMixer C (Eppendorf, Enfield, CT, USA) and allowed to cool to room temperature for 10 min prior to colorimetric assay, according to the manufacturer's instructions (Pierce™ Chromogenic Endotoxin Quant Kit; Thermo Fisher Scientific, Waltham, MA, USA).

All materials used in the assay (e.g., pipette tips, glass tubes, microcentrifuge tubes, and disposable 96-well microplates) were endotoxin-free.

4.10. Tissue Levels of MRI-1867

Tissues were extracted as described previously [45]. MRI-1867 concentration was determined by stable isotope dilution liquid chromatography/tandem mass spectrometry (LC-MS/MS) using an Agilent 6410 triple quadrupole mass spectrometer (Agilent Technologies, Santa Clara, CA, USA) coupled to an Agilent 1200 LC system (Agilent Technologies, USA). Chromatographic and mass spectrometer conditions were set as described previously [45]. The molecular ion and fragments for MRI-1867 were as follows: m/z 548.1→145 and 548.1→257.1 (CID energy: 56 V and 24 V, respectively). The amounts of MRI-1867 in the samples were determined against standard curves. Values are expressed as μmol/gwet tissue weight.

4.11. Drugs

The synthesis, purification, and verification of the MRI-1867 and JD5037 structures were performed as described [34]. Rimonabant was obtained through the National Institute on Drug Abuse Drug Supply Program (ref number NOCD-082). CP 55,940 was from Tocris Bioscience (Minneapolis, MN USA). Cell culture-grade DMSO (ATCC, Manassas, VA, USA) was used as a solvent mix for intracerebroventricular microinfusions. Ethanol (ethyl alcohol, U.S.P. 200 proof, anhydrous; The Warner Graham Company, Cockeysville, MD, USA) was obtained through NIH Supply Center. A standard rodent chow (Teklad laboratory animal diet) was purchased from Envigo (Indianapolis, IN, USA). Pierce™ Chromogenic Endotoxin Quant Kit was from Thermo Fisher Scientific (Waltham, MA, USA). All other chemicals were from MilliporeSigma (Rockville, MD, USA). Drugs were aliquoted and stored at −80 °C. For intragastric administration, chemicals were dissolved in DMSO:Tween 80:water (5:2:93). For intracerebroventricular (i.c.v.) microinfusion, drugs were suspended in DMSO:Tween 80:PEG400:saline (solvent composition for JD5037 3:8:30:59 and rimonabant 1:5:30:64).

4.12. Quantification and Statistical Analysis

Values are presented as mean ± s.e.m., with the number of replicates and the level of significance reported in figures and figure legends. Statistical data analysis was performed using GraphPad Prism 8 for Windows (version 8.0.1, GraphPad Software, San Diego, CA, USA). A two-tailed Student's t-test for paired or unpaired data was used for comparison of values between two groups. For multiple groups, ordinary one-way ANOVA followed by Dunnett's post hoc test was applied. Time-dependent variables were analyzed by two-way ANOVA followed by Tukey's multiple comparisons test. Differences were considered significant when $p < 0.05$.

Author Contributions: G.G. and G.K. conceived the project, conceptualized the experiments and methodology, analyzed the data, and wrote the paper with input from all the authors; O.G.-C., L.S.-M., A.H. and C.N.Z. ran behavioral tests and analyzed biological samples; C.M.W. generated synthetic intermediates of MRI-1867; M.R.I. synthesized and performed chemical purity assessments of MRI-1867 enantiomers; R.C. and Y.L. conducted the mass spectrometry experiments; G.G., L.S.-M.,

A.H., C.N.Z. contributed to data acquisition and analyses; B.G. did a critical review of the manuscript. All authors have read and agreed to the published version of the manuscript.

Funding: This work was supported by intramural funds from the National Institute on Alcohol Abuse and Alcoholism to G.K.

Institutional Review Board Statement: All animal procedures were approved by the Institutional Animal Care and Use Committee of NIAAA, NIH (Animal Experimentation permit number LPS-GK-1), and the experiments were carried out in accordance with its guidelines.

Data Availability Statement: Not applicable.

Acknowledgments: We thank Judith Harvey-White for her technical assistance with tissue extractions for the HPLC mass spectrometry, Luo Guoxiang for genotyping CB1KO mouse line.

Conflicts of Interest: R.C., M.R.I. and G.K. are listed as coinventors on a US patent (US 9,765,031 B2) covering MRI-1867 and related compounds.

Sample Availability: Samples of MRI-1867 are available from R.C., M.R.I. and G.K.

References

1. Murray, C.J.L.; Aravkin, A.Y.; Zheng, P.; Abbafati, C.; Abbas, K.M.; Abbasi-Kangevari, M.; Abd-Allah, F.; Abdelalim, A.; Abdollahi, M.; Abdollahpour, I.; et al. Global burden of 87 risk factors in 204 countries and territories, 1990-2019: A systematic analysis for the Global Burden of Disease Study 2019. *Lancet* **2020**, *396*, 1223–1249. [CrossRef]
2. McGinn, M.A.; Pantazis, C.B.; Tunstall, B.J.; Marchette, R.C.N.; Carlson, E.R.; Said, N.; Koob, G.F.; Vendruscolo, L.F. Drug addiction co-morbidity with alcohol: Neurobiological insights. *Int. Rev. Neurobiol.* **2021**, *157*, 409–472. [CrossRef]
3. Nordeck, C.D.; Riehm, K.E.; Smail, E.J.; Holingue, C.; Kane, J.C.; Johnson, R.M.; Veldhuis, C.B.; Kalb, L.G.; Stuart, E.A.; Kreuter, F.; et al. Changes in drinking days among US adults during the COVID-19 pandemic. *Addiction* **2021**. [CrossRef] [PubMed]
4. Koob, G.F.; Volkow, N.D. Neurobiology of addiction: A neurocircuitry analysis. *Lancet Psychiatry* **2016**, *3*, 760–773. [CrossRef]
5. Morley, K.C.; Perry, C.J.; Watt, J.; Hurzeler, T.; Leggio, L.; Lawrence, A.J.; Haber, P. New approved and emerging pharmacological approaches to alcohol use disorder: A review of clinical studies. *Expert Opin. Pharmacother.* **2021**, *22*, 1291–1303. [CrossRef] [PubMed]
6. Gupta, A.; Khan, H.; Kaur, A.; Singh, T.G. Novel Targets Explored in the Treatment of Alcohol Withdrawal Syndrome. *CNS Neurol. Disord. Drug Targets* **2021**, *20*, 158–173. [CrossRef]
7. Leclercq, S.; De Timary, P.; Delzenne, N.M.; Starkel, P. The link between inflammation, bugs, the intestine and the brain in alcohol dependence. *Transl. Psychiatry* **2017**, *7*, e1048. [CrossRef]
8. Llopis, M.; Cassard, A.M.; Wrzosek, L.; Boschat, L.; Bruneau, A.; Ferrere, G.; Puchois, V.; Martin, J.C.; Lepage, P.; Le Roy, T.; et al. Intestinal microbiota contributes to individual susceptibility to alcoholic liver disease. *Gut* **2016**, *65*, 830–839. [CrossRef]
9. Mason, C.M.; Dobard, E.; Kolls, J.; Nelson, S. Effect of alcohol on bacterial translocation in rats. *Alcohol. Clin. Exp. Res.* **1998**, *22*, 1640–1645. [CrossRef] [PubMed]
10. Liangpunsakul, S.; Toh, E.; Ross, R.A.; Heathers, L.E.; Chandler, K.; Oshodi, A.; McGee, B.; Modlik, E.; Linton, T.; Mangiacarne, D.; et al. Quantity of alcohol drinking positively correlates with serum levels of endotoxin and markers of monocyte activation. *Sci. Rep.* **2017**, *7*, 4462. [CrossRef] [PubMed]
11. Chen, M.T.; Hou, P.F.; Zhou, M.; Ren, Q.B.; Wang, X.L.; Huang, L.; Hui, S.C.; Yi, L.; Mi, M.T. Resveratrol attenuates high-fat diet-induced non-alcoholic steatohepatitis by maintaining gut barrier integrity and inhibiting gut inflammation through regulation of the endocannabinoid system. *Clin. Nutr.* **2020**, *39*, 1264–1275. [CrossRef]
12. Leclercq, S.; Le Roy, T.; Furgiuele, S.; Coste, V.; Bindels, L.B.; Leyrolle, Q.; Neyrinck, A.M.; Quoilin, C.; Amadieu, C.; Petit, G.; et al. Gut Microbiota-Induced Changes in beta-Hydroxybutyrate Metabolism Are Linked to Altered Sociability and Depression in Alcohol Use Disorder. *Cell Rep.* **2020**, *33*, 108238. [CrossRef]
13. Bala, S.; Marcos, M.; Gattu, A.; Catalano, D.; Szabo, G. Acute Binge Drinking Increases Serum Endotoxin and Bacterial DNA Levels in Healthy Individuals. *PLoS ONE* **2014**, *9*, e96864. [CrossRef]
14. Airapetov, M.; Eresko, S.; Lebedev, A.; Bychkov, E.; Shabanov, P. The role of Toll-like receptors in neurobiology of alcoholism. *Biosci. Trends* **2021**, *15*, 74–82. [CrossRef]
15. Qin, L.; He, J.; Hanes, R.N.; Pluzarev, O.; Hong, J.S.; Crews, F.T. Increased systemic and brain cytokine production and neuroinflammation by endotoxin following ethanol treatment. *J. Neuroinflamm.* **2008**, *5*, 10. [CrossRef]
16. Lowe, P.P.; Gyongyosi, B.; Satishchandran, A.; Iracheta-Vellve, A.; Cho, Y.; Ambade, A.; Szabo, G. Reduced gut microbiome protects from alcohol-induced neuroinflammation and alters intestinal and brain inflammasome expression. *J. Neuroinflamm.* **2018**, *15*, 1–12. [CrossRef] [PubMed]
17. Donnadieu-Rigole, H.; Pansu, N.; Mura, T.; Pelletier, S.; Alarcon, R.; Gamon, L.; Perney, P.; Apparailly, F.; Lavigne, J.P.; Dunyach-Remy, C. Beneficial Effect of Alcohol Withdrawal on Gut Permeability and Microbial Translocation in Patients with Alcohol Use Disorder. *Alcohol. Clin. Exp. Res.* **2018**, *42*, 32–40. [CrossRef]

18. Leclercq, S.; Matamoros, S.; Cani, P.D.; Neyrinck, A.M.; Jamar, F.; Starkel, P.; Windey, K.; Tremaroli, V.; Backhed, F.; Verbeke, K.; et al. Intestinal permeability, gut-bacterial dysbiosis, and behavioral markers of alcohol-dependence severity. *Proc. Natl. Acad. Sci. USA* **2014**, *111*, E4485–E4493. [CrossRef] [PubMed]

19. Han, X.N.; Fink, M.P.; Yang, R.K.; Delude, R.L. Increased iNOS activity is essential for intestinal epithelial tight junction dysfunction in endotoxemic mice. *Shock* **2004**, *21*, 261–270. [CrossRef]

20. Tang, Y.; Forsyth, C.B.; Farhadi, A.; Rangan, J.; Jakate, S.; Shaikh, M.; Banan, A.; Fields, J.Z.; Keshavarzian, A. Nitric Oxide Mediated Intestinal Injury Is Required for Alcohol-Induced Gut Leakiness and Liver Damage. *Alcohol. Clin. Exp. Res.* **2009**, *33*, 1220–1230. [CrossRef] [PubMed]

21. Forsyth, C.B.; Tang, Y.; Shaikh, M.; Zhang, L.; Keshavarzian, A. Role of snail activation in alcohol-induced iNOS-mediated disruption of intestinal epithelial cell permeability. *Alcohol. Clin. Exp. Res.* **2011**, *35*, 1635–1643. [CrossRef] [PubMed]

22. Lambert, J.C.; Zhou, Z.; Wang, L.; Song, Z.; McClain, C.J.; Kang, Y.J. Preservation of intestinal structural integrity by zinc is independent of metallothionein in alcohol-intoxicated mice. *Am. J. Pathol.* **2004**, *164*, 1959–1966. [CrossRef]

23. McKim, S.E.; Gabele, E.; Isayama, F.; Lambert, J.C.; Tucker, L.M.; Wheeler, M.D.; Connor, H.D.; Mason, R.P.; Doll, M.A.; Hein, D.W.; et al. Inducible nitric oxide synthase is required in alcohol-induced liver injury: Studies with knockout mice. *Gastroenterology* **2003**, *125*, 1834–1844. [CrossRef]

24. Oppong-Damoah, A.; Gannon, B.M.; Murnane, K.S. The Endocannabinoid System and Alcohol Dependence: Will Cannabinoid Receptor 2 Agonism be More Fruitful than Cannabinoid Receptor 1 Antagonism? *CNS Neurol. Disord. Drug Targets* **2021**. [CrossRef] [PubMed]

25. Lavanco, G.; Castelli, V.; Brancato, A.; Tringali, G.; Plescia, F.; Cannizzaro, C. The endocannabinoid-alcohol crosstalk: Recent advances on a bi-faceted target. *Clin. Exp. Pharmacol. Physiol.* **2018**, *45*, 889–896. [CrossRef]

26. Kunos, G. Interactions Between Alcohol and the Endocannabinoid System. *Alcohol. Clin. Exp. Res.* **2020**, *44*, 790–805. [CrossRef] [PubMed]

27. Wang, L.; Liu, H.; Harvey-White, J.; Zimmer, A.; Kunos, G. Endocannabinoid signaling via cannabinoid receptor 1 is involved in ethanol preference and its age-dependent decline in mice. *Proc. Natl. Acad. Sci. USA* **2003**, *100*, 1393–1398. [CrossRef] [PubMed]

28. Balla, A.; Dong, B.; Shilpa, B.M.; Vemuri, K.; Makriyannis, A.; Pandey, S.C.; Sershen, H.; Suckow, R.F.; Vinod, K.Y. Cannabinoid-1 receptor neutral antagonist reduces binge-like alcohol consumption and alcohol-induced accumbal dopaminergic signaling. *Neuropharmacology* **2018**, *131*, 200–208. [CrossRef] [PubMed]

29. Silva, A.A.F.; Barbosa-Souza, E.; Confessor-Carvalho, C.; Silva, R.R.R.; De Brito, A.C.L.; Cata-Preta, E.G.; Silva Oliveira, T.; Berro, L.F.; Oliveira-Lima, A.J.; Marinho, E.A.V. Context-dependent effects of rimonabant on ethanol-induced conditioned place preference in female mice. *Drug Alcohol Depend.* **2017**, *179*, 317–324. [CrossRef]

30. Garcia-Baos, A.; Alegre-Zurano, L.; Cantacorps, L.; Martin-Sanchez, A.; Valverde, O. Role of cannabinoids in alcohol-induced neuroinflammation. *Prog. Neuropsychopharmacol. Biol. Psychiatry* **2021**, *104*, 110054. [CrossRef]

31. Karwad, M.A.; Couch, D.G.; Theophilidou, E.; Sarmad, S.; Barrett, D.A.; Larvin, M.; Wright, K.L.; Lund, J.N.; O'Sullivan, S.E. The role of CB1 in intestinal permeability and inflammation. *FASEB J.* **2017**, *31*, 3267–3277. [CrossRef]

32. Nathan, P.J.; O'Neill, B.V.; Napolitano, A.; Bullmore, E.T. Neuropsychiatric adverse effects of centrally acting antiobesity drugs. *CNS Neurosci. Ther.* **2011**, *17*, 490–505. [CrossRef] [PubMed]

33. Godlewski, G.; Cinar, R.; Coffey, N.J.; Liu, J.; Jourdan, T.; Mukhopadhyay, B.; Chedester, L.; Liu, Z.; Osei-Hyiaman, D.; Iyer, M.R.; et al. Targeting Peripheral CB1 Receptors Reduces Ethanol Intake via a Gut-Brain Axis. *Cell Metab.* **2019**, *29*, 1320–1333.e8. [CrossRef] [PubMed]

34. Cinar, R.; Iyer, M.R.; Liu, Z.Y.; Cao, Z.X.; Jourdan, T.; Erdelyi, K.; Godlewski, G.; Szanda, G.; Liu, J.; Park, J.K.; et al. Hybrid inhibitor of peripheral cannabinoid-1 receptors and inducible nitric oxide synthase mitigates liver fibrosis. *JCI Insight* **2016**, *1*, e87336. [CrossRef]

35. Roger, C.; Buch, C.; Muller, T.; Leemput, J.; Demizieux, L.; Passilly-Degrace, P.; Cinar, R.; Iyer, M.R.; Kunos, G.; Verges, B.; et al. Simultaneous Inhibition of Peripheral CB1R and iNOS Mitigates Obesity-Related Dyslipidemia Through Distinct Mechanisms. *Diabetes* **2020**, *69*, 2120–2132. [CrossRef]

36. Martin, B.R.; Compton, D.R.; Thomas, B.F.; Prescott, W.R.; Little, P.J.; Razdan, R.K.; Johnson, M.R.; Melvin, L.S.; Mechoulam, R.; Ward, S.J. Behavioral, Biochemical, and Molecular Modeling Evaluations of Cannabinoid Analogs. *Pharmacol. Biochem. Behav.* **1991**, *40*, 471–478. [CrossRef]

37. Tam, J.; Vemuri, V.K.; Liu, J.; Batkai, S.; Mukhopadhyay, B.; Godlewski, G.; Osei-Hyiaman, D.; Ohnuma, S.; Ambudkar, S.V.; Pickel, J.; et al. Peripheral CB1 cannabinoid receptor blockade improves cardiometabolic risk in mouse models of obesity. *J. Clin. Investig.* **2010**, *120*, 2953–2966. [CrossRef] [PubMed]

38. Basavarajappa, B.S.; Yalamanchili, R.; Cravatt, B.F.; Cooper, T.B.; Hungund, B.L. Increased ethanol consumption and preference and decreased ethanol sensitivity in female FAAH knockout mice. *Neuropharmacology* **2006**, *50*, 834–844. [CrossRef]

39. Hungund, B.L.; Szakall, I.; Adam, A.; Basavarajappa, B.S.; Vadasz, C. Cannabinoid CB1 receptor knockout mice exhibit markedly reduced voluntary alcohol consumption and lack alcohol-induced dopamine release in the nucleus accumbens. *J. Neurochem.* **2003**, *84*, 698–704. [CrossRef] [PubMed]

40. Hansson, A.C.; Bermudez-Silva, F.J.; Malinen, H.; Hyytia, P.; Sanchez-Vera, I.; Rimondini, R.; Rodriguez de Fonseca, F.; Kunos, G.; Sommer, W.H.; Heilig, M. Genetic impairment of frontocortical endocannabinoid degradation and high alcohol preference. *Neuropsychopharmacology* **2007**, *32*, 117–126. [CrossRef] [PubMed]

41. Perra, S.; Pillolla, G.; Melis, M.; Muntoni, A.L.; Gessa, G.L.; Pistis, M. Involvement of the endogenous cannabinoid system in the effects of alcohol in the mesolimbic reward circuit: Electrophysiological evidence in vivo. *Psychopharmacology* **2005**, *183*, 368–377. [CrossRef]

42. Pava, M.J.; Blake, E.M.; Green, S.T.; Mizroch, B.J.; Mulholland, P.J.; Woodward, J.J. Tolerance to cannabinoid-induced behaviors in mice treated chronically with ethanol. *Psychopharmacology* **2012**, *219*, 137–147. [CrossRef]

43. Leclercq, S.; De Saeger, C.; Delzenne, N.; De Timary, P.; Starkel, P. Role of Inflammatory Pathways, Blood Mononuclear Cells, and Gut-Derived Bacterial Products in Alcohol Dependence. *Biol. Psychiatry* **2014**, *76*, 725–733. [CrossRef]

44. Souza-Smith, F.M.; Lang, C.H.; Nagy, L.E.; Bailey, S.M.; Parsons, L.H.; Murray, G.J. Physiological processes underlying organ injury in alcohol abuse. *Am. J. Physiol. Endocrinol. Metab.* **2016**, *311*, E605–E619. [CrossRef]

45. Cinar, R.; Gochuico, B.R.; Iyer, M.R.; Jourdan, T.; Yokoyama, T.; Park, J.K.; Coffey, N.J.; Pri-Chen, H.; Szanda, G.; Liu, Z.; et al. Cannabinoid CB1 receptor overactivity contributes to the pathogenesis of idiopathic pulmonary fibrosis. *JCI Insight* **2017**, *2*, e92281. [CrossRef] [PubMed]

46. Alhamoruni, A.; Lee, A.C.; Wright, K.L.; Larvin, M.; O'Sullivan, S.E. Pharmacological effects of cannabinoids on the Caco-2 cell culture model of intestinal permeability. *J. Pharmacol. Exp. Ther.* **2010**, *335*, 92–102. [CrossRef] [PubMed]

47. Alhamoruni, A.; Wright, K.L.; Larvin, M.; O'Sullivan, S.E. Cannabinoids mediate opposing effects on inflammation-induced intestinal permeability. *Br. J. Pharmacol.* **2012**, *165*, 2598–2610. [CrossRef] [PubMed]

48. Muccioli, G.G.; Naslain, D.; Backhed, F.; Reigstad, C.S.; Lambert, D.M.; Delzenne, N.M.; Cani, P.D. The endocannabinoid system links gut microbiota to adipogenesis. *Mol. Syst. Biol.* **2010**, *6*, 392. [CrossRef]

49. Cinar, R.; Park, J.K.; Zawatsky, C.N.; Coffey, N.J.; Bodine, S.P.; Abdalla, J.; Yokoyama, T.; Jourdan, T.; Lay, L.; Zuo, M.X.G.; et al. CB1R and iNOS are distinct players promoting pulmonary fibrosis in Hermansky–Pudlak syndrome. *Clin. Transl. Med.* **2021**, *11*, e471. [CrossRef] [PubMed]

50. Udi, S.; Hinden, L.; Ahmad, M.; Drori, A.; Iyer, M.R.; Cinar, R.; Herman-Edelstein, M.; Tam, J. Dual inhibition of cannabinoid CB1 receptor and inducible NOS attenuates obesity-induced chronic kidney disease. *Br. J. Pharmacol.* **2020**, *177*, 110–127. [CrossRef] [PubMed]

51. Cinar, R.; Iyer, M.R.; Kunos, G. The therapeutic potential of second and third generation CB1R antagonists. *Pharmacol. Ther.* **2020**, *208*, 107477. [CrossRef]

52. Zimmer, A.; Zimmer, A.M.; Hohmann, A.G.; Herkenham, M.; Bonner, T.I. Increased mortality, hypoactivity, and hypoalgesia in cannabinoid CB1 receptor knockout mice. *Proc. Natl. Acad. Sci. USA* **1999**, *96*, 5780–5785. [CrossRef] [PubMed]

53. Walf, A.A.; Frye, C.A. The use of the elevated plus maze as an assay of anxiety-related behavior in rodents. *Nat. Protoc.* **2007**, *2*, 322–328. [CrossRef] [PubMed]

54. Rhodes, J.S.; Best, K.; Belknap, J.K.; Finn, D.A.; Crabbe, J.C. Evaluation of a simple model of ethanol drinking to intoxication in C57BL/6J mice. *Physiol. Behav.* **2005**, *84*, 53–63. [CrossRef] [PubMed]

55. Jin, S.; Cao, Q.; Yang, F.; Zhu, H.; Xu, S.; Chen, Q.; Wang, Z.; Lin, Y.; Cinar, R.; Pawlosky, R.J.; et al. Brain ethanol metabolism by astrocytic ALDH2 drives the behavioural effects of ethanol intoxication. *Nat. Metab.* **2021**, *3*, 337–351. [CrossRef]

Article

Three of a Kind: Control of the Expression of Liver-Expressed Antimicrobial Peptide 2 (LEAP2) by the Endocannabinoidome and the Gut Microbiome

Mélissa Shen [1], Claudia Manca [1,2,†], Francesco Suriano [3,‡], Nayudu Nallabelli [1], Florent Pechereau [4], Bénédicte Allam-Ndoul [4], Fabio Arturo Iannotti [5], Nicolas Flamand [1], Alain Veilleux [4], Patrice D. Cani [3], Cristoforo Silvestri [1,4,*] and Vincenzo Di Marzo [1,2,4,5,6,*]

[1] Quebec Heart and Lung Institute Research Centre, Department of Medicine, Faculty of Medicine, Université Laval, Quebec City, QC G1V 0A6, Canada; melissa.shen.1@ulaval.ca (M.S.); claudia.manca.1@ulaval.ca (C.M.); nayudu.nallabelli.1@ulaval.ca (N.N.); Nicolas.Flamand@criucpq.ulaval.ca (N.F.)

[2] Unité Mixte Internationale en Recherche Chimique et Biomoléculaire du Microbiome et son Impact sur la Santé Métabolique et la Nutrition, Université Laval, Quebec City, QC G1V 0A6, Canada

[3] Metabolism and Nutrition Research Group, Louvain Drug Research Institute (LDRI), Walloon Excellence in Life Sciences and BIOtechnology (WELBIO), UCLouvain, Université Catholique de Louvain, 1200 Brussels, Belgium; francesco.suriano@uclouvain.be (F.S.); patrice.cani@uclouvain.be (P.D.C.)

[4] Centre Nutrition, Santé et Société (NUTRISS), Institut sur la Nutrition et les Aliments Fonctionnels (INAF), École de Nutrition (FSAA), Université Laval, Quebec City, QC G1V 0A6, Canada; florent.pechereau.1@ulaval.ca (F.P.); benedicte.allam-ndoul@criucpq.ulaval.ca (B.A.-N.); Alain.Veilleux@fsaa.ulaval.ca (A.V.)

[5] Endocannabinoid Research Group, Institute of Biomolecular Chemistry, Consiglio Nazionale delle Ricerche, 80078 Pozzuoli, Italy; fabio.iannotti@icb.cnr.it

[6] Canada Excellence Research Chair on the Microbiome-Endocannabinoidome Axis in Metabolic Health, Université Laval, Quebec City, QC G1V 0A6, Canada

[*] Correspondence: cristoforo.silvestri@criucpq.ulaval.ca (C.S.); vincenzo.di-marzo.1@ulaval.ca (V.D.); Tel.: +1-418-656-8711 (ext. 7229) (C.S.); +1-418-656-8711 (ext. 7263) (V.D.)

[†] Current Address: Department of Biomedical Science, University of Cagliari, 09042 Monserrato, Italy.

[‡] Current Address: Mucin Biology Groups, Institute of Biomedicine, Department of Medical Biochemistry, University of Gothenburg, 413 90 Gothenburg, Sweden.

Citation: Shen, M.; Manca, C.; Suriano, F.; Nallabelli, N.; Pechereau, F.; Allam-Ndoul, B.; Iannotti, F.A.; Flamand, N.; Veilleux, A.; Cani, P.D.; et al. Three of a Kind: Control of the Expression of Liver-Expressed Antimicrobial Peptide 2 (LEAP2) by the Endocannabinoidome and the Gut Microbiome. *Molecules* **2022**, 27, 1. https://doi.org/10.3390/molecules27010001

Academic Editor: Mauro Maccarrone

Received: 30 October 2021
Accepted: 16 December 2021
Published: 21 December 2021

Publisher's Note: MDPI stays neutral with regard to jurisdictional claims in published maps and institutional affiliations.

Abstract: The endocannabinoidome (expanded endocannabinoid system, eCBome)-gut microbiome (mBIome) axis plays a fundamental role in the control of energy intake and processing. The liver-expressed antimicrobial peptide 2 (LEAP2) is a recently identified molecule acting as an antagonist of the ghrelin receptor and hence a potential effector of energy metabolism, also at the level of the gastrointestinal system. Here we investigated the role of the eCBome-gut mBIome axis in the control of the expression of LEAP2 in the liver and, particularly, the intestine. We confirm that the small intestine is a strong contributor to the circulating levels of LEAP2 in mice, and show that: (1) intestinal *Leap2* expression is profoundly altered in the liver and small intestine of 13 week-old germ-free (GF) male mice, which also exhibit strong alterations in eCBome signaling; fecal microbiota transfer (FMT) from conventionally raised to GF mice completely restored normal *Leap2* expression after 7 days from this procedure; in 13 week-old female GF mice no significant change was observed; (2) *Leap2* expression in organoids prepared from the mouse duodenum is elevated by the endocannabinoid noladin ether, whereas in human Caco-2/15 epithelial intestinal cells it is elevated by PPARγ activation by rosiglitazone; (3) *Leap2* expression is elevated in the ileum of mice with either high-fat diet—or genetic leptin signaling deficiency—(i.e., *ob/ob* and *db/db* mice) induced obesity. Based on these results, we propose that LEAP2 originating from the small intestine may represent a player in eCBome- and/or gut mBIome-mediated effects on food intake and energy metabolism.

Keywords: endocannabinoid; PPARs; gut microbiome; intestine; ghrelin; LEAP2

1. Introduction

The liver-expressed antimicrobial peptide 2 (LEAP2) was originally isolated from human hemofiltrates [1]. Its encoding gene is composed of three exons and two introns located at chromosome 5q31 in humans and chromosome 11 in mice. There are two different transcripts of LEAP2 in humans; the one mainly found in the liver, small intestine, kidney, colon, and gastric antrum is the 350bp transcript, whereas, the lung, heart, and trachea express the 550bp transcript [1,2]. The murine *Leap2* RNA encodes for a 76 amino acid protein, although the mature LEAP2 peptide is composed of 40 amino acid residues [1]. The peptide is positively charged and shows similar characteristics to other cationic peptides with antimicrobial activity in vitro. It is composed of an N-terminal (1–14 amino acids) and a C-terminal (15–40 amino acids) domain and two disulfide bonds. LEAP2 shares structural characteristics with antimicrobial peptides such as defensins and LEAP1/hepcidin [1]. Several Gram-positive bacteria, such as *Bacillus megaterium* and *Bacillus subtilis*, are sensitive to treatment with a synthetic LEAP2 peptide, which did not affect the growth of Gram-negative *Escherichia coli* and *Pseudomonas* [1]. The antimicrobial activity of LEAP2 was suggested to be dependent on the amino-terminal domain and not on the induction of membrane destabilization and/or pore formation under physiological conditions [3], indicating that the peptide may bind to an intracellular target to exert this activity.

The strong sequence conservation of LEAP2 in vertebrates might indicate a physiological role unrelated to antimicrobial activity [3,4]. Whilst it was shown that the peptide is not mitogenic for epithelial cells and does not function directly to link the innate and adaptive immune systems [2,5], regulation of the action of ghrelin, a key hormone produced from the stomach and acting in the brain to regulate food intake, reward, and other fundamental central nervous system (CNS) functions, was recently suggested as an additional function of LEAP2. Ge et al. [6] used a mouse model of vertical sleeve gastrectomy (VSG) as a tool to identify secreted proteins and peptides that might act as metabolic regulators. They analyzed various genes that encode for secreted proteins and peptides in the stomach and intestines and found that one set of genes exhibited inverse regulation between the stomach and duodenum. Among these genes, *Leap2* expression increased by 52-fold in the stomach and decreased by 94% in the duodenum following VSG [6]. Subsequently, it was found that LEAP2 acts as an endogenous antagonist of the ghrelin receptor by specifically inhibiting ghrelin binding to its receptor, the GHSR, in a non-competitive manner, thus blocking ghrelin-mediated GH release, food intake, and glucose mobilization [6]. This discovery led to the proposal of LEAP2 as a new potential therapeutic target for uncontrolled ghrelin signaling-related diseases, such as obesity and diabetes, cachexia, anorexia, alcohol abuse, and Prader-Willi Syndrome.

Two other reciprocally interacting players in energy metabolism and its pathological disturbances are: (1) the endocannabinoidome (eCBome), which includes: (a) the endocannabinoids anandamide (AEA), 2-arachidonoylglycerol (2-AG), and noladin ether; (b) the cannabinoid CB1 and CB2 receptors; (c) endocannabinoid-related mediators, such as the *N*-acylethanolamines (NAE), like AEA, and other long-chain fatty acid amides and esters in general, such as the 2-monoacyl-glycerols (2-MAGs), like 2-AG, and the *N*-acyl amino acids, the *N*-acyl-glycines, as well as their receptors (which encompass peroxisome proliferator-activated receptors [PPARs], transient receptor potential [TRP] channels and orphan G-protein-coupled receptors [GPRs]) and anabolic/catabolic enzymes [7]; and (2) the gut microbiome (mBIome), which encompasses thousands of commensal intestinal microorganism species with their armamentarium of genes, proteins and metabolites signaling to the host [8]. Both these systems have been related to ghrelin function, for example through the following mechanisms: (1) ghrelin has been shown to enhance food intake partly via endocannabinoid biosynthesis and CB1 receptor activation, e.g., in the hypothalamus [9,10], while, vice versa, some central effects of CB1 receptors have been suggested to be mediated by activation of the ghrelin receptor, GHSR [11]; additionally, the anorexigenic NAE, *N*-oleoyl-ethanolamine (OEA), which is inactive at cannabinoid receptors and activates instead PPARα, TRPV1 and GPR119, or CB1 activation by endocannabinoids, were suggested to inhibit or stimulate, respectively, ghrelin release from the stomach [12–14]; interestingly, in human plasma, similar or opposing alterations in the levels of ghrelin and 2-AG or non-endocannabinoid NAEs, respectively, occur following exposure to palatable food of lean volunteers [15], whereas in obese individuals exposed to chocolate, the plasma levels of both orexigenic and anorexigenic eCBome mediators (i.e., AEA and OEA, respectively) were directly correlated to ghrelin levels [16]; and (2) through some of its specific metabolites, such as the short chain fatty acids (SCFAs), the gut microbiome impacts on ghrelin action at GHSR, whereas concomitant changes in circulating ghrelin levels and specific gut microbiota taxa are also known to occur under different experimental conditions; however, the mechanisms by which the gut microbiota interacts with ghrelin secretion and signaling are still largely unknown [17]. These, and many other previously published data, strongly suggest that both the eCBome and the gut mBIome, through their multiple signaling mechanisms, are likely to regulate energy metabolism also via interactions with ghrelin.

Based on this background, we hypothesized that some of the effects of the eCBome and the gut mBIome on ghrelin action may occur via changes in *Leap2* expression in the liver and, particularly, the intestine, and investigated this hypothesis either in vivo or in vitro. In particular, we first aimed at identifying a direct effect of the gut mBIome by studying *Leap2* expression in germ-free (GF) mice before and after the reinstatement of functional gut microbiota by fecal microbiota transfer (FMT) from conventionally raised (CR) mice. Next, we investigated if the effect of the gut mBIome on *Leap2* expression in the gut was due to its previous effect on the gut eCBome, by looking at how several eCBome mediators, or eCBome receptor activating and inactivating pharmacological tools, affected *Leap2* expression in differentiated intestinal epithelial CaCo-2/15 cells or organoids prepared from the mouse small intestine, in the presence or absence of a gut microbiota-derived pro-inflammatory signal, i.e., lipopolysaccharide (LPS). Finally, we investigated whether, and how, intestinal *Leap2* expression is altered in mouse models of obesity/type 2 diabetes and how such alterations relate to those of the eCBome in the same models, where profound gut microbiota perturbations, sometimes referred to as dysbiosis, also exist.

2. Results

2.1. Germ-Free Mice Exhibit Altered Levels of Leap2 in the Small Intestine

We investigated the impact of the gut microbiota on *Leap2* expression in the liver and intestinal sections of both male and female GF mice at 4 weeks and 13 weeks of age as compared to conventionally raised CR mice of the same sex and age (Figure 1). In 4 week-old male mice, we observed a significant decrease in *Leap2* mRNA expression in the liver, and a non-significant increase in the jejunum and ileum, in GF vs. CR mice, whereas there were no trends in the duodenum (Figure 1A). In 13 week-old males, we observed again a significant decrease in *Leap2* mRNA levels in the liver and a significant increase in expression in the duodenum in GF vs. CR mice, whereas, in the jejunum and ileum, there were no changes (Figure 1A). Levels of circulating LEAP2 protein showed a decrease at 4 weeks (Figure 1B), whereas there was a significant increase in 13 week-old male mice (Figure 1B). This suggests that the liver and duodenum, respectively, may contribute the most to LEAP2 circulating plasma levels in 4 and 13-week old male GF mice.

Concerning female mice, *Leap2* mRNA expression showed a tendency to increase at 4 weeks old in GF vs. CR mice in the liver and duodenum, and the same pattern was observed in the jejunum, where there was a significant increase (Figure 1C). In 13 week-old females, there was a non-significant decrease in *Leap2* mRNA expression in the liver and ileum of GF vs. CR mice, whereas there were no observable changes in the other analyzed tissues (Figure 1C). Protein levels of LEAP2 in the blood did not change in either 4 or 13-week old female GF mice (Figure 1D).

We next performed a fecal microbiota transfer (FMT) from CR male mice of 13 weeks of age into GF male mice to evaluate whether the reintroduction of the gut microbiota is able to reverse the *Leap2* expression changes observed in GF male mice. For this, we privileged male 12 week-old GF mice, since we saw the most interesting results and changes in older male mice compared to female mice. This procedure was previously reported by us in these same mice to successfully reinstate gut microbiota in GF mice [18]. We first checked if we could repeat the results of the previous experiment, for instance, the decrease in the liver and increase in the duodenum of *Leap2* mRNA expression in GF vs. CR mice, and this was indeed the case. More importantly, in both the liver and duodenum, we found that the FMT was able to revert the changes in *Leap2* expression found in male GF mice, with an increase in the liver and a decrease in the duodenum following FMT (Figure 2A). Circulating protein levels of LEAP2 that were upregulated in 13 week-old male GF mice were likewise reduced to baseline CR mouse levels by FMT (Figure 2B).

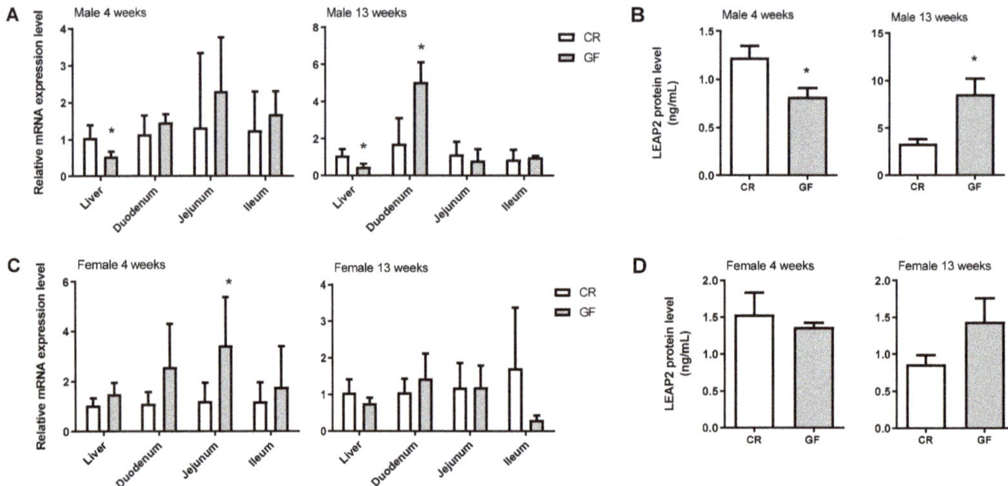

Figure 1. *Leap2* expression in germ-free mice. Gene expression of *Leap2* was measured in the liver and small intestinal regions of conventionally raised (CR) and germ-free (GF) male mice at 4 and 13 (**A**) weeks of age by qPCR and r plotted relative to CR mice for each age and each tissue. LEAP2 protein levels (ng/mL) were measured in the plasma of CR and GF male mice at 4 and 13 (**B**) weeks of age by ELISA and represented as means $\pm$ S.D. n = 5–6. * $p \leq 0.05$. Gene expression of *Leap2* was measured in the liver and small intestinal regions of conventionally raised (CR) and germ-free (GF) female mice at 4 and 13 (**C**) weeks of age by qPCR and plotted relative to CR mice for each age and each tissue. LEAP2 protein levels (ng/mL) were measured in the plasma of CR and GF female mice at 4 and 13 (**D**) weeks of age by ELISA and represented as means $\pm$ S.D. n = 5–6. * $p \leq 0.05$.

These data indicate that the GF status results in sex-dependent significant changes in *Leap2* mRNA expression in the liver and the duodenum, with the latter tissue producing the predominant effect on circulating LEAP2 protein levels in adult male mice, and that such changes are directly due to the lack of the gut microbiota. Since the small intestine of GF mice also exhibits strong FMT-reversible alterations in the levels of eCBome mediators (namely NAEs such as OEA and LEA, which are increased) and receptor mRNAs (namely *Cnr1* and *Ppara*, which are increased, and *Gpr55* and *Gpr18*, which are decreased) [18], we next tested the effect of drugs activating CB1, PPARα, GPR55 and GPR18 receptors on *Leap2* mRNA expression in two different in vitro models of intestinal tissue.

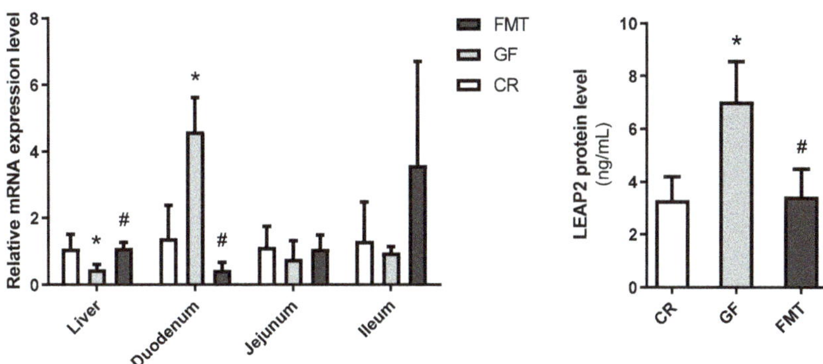

Figure 2. *Leap2* expression in germ-free male mice after a fecal microbiota transfer. Relative *Leap2* gene expression was measured by qPCR (**left**) in the liver and small intestinal regions and LEAP2 protein levels ((**right**); ng/mL) were measured in the plasma by ELISA of 13 week-old conventionally raised (CR) (n = 10), GF (n = 5) and in GF mice after fecal microbiota transfer (FMT) (n = 6). Relative mRNA expression is plotted relative to CR mice for each tissue. Plots represent means $\pm$ S.D. Post-hoc analysis was carried out between GF and CR mice to determine the impact of the lack of microbiota or GF and FMT mice to determine the impact of the introduction of microbiota. * GF vs. CR; # FMT vs. GF. * $p \leq 0.05$, # $p \leq 0.05$.

2.2. Leap2 mRNA Expression in Organoids from the Mouse Duodenum Is Increased by Noladin Ether

We have previously reported that several components of the eCBome are modified within the intestinal sections of GF mice, especially within the duodenum [18]. As we observed changes in the expression of *Leap2* in the duodenum of GF mice, we investigated *Leap2* mRNA expression in small intestine organoids isolated from the mouse duodenum and treated with different concentrations of eCBome mediator inactivating enzyme inhibitors, in the absence or presence of a pro-inflammatory cocktail (LPS, TNFα, IL-1β and IFNγ [LC]). In the absence of LC, organoids treated with: (1) MAGL and FAAH inhibitors, i.e., URB597 and JZL184, respectively, which can indirectly activate the targets of NAEs and 2 MAGs; or (2) CB_1 and CB_2 antagonists/inverse agonists (rimonabant and SR144528, respectively) at 10 nM, did not exhibit significant alterations in *Leap2* expression (Figure 3A). Higher concentrations of the enzyme inhibitors (100 nM and 1 µM) similarly failed to modify *Leap2* expression (Figure 3B). These data may suggest that MAGL and FAAH do not play a major role in inactivating NAEs and 2-MAGs in small intestine epithelial cells, and/or that there is no endogenous tone by AEA and 2-AG at CB1/CB2 receptors regulating *Leap2* expression within duodenum-derived organoids. In order to confirm this hypothesis, we next tested the effects of the hydrolysis-resistant AEA and 2-AG analogs, ACEA, and noladin ether, respectively, to target CB_1/CB_2 receptors directly. Importantly, ACEA and noladin ether, at higher concentrations, are also known to activate TRPV1 and PPARα, respectively [19–21]. Treatment with ACEA at different concentrations, 0.1, 1.0, and 10 µM did not change *Leap2* expression (Figure 3C). However, while the two lowest concentrations of noladin ether (0.1 and 1 µM) also did not change *Leap2* expression, the highest concentration (10 µM) resulted in a statistically significant upregulation of more than 2-fold (Figure 3C).

The LC inflammatory cocktail did not affect *Leap2* expression, nor did co-incubation of LC with URB597, JZL184, SR144528 or rimonabant, all at a 0.1 µM concentration (Figure 4D).

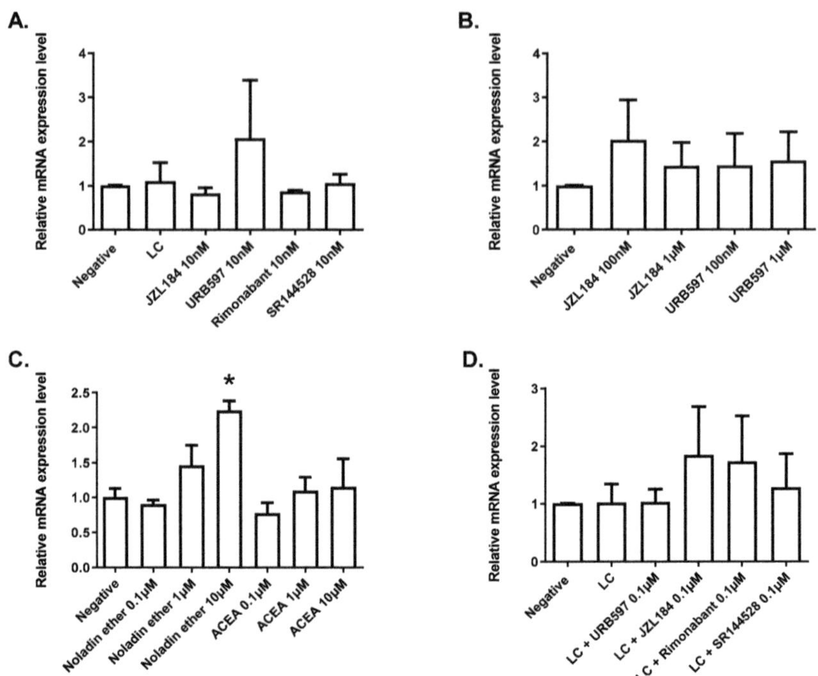

Figure 3. Expression in the duodenum-derived intestinal organoids. Relative *Leap2* gene expression was measured by qPCR in organoids treated with the indicated compounds at the indicated concentrations to test responsiveness to cannabinoid receptor activity through pharmacological manipulation ((**A–C**), Rimonabant; CB1 antagonist, SR144528; CB2 antagonist, URB597; FAAH inhibitor, JZL184; MGLL inhibitor) and in response to an inflammatory lipopolysaccharide/cytokine cocktail ((**D**), LC; lipopolysaccharide [LPS, 10 μg/mL], tumor necrosis factor-alpha [TNFα, 100 ng/mL], interleukin 1 beta [IL-1β; 100 ng/mL] and interferon-gamma [IFNγ, 100 ng/mL]) for 24 h and then co-treated with the indicated compound for a further 24 h. Experiments were performed in duplicate or in triplicate with organoids from at least 3 independent organoid cultures (n = 3–5). Each culture was isolated from the duodenum of different mice. Values are represented as mean ± S.E.M and were compared relative to control (Negative; without DMSO). * $p \leq 0.005$ vs. DMSO.

2.3. LEAP2 mRNA Expression in Human Intestinal Epithelial Caco-2/15 Cells Is Increased following PPARγ Activation

Duodenal organoids contain all cellular elements of the duodenum, and the epithelial layer in general (except for myenteric neurons). Additionally, the geometry of the organoids is inverse to that of the normal intestinal mucosa, since their basolateral side is outside and directly exposed to treatment, and the apical side is inside and less exposed. Therefore, we wanted to investigate the potential for the eCBome to modulate *LEAP2* expression more selectively in epithelial intestinal cells. To do so, we used the human Caco-2/15 intestinal cell line, an easy model to culture that could give us an insight of how LEAP2 expression might behave in humans, which we first fully differentiated into enterocytes to be treated on their apical side. Differentiated Caco-2/15 cells in monolayers were treated with a 1 μM concentration of agonists and antagonists of eCBome receptors, i.e., ACEA (CB1 and TRPV1 agonist), noladin ether (CB1, GPR55, and PPARα agonist), rimonabant (CB1 antagonist/inverse agonist), fenofibrate (PPARα agonist), GW6471 (PPARα antagonist), capsaicin (TRPV1 agonist), capsazepine (TRPV1 antagonist), rosiglitazone (PPARγ agonist) and GW9662 (PPARγ antagonist), as well as with inhibitors of catabolic enzymes, i.e., URB597 and JZL184. We observed no changes in *LEAP2* expression after treatment with any of these compounds at 1 µM for 24 h, except for a 4-fold increase in enterocytes treated with rosiglitazone (Figure 4A), indicating a possible role of PPARγ. In view of this latter result, we investigated if the effect of rosiglitazone could be abolished using a PPARγ antagonist (GW9662) at 1 µM concentration. We tested two doses (0.1 µM and 1 µM) of rosiglitazone, which dose-dependently increased *LEAP2* expression (Figure 4B), supporting our initial observation (Figure 4A). The PPARγ antagonist alone produced no effect on LEAP2 expression; however, co-treatment with rosiglitazone and GW9662 significantly inhibited the rosiglitazone-mediated increase in LEAP2 expression (Figure 4B). These results indicate that PPARγ is involved in the up-regulation of LEAP2. Since AEA, unlike ACEA, has been suggested to activate PPARγ, at concentrations higher than those required to activate cannabinoid receptors, we tested this compound and found that up to a 10 µM concentration, it did not affect LEAP2 expression, though a trend towards increase was observed (Figure 4C). Finally, *N*-arachidonoyl-glycine (NAGly), which has been proposed as both a PPARα and GPR18 agonist (see above), was also tested at 10 µM and similarly found not to stimulate LEAP2 expression (Figure 4C).

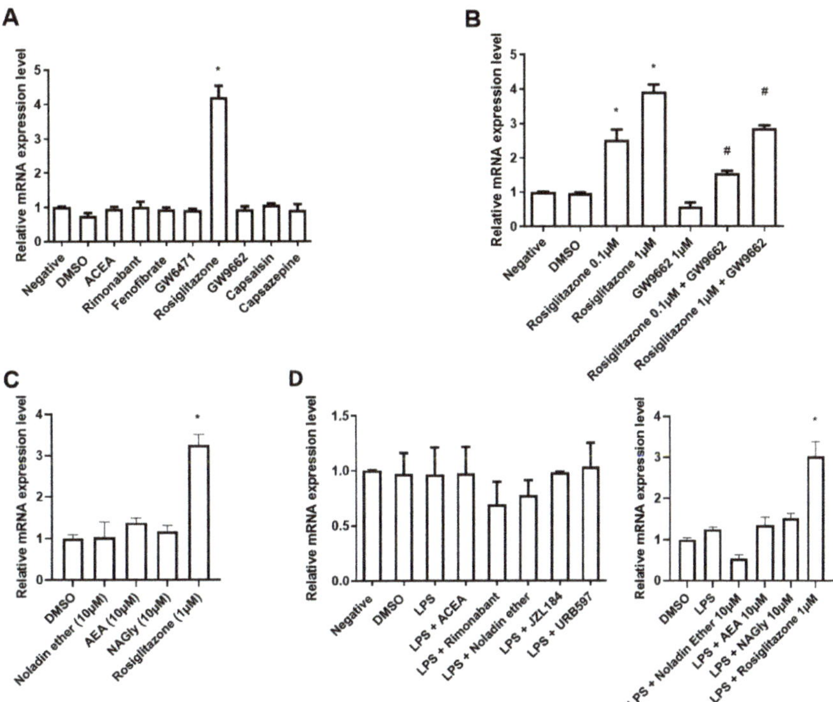

Figure 4. *LEAP2* expression in Caco-2/15 cells. Differentiated cells were treated with the indicated compounds at 1 µM (unless otherwise indicated) for 24 h. Effect of modulation of receptor activity through pharmacological manipulation. ((**A**), ACEA; CB1 and TRPV1 agonist, Rimonabant; CB1 antagonist, Fenofibrate; PPARα agonist, GW6471; PARA antagonist, Rosiglitazone; PPARγ agonist and GW9662; PPARγ antagonist, Capsaicin; TRPV1 agonist and Capsazepine; TRPV1 antagonist). Specificity of *LEAP2* gene expression responsiveness to PPARγ activation (**B**). Effects of noladin ether *N*-arachidonoylethanolamine (AEA) and *N*-arachidonoyl-glycine (NAGly) (**C**). Effects of inflammatory stimulation by lipopolysaccharide (LPS at 10 µg/mL) without and with eCBome pharmacological manipulation ((**D**), URB597; FAAH inhibitor, JZL184; MAGL inhibitor). Values are represented as mean ± S.E.M and were compared relative to DMSO. n = 3. * $p \leq 0.05$ vs. DMSO; # $p \leq 0.05$ vs. relevant Rosiglitazone control.

Similar to what was done with organoids, we also tested the impact of inflammation on *LEAP2* expression in enterocytes, and the effect thereupon of pharmacological manipulation of eCBome receptors and enzymes. In this case, the inflammatory state was induced by incubation with only LPS, concomitantly with agonists and antagonists/inverse agonists of eCBome receptors (CB1, PPARγ, PPARα, GPR55, GPR18, TRPV1) and inhibitors (URB597 and JZL184) of the catabolic enzymes (FAAH and MAGL) for 24 h. LPS did not alter LEAP2 expression, either alone or in the presence of any of the compounds, except again for rosiglitazone (Figure 4D). In particular, also, in this case, AEA and NAGly only tended to increase expression in a non-statistically significant manner.

2.4. Effect of Noladin Ether on PPARα and PPARγ in a Luciferase Functional Assay

Due to the stimulatory effect of noladin ether on *Leap2* expression in intestinal organoids and its lack of effect in Caco-2/15 cells, where PPARγ, but not PPARα, activation instead produces elevation of *LEAP2* mRNA levels, and in view of the previous reports of noladin ether as a PPARα agonist, we next wondered if high noladin ether concentrations may be acting through PPARs to affect *LEAP2* mRNA expression in cells other than epithelial enterocytes. In fact, the possibility of PPARα being involved in the noladin ether effect in organoids still exists, since we did not test here PPARα agonists in organoids. To test if high concentrations of noladin ether can activate PPAR-mediated transcription, we performed luciferase assays with this compound at concentrations from 1 to 25 μM. As expected, PPARα/HEK293 and PPARγ/HEK293 cells showed a massive luciferase induction upon exposure to the selective agonists GW7647 and rosiglitazone, respectively. However, no changes were found in either case with noladin ether (Figure 5A,B). Therefore, the mechanism by which noladin ether upregulates *Leap2* expression in duodenal organoids remains to be determined.

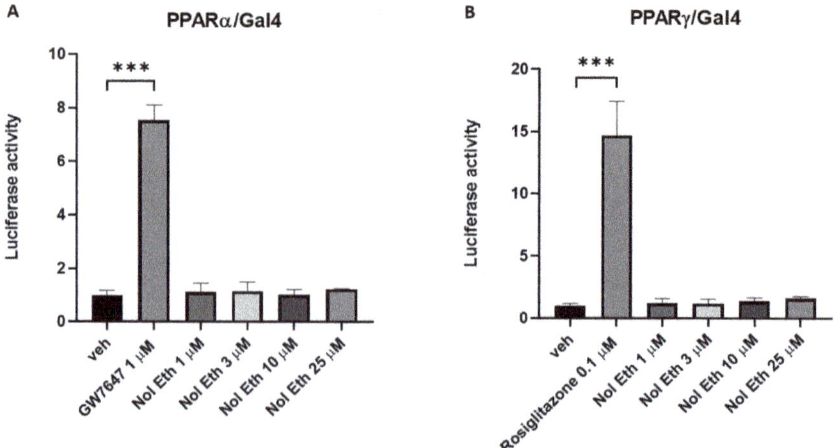

Figure 5. Effect of noladin ether (Nol Eth) in PPARα- or PPARγ reporter cell lines. Luciferase assay was performed in HEK293 cells transiently expressing the ligand-binding domain (LBD) of human PPARα (**A**) and/or PPARγ (**B**) fused to the yeast GAL4 DNA binding domain (DBD). Bar graphs showing the ratio between the firefly and *Renilla* luciferase in response to increasing concentrations of noladin ether. GW7647 (1 μM) and rosiglitazone (0.1 μM) were used as positive controls for PPARα (**A**) and PPARγ (**B**), respectively. The vehicle group value was set to 1. Each point is the mean ± SEM of four separate determinations performed in duplicate. *** $p \leq 0.0005$ versus the vehicle (dimethyl sulfoxide, DMSO) group.

2.5. Diet-Induced and Genetically Obese Mice Express Altered Intestinal Levels of Leap2 in Relation with Altered Endocannabinoidome Signaling

It was previously shown that ghrelin levels are altered in a mouse model of diet-induced obesity (DIO) [22]. Here we wanted to explore how DIO affects *Leap2* expression in mice fed with high-fat high sucrose (HFHS) diet at different time points following the beginning of the diet (0 days, 10 days, 21 days, and 56 days). We have previously shown [23] that these same mice progressively gained weight and become glucose intolerant over the duration of the protocol while developing time-dependent alterations in their circulating and intestinal eCBome mediators, enzymes, and receptors [23]. We report a gradual increase in *Leap2* expression in the liver, from day 0 to day 56 that did not reach statistical significance (Figure 6A). Then, we looked at the duodenum and saw no changes in *Leap2* expression, although we observed a non-significant increase on day 21 (Figure 6B). In the jejunum, instead, we observed a significant decrease in *Leap2* expression on day 21, and no changes on days 3, 10, and 56 compared to the baseline (Figure 6C). Finally, in the ileum, we observed a significant increase in *Leap2* expression at day 3, when the mice were already glucose intolerant [23], and a gradual decrease from day 10 to day 56 (Figure 6D), when the mice gradually became obese [23].

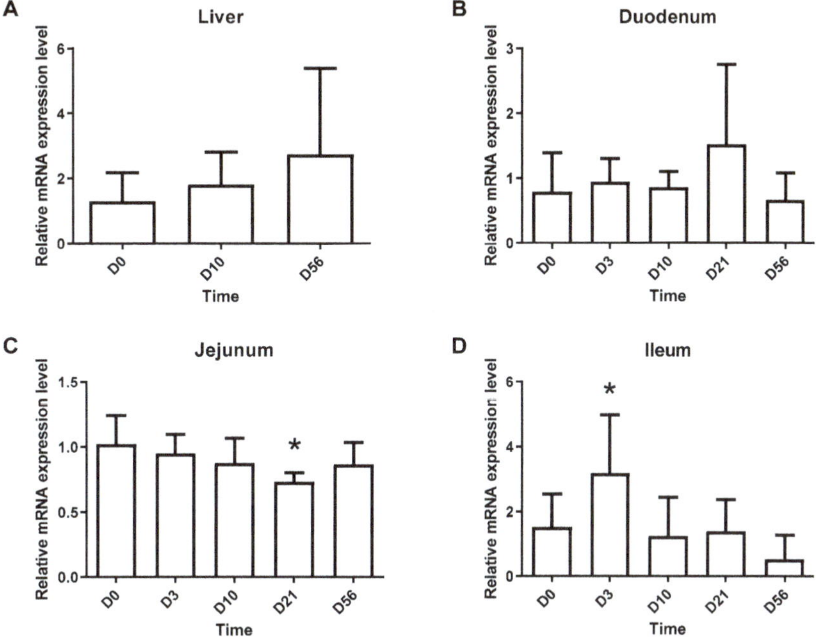

Figure 6. *Leap2* mRNA expression in the liver and small intestine following the initiation of an HFHS diet. Gene expression of *Leap2* in the liver (**A**), duodenum (**B**), jejunum (**C**), and ileum (**D**) at different HFHS diet time points as measured by qPCR. Relative mRNA expression relative to mice at day 0 is represented as mean $\pm$ S.D.). n = 10–12. * $p \leq 0.05$.

We have reported previously [23] that in the ileum of the same HFHS mice used in the present study the levels of some eCBome mediators, i.e., OEA, 2-oleoyl-glycerol (2-OG), and 2-linoleoyl-glycerol (2-LG), which may act on PPARα, GPR119 and TRPV1 receptors [7], change in a very similar way to what found here for *Leap2* mRNA levels, i.e., they peak at day 3 and then significantly decrease, whereas the levels of the eCB and weak PPARγ agonist, AEA, instead start increasing after 3 days of the HFHS diet. Likewise, the mRNA levels of eCBome genes such as *Faah*, *Abhd6*, and *Ppara* peaked at day 3 before decreasing, whereas *Cnr2*, *Pparg*, and *Plcb1* mRNA levels showed a negative peak at day 3 and then increased with time [23]. Therefore, we wanted to see if the changes we saw in *Leap2* expression in the ileum were correlated with changes in eCBome genes. Interestingly, *Leap2* expression was significantly and positively correlated with the ileal mRNA levels of *Ppara*, *Faah*, and *Gde1*, but not *Abhd6*, and negatively with those of *Plcb1* and *Trpv2*, but not *Cnr2* and *Pparg* (Figure 7). These correlations might suggest the implication of the eCBome, and in particular, of NAEs—which are biosynthesized through GDE1 (among other enzymes) and (like *N*-acyl-glycines) are degraded by FAAH, and act as agonists at PPARα (as in the case of OEA and PEA, similar to *N*-acyl-glycines) and PPARγ (as in the case of AEA) and/or antagonists at TRPV2 (as in the case of OEA and *N*-linoleoyl-ethanolamine [LEA]) [7,24,25]—in *Leap2* up-regulation during the development of HFHS-induced glucose intolerance and obesity. The negative correlation with *Plcb1* might suggest instead the existence of a negative correlation with 2-MAGs, of whose biosynthesis this enzyme catalyzes the rate-limiting step. However, ileal *Leap2* mRNA levels did not correlate with any eCBome mediator levels measured in the ileum (data not shown).

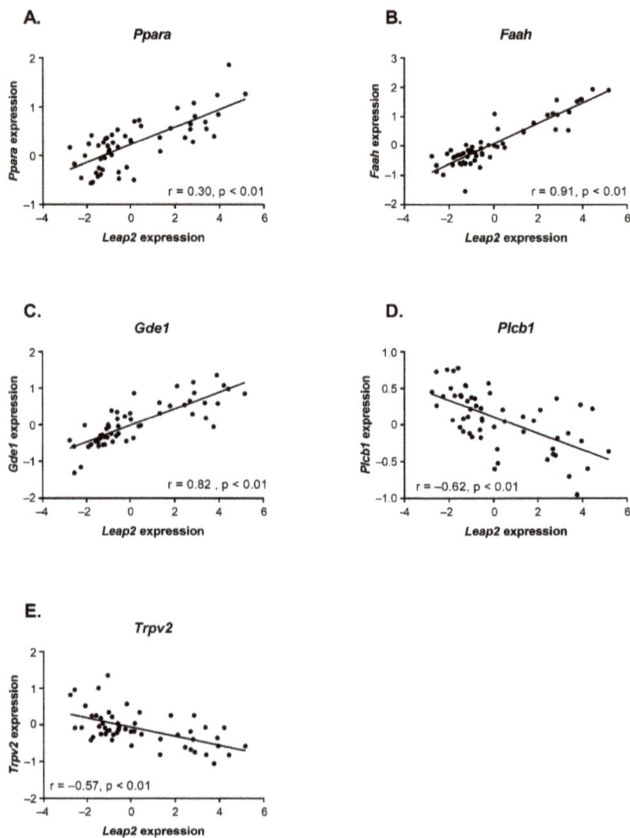

Figure 7. Correlation between the mRNA expression levels of *Leap2* and those of (**A**) *Ppara*, (**B**) *Faah*, (**C**) *Gde1*, (**D**) *Plcb1* and (**E**) *Trpv2* in the ileum of HFHS-fed mice. The figure shows scatterplots with regression lines. Pearson correlation coefficients (r) and *p*-values are shown in graphs (n = 54–56). Data were obtained using the 2-ΔΔCt method relative to the *Tbp* (TATA-binding protein) housekeeping gene (RT-qPCR).

We next analyzed two genetic models of obesity: leptin gene mutant (*ob/ob*) mice and leptin receptor mutant (*db/db*) mice. Although there were trends for a decrease in the liver and for increases in the jejunum and duodenum of *ob/ob* mice, the only significant difference was found in the ileum, where there was a strong increase in *Leap2* expression in both *ob/ob* and in *db/db* mice compared to the corresponding controls (Figure 8A). The results observed in these and DIO mice may suggest that *Leap2* expression in the small intestine, and the ileum, in particular, maybe due to glucose intolerance rather than obesity *per se*, since this dysmetabolic feature was common to both genetically obese mice and HFHS mice when they showed increase *Leap2* mRNA levels.

Finally, also in the case of *ob/ob* and *db/db* mice we looked at possible correlations between the mRNA expression of *Leap2* and eCBome signaling in the ileum. First, we observed that some mediators, such as 2-PG, were also higher in *ob/ob* and *db/db* mice as compared to their respective controls. Other mediators, i.e., NAGly, LEA, and *N*-docosahexaenoylethanolamine (DHEA), instead were significantly, or tended to be, lower in the *ob/ob* and *db/db* mice when compared to the respective controls (Figure 8B). Accordingly, we found that the levels of 2-PG, a PPARα agonist [26], positively correlated with *Leap2* mRNA expression (Figure 8C), which however did not correlate with any of the mRNAs of eCBome genes (data not shown).

In summary, the results obtained in animal models of glucose intolerance and obesity suggest that ileal *Leap2* expression may represent an early adaptive response aimed at counteracting dysmetabolism. This response might be under the indirect positive control of PPARα and its eCBome agonists (as suggested by the positive correlations with *Ppara*, the NAE synthesizing enzyme *Gde1*, and AEA and *N*-acyl-glycine degrading enzyme *Faah*, as well as 2-PG, which is also a good PPARα agonist). Based on the results of the previous section, however, such control would not occur in enterocytes, where instead PPARγ activation may play a role.

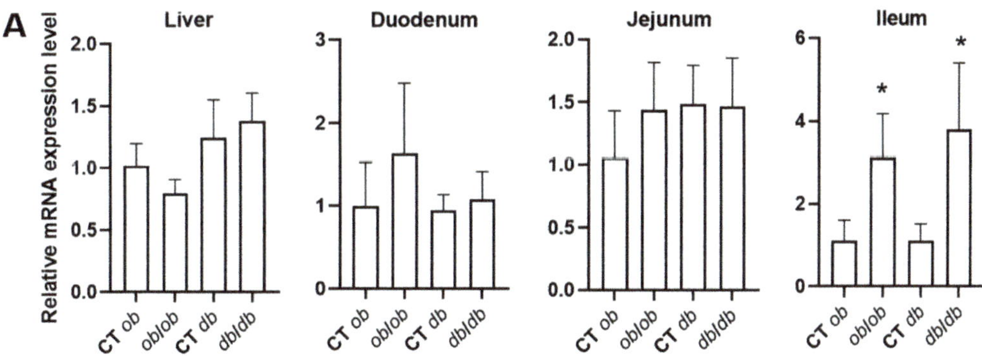

Figure 8. *Cont.*

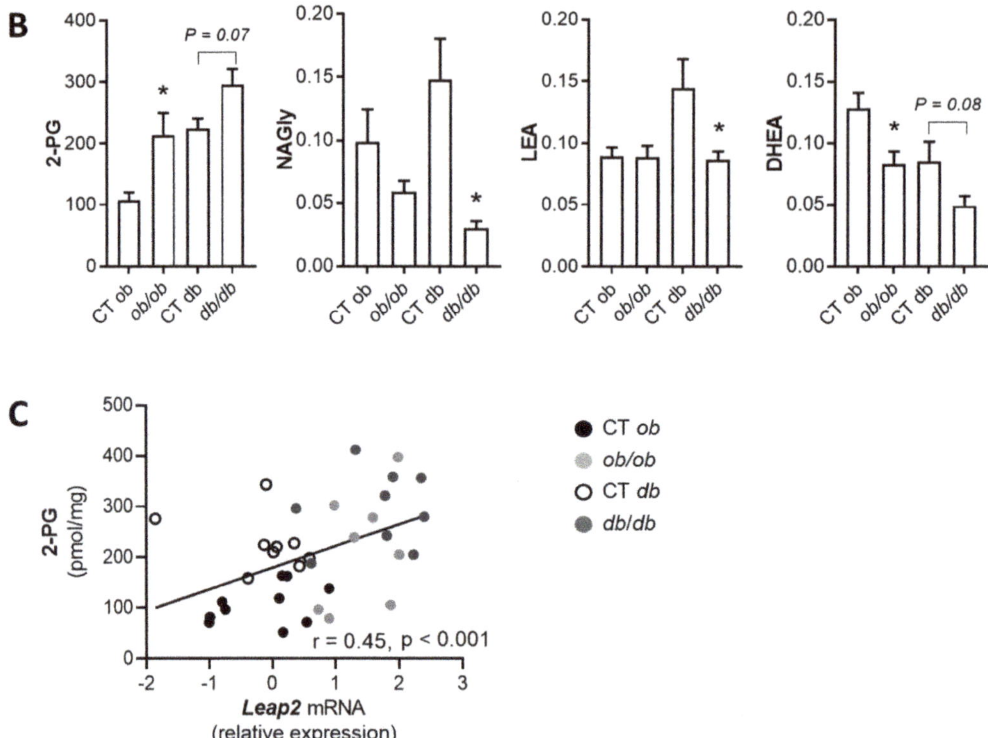

Figure 8. mRNA expression (**A**) in the liver, duodenum, jejunum, and ileum of *ob/ob* and *db/db* mice as measured by qPCR. Relative mRNA expression is represented as mean ± S.D. Values were compared relative to control (i.e., CT *ob*; CT *db*) mice for each tissue. Levels of select eCBome mediators within the ileum of *ob/ob* and *db/db* mice and their respective controls ((**B**); pmol/mg wet tissue weight; 2-palmitoylglycerol; 2-PG, *N*-Arachidonoyl-glycine; NAGly; *N*-linoleoylethanolamine; LEA, *N*-docosahexaenoylethanolamine; DHEA), as measured by HPLC-MS/MS. Data are presented as the mean ± S.E.M. of n = 8–10 and were analyzed by one-way ANOVA followed by Tukey's post hoc test. * $p \leq 0.05$ vs. the respective control. (**B**). Pearson correlation analysis between the mRNA expression of *Leap2* and the concentration of 2-PG both measured/quantified in the ileum of *ob/ob* and *db/db* mice, and their respective controls (**C**). n = 8–10 mice.

3. Discussion

LEAP2 is a recently discovered, a potential endogenous regulator of food intake and energy metabolism through antagonism of ghrelin action [6]. In agreement with these properties of this peptide, originally isolated from the liver as an antimicrobial agent, are the following previous findings: (1) selectively in female mice subjected to a high-fat diet, *Leap2* deletion raises body weight, food intake, lean mass, and hepatic fat, and reduces O_2 consumption, heat production, and locomotor activity during the first part of the dark period [27]; (2) in both female and male lean mice, *Leap2* deletion renders the animals more sensitive to the hyperphagic actions of ghrelin [27]; LEAP2 administration also reduces blood glucose levels in lean mice [6]; (3) in humans and mice, circulating LEAP2 levels display an inverse pattern compared to ghrelin, by increasing with food intake and obesity, and decreasing upon fasting and weight loss [28,29]; refeeding decreases circulating levels of ghrelin, while LEAP2 goes up to baseline levels; and (4) like other anorectic signals, LEAP2 is also produced by the small intestine [2]. Others have reported that all regions of the mouse small intestine express *Leap2*, with the highest expression having been found in the jejunum (whereas we detected slightly higher levels in the duodenum), with next to no detectable levels expressed in the stomach [6]. Regardless, after bariatric surgery (vertical sleeve gastrectomy), the expression in the jejunum decreased significantly, while expression in the stomach increased over 50-fold. This shows that *Leap2* expression can respond to various experimental paradigms in an intestinal-region-specific manner. However, LEAP2 regulation in this tissue has not yet been investigated, nor have its potential interactions therein with two major and inter-related players in obesity, i.e., the eCBome and the gut mBIome. Here we have reported evidence suggesting that: (1) the presence or absence of the gut microbiota strongly and directly affects *Leap2* mRNA expression in the liver and duodenum, and LEAP2 circulating levels, in 13 week-old male, but not female, mice (gender differences are known to influence gene expression [30] and their targeting by transcription [31] and translation regulators that ultimately lead to different mRNAs or protein products [32]); (2) despite the fact that the gut microbiota also strongly and directly affects small intestinal eCBome signaling, generalized modulation of eCBome receptors does not result in significant changes of *Leap2* expression in a model of human enterocytes, except for the activation by the synthetic (i.e., rosiglitazone) agonist of PPARγ, which caused a strong upregulation; additionally, noladin ether, another endocannabinoid discovered by Raphael Mechoulam and his group in 2001 [33], caused a significant elevation of *Leap2* expression in organoids from the mouse duodenum, through yet to be investigated mechanisms; and (3) *Leap2* expression in the ileum of HFHS- and genetically (leptin signaling deficiency)-induced obesity is increased concomitantly with glucose intolerance rather than obesity, and in a manner variedly and strongly correlated with the ileal levels of either an eCBome mediator (2-PG, positive) or mRNAs of proteins that participate in either NAE or 2-MAG biosynthesis (*Gde1*, positive; *Plcb1*, negative) or actions (*Ppara*, positive; *Trpv2*, negative), or in NAE and *N*-acyl-glycine inactivation (*Faah*, positive).

The gut microbiota is known to affect host physiology, and metabolism in particular, through a plethora of microorganism-derived molecules, of which eCB-like mediators different from those produced by the host are also part [34]. Therefore, it was not surprising to find previously that the gut microbiota is a strong determinant of eCBome signaling in the mouse intestine [18], although this action of intestinal microorganisms could also be mediated by other, non-eCBome-related molecules, such as, for example, short-chain fatty acids, tryptophan metabolites and secondary bile acids [35,36]. Likewise, the tonic inhibitory effect of the gut microbiota on duodenal *Leap2* expression, described here for the first time in adult male mice, might be both eCBome- and non-eCBome-mediated. It is noteworthy that the intestinal mBIome is markedly varied along the length of the gastrointestinal tract, with there being relatively few bacteria in the duodenum, and the number and diversity of bacteria increasing distally [37,38]. This of course does not discount the ability of the duodenal microbiome, by being significantly modified with obesity [38,39], to impact both the host physiology and small intestinal bacterial overgrowth [40]. Indeed, we have reported that fecal microbiota transfer into germ-free mice reconstitutes the small intestinal bacterial communities and significantly modulates the eCBome of the small intestine, including the duodenum [18].

We found that at least one eCBome mediator, noladin ether, can stimulate *Leap2* expression in duodenal organoids, which, together with the previous observation that the gut microbiota tonically reduces duodenal PPARα expression in the small intestine of adult male mice [18], may suggest that this receptor and its endogenous ligands might mediate in part gut microbiota tonic inhibition of duodenal *Leap2* expression. However, we could not confirm the proposed mechanism of action of noladin ether in organoids, as this compound did not activate recombinant PPARγ or PPARα in a functional assay, nor could we show that it stimulates *Leap2* expression in a model of human enterocytes. Additionally: (1) PPARγ (an alternative AEA target), although involved in stimulating *Leap2* expression in vitro, particularly when rosiglitazone was used as an agonist, is not significantly altered by the lack of presence of the gut microbiota in terms of mRNA expression in the duodenum [18];

(2) GPR55 (another proposed target for AEA and noladin ether) and GPR18 (a proposed target for NAGly) are tonically stimulated, rather than inhibited, in the duodenum by the gut microbiota [18], and hence cannot partake in microbiota tonic inhibition of *Leap2* expression; and (3) also the duodenal expression of *Cnr1* or *Cnr2*, the two preferential targets for eCBs, is not significantly altered by the gut microbiota [18]. In sum, present and previous data suggest that the effects of the gut microbiota on *Leap2* expression and eCBome signaling might be two unrelated, or only very partially related, phenomena. Nevertheless, the increase in *Leap2* expression observed in GF mice might still account, at least in part, for their more favorable metabolic and glycemic profile under conditions of DIO [41].

Indeed, we observed here, again for the first time, that increased RNA levels encoding for this peptide are produced in the ileum 3 days following an HFHS diet, concomitantly with the appearance of hyperglycemia, and in both *ob/ob* and *db/db* mice that, apart from being obese, are also glucose intolerant. This finding may suggest that LEAP2 is overproduced by the ileum to counteract glucose intolerance induced by an HFHS diet or hyperphagia. Since the ileum of HFHS and of *ob/ob* and *db/db* mice presents with inter-related alterations in gut eCBome signaling (present results and [23]) and microbiota composition [23,42,43], we hypothesized a role of either system in the modulation of *Leap2* expression. However, also, in this case, the only potential eCBome signaling pathway that changed in a manner similar to *Leap2* expression following HFHS and in *ob/ob* and *db/db* mice was that mediated by PPARα and/or its ligands. Conversely, as mentioned above, the only eCBome signaling pathway that was clearly implicated in *Leap2* expression stimulation in human enterocytes was that mediated by PPARγ, whose expression, however, peaks negatively in HFHS mice when *Leap2* expression peaks positively. This may suggest that also in the case of obesity and hyperglycemia the regulation of *Leap2* expression and eCBome signaling might be two unrelated phenomena, and not controlled in a coordinated manner by gut dysbiosis.

A limitation of this study is that the selected in vitro systems for the small intestine do not necessarily reflect what is going on in vivo in the ileum. The organoids were prepared from the duodenum and not from the ileum, and therefore results obtained in this system may only be relevant to those in GF mice, where the strongest reduction in *Leap2* expression was indeed found in this small intestinal section. Yet, the geometry of organoids is such that their basolateral side, rather than the apical one, is the one that is exposed to treatments, which may have altered the effects of the latter. Differentiated Caco-2 cells, instead, model only one cell type (enterocytes) among the several ones that are found in the small intestinal mucosa. Therefore, the findings in these two in vitro systems may not necessarily be relevant to the actual in vivo regulation of *Leap2* expression by the gut microbiota in the dysbiosis typical of GF and hyperglycemic/obese mice, thus leaving still open the possibility that the eCBome, *per se* or following its modulation by the gut microbiome, might be an important determinant of LEAP2 production. In this sense, it is noteworthy that the mRNA expression of this potentially metabolically beneficial endogenous antagonist of ghrelin action is stimulated in vitro by, and correlates in vivo with, non-CB1-mediated (and hence non-metabolically "noxious") eCBome signaling, i.e., respectively: (1) activation of PPARγ, which is known to be an intermediate in anti-diabetic and anti-glucose intolerance drugs, and (2) expression of PPARα and the levels of one of its endogenous agonists, i.e., 2-PG, with potential anorectic and anti-dyslipidemic actions.

In summary, we have provided here unprecedented evidence for the existence of gut microbiota tonic in vivo control over intestinal *Leap2* expression, which persists into adulthood, at least in the duodenum, of male mice. We also provided preliminary in vitro data suggesting that stimulation of the levels of the mRNA encoding for this metabolically beneficially peptide may be exerted by non-CB1-mediated eCBome signaling. It remains to be clarified whether the gut microbiota-eCBome axis is involved in the control of intestinal LEAP2 levels, especially during obesity and hyperglycemia, where, as shown previously and confirmed here, all the members of this triangle undergo adaptive regulation.

4. Materials and Methods

4.1. Animals and Housing

Conventionally Raised (CR) and Germ-free (GF) C57BL/6NTac mice were purchased from Taconic (Taconic Bioscience, NY, USA) and maintained in the animal facility of the Institut Universitaire de Cardiologie et Pneumologie de Québec (IUCPQ, QC, Canada). All animals have grouped 3–4 mice per cage under a 12 h:12 h light-dark cycle with ad libitum access to NIH-31 Open Formula Autoclavable Diet (Zeigler, PA, USA) and water. GF mice were housed in axenic status and fecal samples were weekly tested for microbes and parasites by the facility's staff to ensure that the GF unit was indeed sterile. Both GF and CR mice were acclimatized for at least one week prior to starting the procedures.

4.2. Animal Experiments and Fecal Microbial Transplant (FMT)

Twelve (6 male and 6 female) CR and GF mice at 4 and 13 weeks of age, were intraperitoneally anesthetized with a cocktail of ketamine/xylazine/acepromazine at a dose of 50/10/1.7 mg/kg body weight and euthanized by cervical dislocation, following an intra-cardiac puncture. Whole blood was collected in K3-EDTA tubes. The abdominal cavity was opened and the whole digestive tract was carefully aligned from the stomach up to the colon. Once the stomach was removed, the small (duodenum, jejunum, ileum) and large (cecum and colon) intestine were carefully excised and separated and the intestinal contents were harvested by flushing with 1ml of sterile PBS without Ca/Mg (Thermo Fisher Scientific, MA, USA) and snap-frozen. The liver was also isolated from the abdominal cavity. Sections of the liver as well as small and large intestine were stored either in RNALater (Thermo Fisher Scientific, MA, USA) for RNA stabilization or immediately snap-frozen and stored at $-80\,^{\circ}$C for further analysis. This common procedure was concluded, for each mouse, within a maximum of 15 min, to ensure the preservation of mRNA and lipid for further analysis.

For fecal microbiota transplant (FMT) experiments only 12 weeks old male mice were utilized. GF mice were randomly divided into two groups at the age of 12 weeks: those gavaged with sterile PBS (SHAM; 5 mice) and those gavaged with fecal material (FMT; 6 mice). Material gavaged for FMT consisted of a cocktail of the intestinal contents and stools of a single and 4 CR donor mice, respectively. Briefly, the intestinal contents of the duodenum, jejunum, ileum, colon, and cecum were collected from one 12-week-old CR donor mouse and mixed with stool pellets from all the CR mice to be used as controls. The mixture was well homogenized, weighed, suspended at 1:10 in sterile PBS, and centrifuged at $805 \times g$ for 10 min at room temperature. The supernatant was used to gavage the mice (200 µL of homogenate per mouse) [18]. The FMT mice were then housed (3 per cage, like for CR mice) for one week in conventional conditions in cages contaminated with used litter coming from donor mouse cages. SHAM mice were submitted to a similar gavage with saline solution, but then kept in the germ-free facility for a week in the same conditions as GF mice. CR mice were euthanized the day of the gavages, while SHAM and FMT mice were sacrificed one week after the gavage; the tissues were collected from all animals as previously described.

4.3. RNA Isolation, Reverse Transcription, and qPCR

RNA was extracted from tissues and cells with the RNeasy Plus Mini Kit (Qiagen, Hilden, Germany) following the manufacturer's instruction and eluted in 50 µL of Ultra-Pure Distilled Water (Invitrogen, CA, USA). The concentration and purity of RNA were determined by measuring the absorbance of the RNA in a Biodrop at 260 nm and 280 nm, and RNA integrity was assessed by an Agilent 2100 Bioanalyzer, using the Agilent RNA 6000 Nano Kit (Agilent Technologies, CA, USA). 1 µg of total RNA was reverse transcribed using the iScript cDNA synthesis kit (Bio-Rad, CA, USA) in a reaction volume of 20 µL. *Leap2* mRNA expression levels were determined using primer pairs (Mm.PT.58.6158455.g, IDT, IA, USA) on a CFX384 touch qPCR System (BioRad) using PowerUp SYBR Green qPCR master mix (Thermo Fisher Scientific, CA, USA) in duplicate reactions. Hprt1

(Mm.PT.39a.22214828, IDT, IA, USA) was used as a reference gene. Because the Ct of the reference gene was the same in all the 13 weeks old control samples (first and FMT experiments), we merged the two control sets into 13 weeks control group. Gene expression levels in mice and cells (see below) were evaluated by the 2-ΔΔCt method and represented as fold increase with respect to the baseline of the relevant control. Data were analyzed by ANOVA followed by Dunnett's multiple comparisons.

4.4. Organoid Preparation and Treatment

Crypt-derived organoids were mechanically separated from sacrificed black male C57BL/6 mouse duodenum. Murine small intestines were opened longitudinally, scratched, and washed with cold phosphate-buffered saline (PBS) (Gibco Life Technologies, Carlsbad, CA, USA). Organoids were incorporated in a solid matrix (Corning® Matrigel®) (Corning, Corning, NY, USA) and were incubated with advanced DMEM media (Dulbecco's Modified Eagle's Medium) (Gibco Life Technologies, Carlsbad, CA, USA) supplemented with HEPES buffer (Thermoscientific, Waltham, MA, USA), Glutamine (GlutaMAX) (Gibco Thermoscientific, Waltham, Ma, USA), antibiotic Pen-Strep (Millipore Sigma, Oakville, ON, Canada), Noggin (Millipore Sigma, Oakville, ON, Canada), mRSPO (Peprotech, Cranbury, NJ, USA), B27 supplement (Gibco Life Technologies, Carlsbad, CA, USA), EGF (Cedarlane, Burlington, ON, Canada, and N-acetyl-cystein (Millipore Sigma, Oakville, ON, Canada) and Ly-27. Organoids were split 1 into 3 every week, in 24 wells plates, and media was changed every 2 days. on Mature organoid cultures, 4 days following plating, were exposed 24 h to treatment or vehicle.

4.5. Caco-2/15Cell Differentiation and Treatment

Human Caco-2/15 intestinal cells (kindly provided by Dr. Jean-François Beaulieu (Université de Sherbrooke, Sherbrooke, Canada) were cultured in high glucose Dulbecco's Modified Eagle Medium (DMEM) supplemented with 10% Fetal Bovine Serum (FBS, Neuromics), 10 mM HEPES, 1X GlutaMAX, and 100 U/mL Penicillin/Streptomycin. Cell lines were grown until confluence, split, and plated 2×10^5 cell/well for differentiation for 21 days to induce the enterocyte phenotype. On the day of the experiment (after 21 days of differentiation), cells were treated for 24 or 48 h with media containing compounds at the indicated concentrations. DMSO controls were at 0.1%. For the LPS induction, 24 h prior to the experiment, cells were treated with 10ug/mL of LPS (Sigma) before being incubated for 24 h with the indicated compounds at the indicated concentrations. After incubation, media was removed, cells were washed with 1X PBS and frozen for RNA extraction as above.

4.6. PPAR-Luciferase Assays

Human embryonic kidney 293 cells (ATCC CRL-1573) were propagated in a growth medium (GM) composed of Dulbecco's modified Eagle's medium (DMEM cat. n. 41966029; Thermo Fisher, Monza, Italy) supplemented with 10% fetal bovine serum (cat. n. 16000044; Thermo Fisher, Monza, Italy) and 1% Pen/Strep (cat. n. 15140122 Thermo Fisher, Monza, Italy) under standard conditions. After plating (in 24 well plate density; 5×104 cells/well), the cells were transfected on the next day with the following plasmids: (a) pM1-hPPARα-Gal4 or pM1-hPPARγ-Gal4; (b) TK-MH100 × 4-Luc containing the UAS enhancer elements and; (c) Renilla luciferase (pRL, Cat. E2231; Promega, Milan, Italy) using lipofectamine 2000 (cat. n. 11668027; Life Technologies; Milan, Italy). The next day, the growth media was replaced with fresh media containing vehicle (Dimethyl sulfoxide, DMSO $\leq$ 0.03%) or compounds of interest (noladin ether Cat. No. 1411, GW7647 Cat. No. 1677/10 and Rosiglitazone Cat. No. 5325/10, purchased from TOCRIS, Abingdon, UK). On day 3, the cells were harvested and processed for analysis of luciferase activity using a GloMax Luminometer instrument (Promega, Milan, Italy) and the Dual-Luciferase Reporter Assay kit (cat. n. E1910 Promega, Milan, Italy) following published procedures [26].

4.7. Lipid Extraction and HPLC-MS/MS for the Analysis of eCBome Mediators

Lipids were extracted from ileum samples as previously described [18,23,26]. Briefly, about 10 mg of ileum was sampled and homogenized in 1 mL of a 1:1 Tris-HCl 50 mM pH 7: methanol solution containing 0.1 M acetic acid and 5 ng of deuterated standards. One ml of chloroform was then added to each sample, which was then vortexed for 30 s and centrifuged at $3000\times g$ for 5 min. The organic phase was collected and another 1 mL of chloroform was added to the inorganic one. This was repeated twice to ensure the maximum collection of the organic phase. The organic phases were pooled and evaporated under a stream of nitrogen and then suspended in 50 μL of mobile phase containing 50% of solvent A (water + 1 mM ammonium acetate +0.05% acetic acid) and 50% of solvent B (acetonitrile/water 95/5 + 1 mM ammonium acetate +0.05% acetic acid). Forty μL of each sample were finally injected onto an HPLC column (Kinetex C8, 150×2.1 mm, 2.6 μm, Phenomenex) and eluted at a flow rate of 400 μL/min using a discontinuous gradient of solvent A and solvent B [18,23,26]. Quantification of eCBome-related mediator, was carried out by HPLC interfaced with the electrospray source of a Shimadzu 8050 triple quadrupole mass spectrometer and using multiple reaction monitoring in positive ion mode for the compounds and their deuterated homologs [18,23,26].

4.8. Experiments in Mice Undergoing a High Fat High Sucrose Diet

As previously described [20], sixty 6-week-old C57BL/6J male mice were fed ad libitum with a low-fat, low-sucrose purified diet (10% fat and 7% sucrose [LFLS]; Research Diet, NJ, USA) for a 10-day acclimatization period in the animal facility of the Institute of Nutrition and Functional Foods. Mice were then randomly assigned to 6 groups (n = 12) fed a high-fat, high-sucrose purified diet (45% fat and 17% sucrose [HFHS]; Research Diet, NJ, USA) for up to 56 days. These dye-free diets harbor comparable fiber contents, and while the HFHS diet had, by design, a higher fatty acid content, the omega-3/omega-6 ratios were comparable. Six-hour-fasted mice were sacrificed by cardiac puncture to retrieve plasma ($1780\times g$, 10 min) at either 0 (baseline), 3, 10, 21, or 56 days following HFHS diet initiation. Duodenum was collected 2 cm of the pylorus, while jejunum and ileum were collected 10 cm and 2 cm, respectively, from the ileocecal junction. All samples were stored at −80 °C until batch analysis.

4.9. Experiments in ob/ob and db/db Mice

As previously described [36], male homozygous *ob/ob* mice (B6.V-Lepob/ob/JRj) were used as a leptin-deficient obese model, and their lean littermates served as controls (CT ob); (n = 9–10 per group). Male homozygous *db/db* mice (BKS-Lepr/db/db/JOrlRj) functionally deficient for the long-form leptin receptor were used as a hyperleptinemic obese type 2 diabetic model, and their lean littermates served as controls (CT db); (n = 9–10 per group). Mice were purchased at the same time and from the same supplier (Janvier Laboratories, Le Genest-Saint-Isle, France) at the age of 6 weeks. Mice were housed in a specific pathogen and opportunistic free (SOPF) controlled environment (room temperature of 22 ± 2 °C, humidity 55 ± 10%, 12 h daylight cycle, lights off at 6 p.m.) in groups of two mice per cage, with free access to sterile food and sterile water. Upon delivery, mice underwent an acclimation period of one week, during which they were fed a standard diet containing 10% calories from fat (D12450Ji; Research Diet; New Brunswick, NJ, USA) and were then kept ad libitum on the same diet for 7 weeks. Milli-Q water filtered by a Millipak® Express 40 with a 0.22 μm membrane filter (Merck Millipore, Burlington, MA, USA) was autoclaved and provided ad libitum. All mouse experiments were approved by and performed in accordance with the guideline of the local ethics committee (Ethics committee of the Université Catholique de Louvain for Animal Experiments specifically approved this study that received the agreement number 2017/UCL/MD/005). Housing conditions were specified by the Belgian Law of 29 May 2013, regarding the protection of laboratory animals (agreement number LA1230314).

Author Contributions: Conceptualization, C.S., V.D.; methodology, M.S., C.M, F.S, N.N., F.P., B.A.-N., F.A.I., N.F. and C.S.; validation, A.V., P.D.C., V.D. and C.S.; formal analysis, C.S. and V.D.; writing—original draft preparation, V.D.; writing—review and editing, M.S., C.M., F.S., C.S. and V.D.; supervision, A.V., P.D.C., C.S. and V.D.; funding acquisition, C.S., A.V., V.D. and P.D.C. All authors have read and agreed to the published version of the manuscript.

Funding: V.D. is the holder of the Canada Excellence Research Chair on the Gut Microbiome-Endocannabinoidome Axis in Metabolic Health (CERC-MEND) at Université Laval, funded by the Federal Tri-Agency of Canada. V.D. is the recipient of two Canada Foundation for Innovation grants (37392 and 37858). P.D.C. is a research director at Fonds de la Recherche Scientifique (FRS-FNRS), Belgium. This work was supported by FRS-FNRS under the grants WELBIO-CR-2017C-02 and WELBIO-CR-2019C02R, the Funds Baillet-Latour under the grant "Grant For Medical Research 2015", and the EOS program no. 30770923. C.M. was the recipient of a post-doctoral bursary from the Joint International Research Unit on Chemical and Biomolecular Studies on the Microbiome and its Impact on Metabolic Health and Nutrition (JIRU-MicroMeNu), which is funded by the Sentinelle Nord-Apogée Program of Université Laval, funded in turn by the Federal Tri-Agency of Canada. A.V is supported by a grant from the Canadian Institutes of Health Research (DOL342964).

Institutional Review Board Statement: For the *ob/ob* and *db/db* mouse study, all mouse experiments were approved by and performed in accordance with the guideline of the local ethics committee (Ethics committee of the Université Catholique de Louvain for Animal Experiments specifically approved this study that received the agreement number 2017/UCL/MD/005). Housing conditions were specified by the Belgian Law of 29 May 2013, regarding the protection of laboratory animals (agreement number LA1230314). All experimental protocols in germ-free mice and mice fed an HFHS diet were validated and approved by the Laval University animal ethics committee (CPAUL 2018010-1 and CPAUL 2017048-1 respectively).

Acknowledgments: The authors are grateful to Cyril Martin for performing the LC-MS-MS analyses of the ileum of *ob/ob* and *db/db* mice.

Conflicts of Interest: PDC is co-founders of A-Mansia Biotech. PDC is owner of patents concerning the use of *Akkermansia muciniphila* on health. The rest of the authors declare no conflict of interest.

Sample Availability: All compounds used are commercially available.

References

1. Krause, A. Isolation and biochemical characterization of LEAP-2, a novel blood peptide expressed in the liver. *Protein Sci.* **2003**, *12*, 143–152. [CrossRef] [PubMed]
2. Howard, A.; Townes, C.; Milona, P.; Nile, C.J.; Michailidis, G.; Hall, J. Expression and functional analyses of liver expressed antimicrobial peptide-2 (LEAP-2) variant forms in human tissues. *Cell. Immunol.* **2010**, *261*, 128–133. [CrossRef] [PubMed]
3. Henriques, S.T.; Tan, C.C.; Craik, D.J.; Clark, R.J. Structural and Functional Analysis of Human Liver-Expressed Antimicrobial Peptide 2. *ChemBioChem.* **2010**, *11*, 2148–2157. [CrossRef] [PubMed]
4. Hocquellet, A.; Odaert, B.; Cabanne, C.; Noubhani, A.; Dieryck, W.; Joucla, G.; Le Senechal, C.; Milenkov, M.; Chaignepain, S.; Schmitter, J.-M.; et al. Structure–activity relationship of human liver-expressed antimicrobial peptide 2. *Peptides* **2010**, *31*, 58–66. [CrossRef] [PubMed]
5. Chanput, W.; Mes, J.J.; Wichers, H.J. THP-1 cell line: An in vitro cell model for immune modulation approach. *Int. Immunopharmacol.* **2014**, *23*, 37–45. [CrossRef] [PubMed]
6. Ge, X.; Yang, H.; Bednarek, M.A.; Galon-Tilleman, H.; Chen, P.; Chen, M.; Lichtman, J.S.; Wang, Y.; Dalmas, O.; Yin, Y.; et al. LEAP2 Is an Endogenous Antagonist of the Ghrelin Receptor. *Cell Metab.* **2018**, *27*, 461–469.e6. [CrossRef] [PubMed]
7. Di Marzo, V.; Silvestri, C. Lifestyle and Metabolic Syndrome: Contribution of the Endocannabinoidome. *Nutrients* **2019**, *11*, 1956. [CrossRef]
8. Cani, P.D.; Moens de Hase, E.; Van Hul, M. Gut Microbiota and Host Metabolism: From Proof of Concept to Therapeutic Intervention. *Microorganisms* **2021**, *9*, 1302. [CrossRef]
9. Tucci, S.A.; Rogers, E.K.; Korbonits, M.; Kirkham, T.C. The cannabinoid CB1 receptor antagonist SR141716 blocks the orexigenic effects of intrahypothalamic ghrelin. *Br. J. Pharmacol.* **2004**, *143*, 520–523. [CrossRef]
10. Kola, B.; Farkas, I.; Christ-Crain, M.; Wittmann, G.; Lolli, F.; Amin, F.; Harvey-White, J.; Liposits, Z.; Kunos, G.; Grossman, A.B.; et al. The Orexigenic Effect of Ghrelin Is Mediated through Central Activation of the Endogenous Cannabinoid System. *PLoS ONE* **2008**, *3*, e1797. [CrossRef]
11. Charalambous, C.; Lapka, M.; Havlickova, T.; Syslova, K.; Sustkova-Fiserova, M. Alterations in Rat Accumbens Dopamine, Endocannabinoids and GABA Content During WIN55,212-2 Treatment: The Role of Ghrelin. *Int. J. Mol. Sci.* **2020**, *22*, 210. [CrossRef]

12. Cani, P.D.; Montoya, M.L.; Neyrinck, A.M.; Delzenne, N.M.; Lambert, D.M. Potential modulation of plasma ghrelin and glucagon-like peptide-1 by anorexigenic cannabinoid compounds, SR141716A (rimonabant) and oleoylethanolamide. *Br. J. Nutr.* **2004**, *92*, 757–761. [CrossRef]

13. Godlewski, G.; Cinar, R.; Coffey, N.J.; Liu, J.; Jourdan, T.; Mukhopadhyay, B.; Chedester, L.; Liu, Z.; Osei-Hyiaman, D.; Iyer, M.R.; et al. Targeting Peripheral CB1 Receptors Reduces Ethanol Intake via a Gut-Brain Axis. *Cell Metab.* **2019**, *29*, 1320–1333.e8. [CrossRef]

14. Coccurello, R.; Maccarrone, M. Hedonic Eating and the "Delicious Circle": From Lipid-Derived Mediators to Brain Dopamine and Back. *Front. Neurosci.* **2018**, *12*, 271. [CrossRef] [PubMed]

15. Monteleone, P.; Piscitelli, F.; Scognamiglio, P.; Monteleone, A.M.; Canestrelli, B.; Di Marzo, V.; Maj, M. Hedonic Eating Is Associated with Increased Peripheral Levels of Ghrelin and the Endocannabinoid 2-Arachidonoyl-Glycerol in Healthy Humans: A Pilot Study. *J. Clin. Endocrinol. Metab.* **2012**, *97*, E917–E924. [CrossRef]

16. Rigamonti, A.E.; Piscitelli, F.; Aveta, T.; Agosti, F.; De Col, A.; Bini, S.; Cella, S.G.; Di Marzo, V.; Sartorio, A. Anticipatory and consummatory effects of (hedonic) chocolate intake are associated with increased circulating levels of the orexigenic peptide ghrelin and endocannabinoids in obese adults. *Food Nutr. Res.* **2015**, *59*, 29678. [CrossRef]

17. Leeuwendaal, N.K.; Cryan, J.F.; Schellekens, H. Gut peptides and the microbiome: Focus on ghrelin. *Curr. Opin. Endocrinol. Diabetes Obes.* **2021**, *28*, 243–252. [CrossRef]

18. Manca, C.; Boubertakh, B.; Leblanc, N.; Deschênes, T.; Lacroix, S.; Martin, C.; Houde, A.; Veilleux, A.; Flamand, N.; Muccioli, G.G.; et al. Germ-free mice exhibit profound gut microbiota-dependent alterations of intestinal endocannabinoidome signaling. *J. Lipid Res.* **2020**, *61*, 70–85. [CrossRef]

19. Ruparel, N.B.; Patwardhan, A.M.; Akopian, A.N.; Hargreaves, K.M. Desensitization of Transient Receptor Potential Ankyrin 1 (TRPA1) by the TRP Vanilloid 1-Selective Cannabinoid Arachidonoyl-2 Chloroethanolamine. *Mol. Pharmacol.* **2011**, *80*, 117–123. [CrossRef] [PubMed]

20. Sun, Y.; Alexander, S.P.H.; Garle, M.J.; Gibson, C.L.; Hewitt, K.; Murphy, S.P.; Kendall, D.; Bennett, A. Cannabinoid activation of PPARα; a novel neuroprotective mechanism. *Br. J. Pharmacol.* **2007**, *152*, 734–743. [CrossRef] [PubMed]

21. Duncan, M.; Millns, P.; Smart, D.; Wright, J.E.; Kendall, D.A.; Ralevic, V. Noladin ether, a putative endocannabinoid, attenuates sensory neurotransmission in the rat isolated mesenteric arterial bed via a non-CB1/CB2 Gi/o linked receptor. *Br. J. Pharmacol.* **2004**, *142*, 509–518. [CrossRef] [PubMed]

22. Uchida, A.; Zechner, J.F.; Mani, B.K.; Park, W.; Aguirre, V.; Zigman, J.M. Altered ghrelin secretion in mice in response to diet-induced obesity and Roux-en-Y gastric bypass. *Mol. Metab.* **2014**, *3*, 717–730. [CrossRef]

23. Lacroix, S.; Pechereau, F.; Leblanc, N.; Boubertakh, B.; Houde, A.; Martin, C.; Flamand, N.; Silvestri, C.; Raymond, F.; Di Marzo, V.; et al. Rapid and Concomitant Gut Microbiota and Endocannabinoidome Response to Diet-Induced Obesity in Mice. *MSystems* **2019**, *4*, e00407–e00419. [CrossRef]

24. Di Marzo, V. New approaches and challenges to targeting the endocannabinoid system. *Nat. Rev. Drug Discov.* **2018**, *17*, 623–639. [CrossRef] [PubMed]

25. Moriello, A.S.; Chinarro, S.L.; Fernández, O.N.; Eras, J.; Amodeo, P.; Canela-Garayoa, R.; Vitale, R.M.; Di Marzo, V.; De Petrocellis, L. Elongation of the Hydrophobic Chain as a Molecular Switch: Discovery of Capsaicin Derivatives and Endogenous Lipids as Potent Transient Receptor Potential Vanilloid Channel 2 Antagonists. *J. Med. Chem.* **2018**, *61*, 8255–8281. [CrossRef]

26. Depommier, C.; Vitale, R.M.; Iannotti, F.A.; Silvestri, C.; Flamand, N.; Druart, C.; Everard, A.; Pelicaen, R.; Maiter, D.; Thissen, J.-P.; et al. Beneficial Effects of Akkermansia muciniphila Are Not Associated with Major Changes in the Circulating Endocannabinoidome but Linked to Higher Mono-Palmitoyl-Glycerol Levels as New PPARα Agonists. *Cells* **2021**, *10*, 185. [CrossRef] [PubMed]

27. Shankar, K.; Metzger, N.P.; Singh, O.; Mani, B.K.; Osborne-Lawrence, S.; Varshney, S.; Gupta, D.; Ogden, S.B.; Takemi, S.; Richard, C.P.; et al. LEAP2 deletion in mice enhances ghrelin's actions as an orexigen and growth hormone secretagogue. *Mol. Metab.* **2021**, *53*, 101327. [CrossRef]

28. Mani, B.K.; Puzziferri, N.; He, Z.; Rodriguez, J.A.; Osborne-Lawrence, S.; Metzger, N.P.; Chhina, N.; Gaylinn, B.; Thorner, M.O.; Thomas, E.L.; et al. LEAP2 changes with body mass and food intake in humans and mice. *J. Clin. Investig.* **2019**, *129*, 3909–3923. [CrossRef] [PubMed]

29. Fittipaldi, A.S.; Hernández, J.; Castrogiovanni, D.; Lufrano, D.; Francesco, P.N.D.; Garrido, V.; Vitaux, P.; Fasano, M.V.; Fehrentz, J.-A.; Fernández, A.; et al. Plasma levels of ghrelin, des-acyl ghrelin and LEAP2 in children with obesity: Correlation with age and insulin resistance. *Eur. J. Endocrinol.* **2020**, *182*, 165–175. [CrossRef]

30. Yang, X.; Schadt, E.E.; Wang, S.; Wang, H.; Arnold, A.P.; Ingram-Drake, L.; Drake, T.A.; Lusis, A.J. Tissue-specific expression and regulation of sexually dimorphic genes in mice. *Genome Res.* **2006**, *16*, 995–1004. [CrossRef]

31. Lopes-Ramos, C.M.; Chen, C.-Y.; Kuijjer, M.L.; Paulson, J.N.; Sonawane, A.R.; Fagny, M.; Platig, J.; Glass, K.; Quackenbush, J.; DeMeo, D.L. Sex Differences in Gene Expression and Regulatory Networks across 29 Human Tissues. *Cell Rep.* **2020**, *31*, 107795. [CrossRef]

32. Zhang, W.; Huang, R.S.; Duan, S.; Dolan, M.E. Gene set enrichment analyses revealed differences in gene expression patterns between males and females. *In Silico Biol.* **2009**, *9*, 55–63. [CrossRef]

33. Hanuš, L.; Abu-Lafi, S.; Fride, E.; Breuer, A.; Vogel, Z.; Shalev, D.E.; Kustanovich, I.; Mechoulam, R. 2-Arachidonyl glyceryl ether, an endogenous agonist of the cannabinoid CB1 receptor. *Proc. Natl. Acad. Sci. USA* **2001**, *98*, 3662–3665. [CrossRef] [PubMed]

34. Iannotti, F.A.; Marzo, V.D. The gut microbiome, endocannabinoids and metabolic disorders. *J. Endocrinol.* **2021**, *248*, R83–R97. [CrossRef]

35. Agus, A.; Clément, K.; Sokol, H. Gut microbiota-derived metabolites as central regulators in metabolic disorders. *Gut* **2021**, *70*, 1174–1182. [CrossRef]

36. Delzenne, N.M.; Rodriguez, J.; Olivares, M.; Neyrinck, A.M. Microbiome response to diet: Focus on obesity and related diseases. *Rev. Endocr. Metab. Disord.* **2020**, *21*, 369–380. [CrossRef] [PubMed]

37. Kastl, A.J.; Terry, N.A.; Wu, G.D.; Albenberg, L.G. The Structure and Function of the Human Small Intestinal Microbiota: Current Understanding and Future Directions. *Cell. Mol. Gastroenterol. Hepatol.* **2020**, *9*, 33–45. [CrossRef] [PubMed]

38. Nardelli, C.; Granata, I.; D'Argenio, V.; Tramontano, S.; Compare, D.; Guarracino, M.R.; Nardone, G.; Pilone, V.; Sacchetti, L. Characterization of the Duodenal Mucosal Microbiome in Obese Adult Subjects by 16S rRNA Sequencing. *Microorganisms* **2020**, *8*, 485. [CrossRef]

39. Granata, I.; Nardelli, C.; D'Argenio, V.; Tramontano, S.; Compare, D.; Guarracino, M.R.; Nardone, G.; Pilone, V.; Sacchetti, L. Duodenal Metatranscriptomics to Define Human and Microbial Functional Alterations Associated with Severe Obesity: A Pilot Study. *Microorganisms* **2020**, *8*, 1811. [CrossRef]

40. Leite, G.; Morales, W.; Weitsman, S.; Celly, S.; Parodi, G.; Mathur, R.; Barlow, G.M.; Sedighi, R.; Millan, M.J.V.; Rezaie, A.; et al. The duodenal microbiome is altered in small intestinal bacterial overgrowth. *PLoS ONE* **2020**, *15*, e0234906. [CrossRef]

41. Rabot, S.; Membrez, M.; Bruneau, A.; Gérard, P.; Harach, T.; Moser, M.; Raymond, F.; Mansourian, R.; Chou, C.J. Germ-free C57BL/6J mice are resistant to high-fat-diet-induced insulin resistance and have altered cholesterol metabolism. *FASEB J.* **2010**, *24*, 4948–4959. [CrossRef] [PubMed]

42. Suriano, F.; Vieira-Silva, S.; Falony, G.; Roumain, M.; Paquot, A.; Pelicaen, R.; Régnier, M.; Delzenne, N.M.; Raes, J.; Muccioli, G.G.; et al. Novel insights into the genetically obese (ob/ob) and diabetic (db/db) mice: Two sides of the same coin. *Microbiome* **2021**, *9*, 147. [CrossRef] [PubMed]

43. Suriano, F.; Manca, C.; Flamand, N.; Depommier, C.; Van Hul, M.; Delzenne, N.M.; Silvestri, C.; Cani, P.D.; Di Marzo, V. Exploring the endocannabinoidome in genetically obese (ob/ob) and diabetic (db/db) mice: Links with inflammation and gut microbiota. *Biochim. Biophys. Acta (BBA)-Mol. Cell Biol. Lipids* **2022**, *1867*, 159056. [CrossRef] [PubMed]

Article

An Evaluation of Understudied Phytocannabinoids and Their Effects in Two Neuronal Models

Alex Straiker [1,*], Sierra Wilson [1], Wesley Corey [1], Michaela Dvorakova [1,2], Taryn Bosquez [1], Joye Tracey [1], Caroline Wilkowski [1], Kathleen Ho [1], Jim Wager-Miller [1] and Ken Mackie [1]

[1] Gill Center for Molecular Bioscience, Program in Neuroscience, Department of Psychological & Brain Sciences, Indiana University, Bloomington, IN 47405, USA; silywils@iu.edu (S.W.); coreyw@iu.edu (W.C.); michdvor@iu.edu (M.D.); tbosquez@iu.edu (T.B.); jxtracey@iu.edu (J.T.); cwilkows@iu.edu (C.W.); katkho@iu.edu (K.H.); jm99@indiana.edu (J.W.-M.); kmackie@iu.edu (K.M.)
[2] Department of Molecular Pharmacology, Institute of Molecular Genetics of the Czech Academy of Sciences, Videnska 1083, 14220 Prague, Czech Republic
[*] Correspondence: straiker@indiana.edu

Abstract: Cannabis contains more than 100 phytocannabinoids. Most of these remain poorly characterized, particularly in neurons. We tested a panel of five phytocannabinoids—cannabichromene (CBC), cannabidiolic acid (CBDA), cannabidivarin (CBDV), cannabidivarinic acid (CBDVA), and Δ^9-tetrahydrocannabivarin (THCV) in two neuronal models, autaptic hippocampal neurons and dorsal root ganglion (DRG) neurons. Autaptic neurons expressed a form of CB1-dependent retrograde plasticity while DRGs expressed a variety of transient receptor potential (TRP) channels. CBC, CBDA, and CBDVA had little or no effect on neuronal cannabinoid signaling. CBDV and THCV differentially inhibited cannabinoid signaling. THCV inhibited CB1 receptors presynaptically while CBDV acted post-synaptically, perhaps by inhibiting 2-AG production. None of the compounds elicited a consistent DRG response. In summary, we find that two of five 'minor' phytocannabinoids tested antagonized CB1-based signaling in a neuronal model, but with very different mechanisms. Our findings highlight the diversity of potential actions of phytocannabinoids and the importance of fully evaluating these compounds in neuronal models.

Keywords: cannabichromene; cannabidiolic acid; cannabidivarin; cannabidivarinic acid; phytocannabinoids; tetrahydrocannabivarin

Citation: Straiker, A.; Wilson, S.; Corey, W.; Dvorakova, M.; Bosquez, T.; Tracey, J.; Wilkowski, C.; Ho, K.; Wager-Miller, J.; Mackie, K. An Evaluation of Understudied Phytocannabinoids and Their Effects in Two Neuronal Models. *Molecules* **2021**, *26*, 5352. https://doi.org/10.3390/molecules26175352

Academic Editor: Mauro Maccarrone

Received: 31 July 2021
Accepted: 28 August 2021
Published: 2 September 2021

Publisher's Note: MDPI stays neutral with regard to jurisdictional claims in published maps and institutional affiliations.

1. Introduction

Cannabis has been used extensively during much of human history. Due to its changing legal status, cannabis and its phytocannabinoid constituents have recently attracted a great deal of commercial and public interest. Specifically, Δ^9-tetrahydrocannabinol (Δ^9-THC) and cannabidiol (CBD) are the most abundant (and studied) phytocannabinoids; however, more than 100 additional phytocannabinoids are present at lower concentrations in cannabis [1], the so-called "minor cannabinoids". Δ^9-THC, the main intoxicating component of cannabis [2], has been the subject of thousands of studies; research shows that it acts on the endocannabinoid signaling system [3]. This signaling system includes receptors, primarily CB1 [4] and CB2 [5], but also other lipid messengers (endocannabinoids) and enzymes to synthesize and metabolize these messengers 'on demand' [6]. CB1 receptors are enriched in the brain and likely mediate many of the CNS effects of THC. The case of CBD is perhaps more interesting. CBD, as the main non-intoxicating constituent of cannabis [7], was long considered inactive; yet, in the space of 10 years, CBD has transitioned from relatively unknown to a wonder-drug in the popular press, due in part to CBD approval as a therapy treatment for seizures associated with Dravet syndrome and Lennox–Gastaut syndrome, two forms of childhood epilepsy [8]. CBD is now readily available over-the-counter, in a variety of preparations, in grocery stores in many US states. Commercial interests have taken note of the renewed interest in phytocannabinoids and are now focusing on minor cannabinoids. Previously, low and variable concentrations of minor cannabinoids in cannabis served as natural limitations on their exploitation. However, these limitations have been overcome with scaled production and improvements in extraction and synthesis; some groups are even harnessing yeast or algae to synthesize specific cannabinoids [9]. Phytocannabinoids, such as cannabidiolic acid (CBDA) and cannabichromene (CBC), are now finding their way into creams, foods, and beverages, with vendors ascribing health benefits to these compounds, and consumers readily embracing all things cannabinoid. Though it is often assumed that phytocannabinoids act via the cannabinoid signaling system [10], there has been little systematic study on how these compounds work in the body. A first question is whether they interact with the cannabinoid signaling system. Binding studies can be misleading since CBD binds poorly to the orthosteric site of CB1 receptors [11], but effectively and potently inhibits CB1 as a negative allosteric modulator [12,13]. Moreover, the cannabinoid signaling system consists not only of cannabinoid receptors but also of lipid messengers, and the enzymes to synthesize, transport, and metabolize these messengers [14], all processes that 'minor' cannabinoids might affect. Research shows that phytocannabinoids activate several members of the transient receptor potential (TRP) family of ion channels [15]. Some of these receptors may be linked to the endocannabinoid signaling system, since anandamide is an efficacious agonist at TRPV1 [16]. We have therefore tested several 'minor' phytocannabinoids—CBC, CBDA, cannabidivarin (CBDV), cannabidivarinic acid (CBDVA), and Δ^9-tetrahydrocannabivarin (THCV) (Figure 1), in two neuronal models of endogenous cannabinoid signaling. These phytocannabinoids were chosen for their range of reported effects (spanning from analgesia to seizures) and because they are some of the most commercially promoted compounds. The neuronal models include the well-characterized autaptic hippocampal neurons that natively express CB1 receptors, the machinery to synthesize and metabolize the endocannabinoid 2-AG, as well as several forms of CB1-mediated plasticity [17–19]. We also tested dorsal root ganglion neurons, which natively express a variety of TRP receptors.

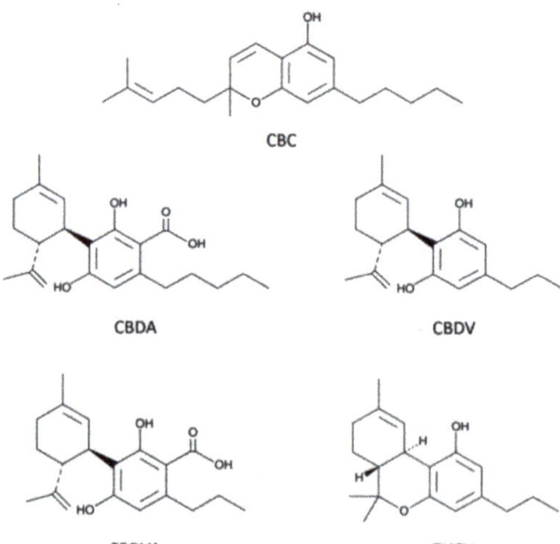

Figure 1. Structures of phytocannabinoids examined in the current study. The figure shows chemical structures of the five phytocannabinoids used in this study; cannabichromene (CBC), cannabidiolic acid (CBDA), cannabidivarin (CBDV), cannabidivarinic acid (CBDVA), and Δ^9-tetrahydrocannabivarin (THCV).

2. Materials and Methods

2.1. Hippocampal Culture Preparation

All procedures used in this study were approved by the Animal Care Committee of Indiana University and conformed to the Guidelines of the National Institutes of Health on the Care and Use of Animals. Mouse hippocampal neurons isolated from the CA1–CA3 region were cultured on microislands as described previously [20,21]. Neurons were obtained from animals (age postnatal day 0–2) and plated onto a feeder layer of hippocampal astrocytes that had been laid down previously [22]. Cultures were grown in high-glucose (20 mM) DMEM containing 10% horse serum, without mitotic inhibitors, and used for recordings after 8 days in culture, and for no more than three hours after removal from culture medium.

2.2. Electrophysiology

When a single neuron is grown on a small island of permissive substrate, it forms synapses or "autapses" onto itself. All experiments were performed on isolated autaptic neurons. Whole cell voltage-clamp recordings from autaptic neurons were carried out at room temperature using an Axopatch 200A amplifier (Molecular Devices, Sunnyvale, CA). The extracellular solution contained (in mM) 119 NaCl, 5 KCl, 2.5 $CaCl_2$, 1.5 $MgCl_2$, 30 glucose, and 20 HEPES. Continuous flow of solution through the bath chamber (~2 mL/min) ensured rapid drug application and clearance. Drugs were typically prepared as stocks, and then diluted into extracellular solution at their final concentration and used on the same day.

Recording pipettes of 1.8–3 MΩ were filled with (in mM) 121.5 K Gluconate, 17.5 KCl, 9 NaCl, 1 $MgCl_2$, 10 HEPES, 0.2 EGTA, 2 MgATP, and 0.5 LiGTP. Access resistance and holding current were monitored and only cells with both stable access resistance and holding current were included for data analysis. Conventional stimulus protocol: the membrane potential was held at –70 mV and excitatory postsynaptic currents (EPSCs) were evoked every 20 s by triggering an unclamped action current with a 1.0 ms depolarizing step. The resultant evoked waveform consisted of a brief stimulus artifact and a large downward spike representing inward sodium currents, followed by the slower EPSC. The size of the recorded EPSCs was calculated by integrating the evoked current to yield a charge value (in pC). Calculating the charge value in this manner yields an indirect measure of the amount of neurotransmitter released while minimizing the effects of cable distortion on currents generated far from the site of the recording electrode (the soma). Data were acquired at a sampling rate of 5 kHz.

DSE stimuli: after establishing a 10–20 s 0.5 Hz baseline, DSE was evoked by depolarizing to 0 mV for 50 ms, 100 ms, 300 ms, 500 ms, 1 s, 3 s and 10 s, followed in each case by resumption of a 0.5 Hz stimulus protocol for 20–80+ seconds, allowing EPSCs to recover to baseline values. This approach allowed us to determine the sensitivity of the synapses to DSE induction. To allow comparison, baseline values (prior to the DSE stimulus) are normalized to one. DSE inhibition values are presented as fractions of 1, i.e., a 50% inhibition from the baseline response is 0.50 $\pm$ standard error of the mean. The x-axis of DSE depolarization response curves are log-scale seconds of the duration of the depolarization used to elicit DSE. Depolarization response curves are obtained to determine pharmacological properties of endogenous 2-AG signaling by depolarizing neurons for progressively longer durations (50 ms, 100 ms, 300 ms, 500 ms, 1 s, 3 s, and 10 s).

2.3. Flamindo cAMP Assay

2.3.1. Cell Culture and Transfection

HEK293 cells were purchased from ATCC. Cells were cultured in high glucose Dulbecco's Modified Eagle Medium (Thermo Fisher Scientific, Waltham, MA, USA) and supplemented with 10% fetal bovine serum and a 1% Pen/Strep solution. Cultures were maintained at 37 °C with an atmosphere of 5% CO_2. For the imaging experiments, the cells were dissociated using trypsin-EDTA (0.05%) and cultured on poly-D-lysine pre-coated

18 mm glass coverslips in 12-well plates. One day post-plating, the cells were transfected with the receptor of interest (rat CB_1), the fluorescent protein EYFP, and the Pink Flamindo cAMP indicator [23], using Lipofectamine 2000 Transfection Reagent (Thermo Fisher Scientific). After 3.5 h, the transfection reagent was replaced with cell culture media and the cells used for experiments within two days of transfection.

2.3.2. Cell Imaging and cAMP Binding Assay

Transfected HEK293 cells, were imaged in an extracellular solution containing (mM) NaCl 119, KCl 5, $CaCl_2$ 2, $MgCl_2$ 1, glucose 30 and HEPES 20, using a Leica TCS SP5 confocal microscope with an oil-immersion $20\times$ objective. Images were acquired using an argon (40%), DPSS 561, with fluorescent wavelength settings set to 488–550 nm (EYFP), and 594–773-nm (Pink Flamindo). Drugs were initially prepared as a stock in DMSO or ethanol, then diluted using extracellular solution to their final concentration shortly before use.

Pink Flamindo is a fluorescent protein cAMP-indicator where increasing magnitudes of brightness in expressing cells, is indicative of elevated levels of cellular cAMP. CB1 agonists inhibit cAMP accumulation. CB1-transfected HEK293 cells were prepared as described above and were used to measure the inhibition of forskolin (Fsk)-induced production of cAMP, caused by the CB1 agonist, 2-AG (2.5 μM). The test compound and 2-AG were co-applied, followed by the potent adenylyl cyclase activator, forskolin (Fsk; 100 μM). Images were acquired every 30 s for 15 min and then analyzed using FIJI software with the 1-click ROI manager plugin [24], to measure the change in fluorescence intensity. Target cells were chosen by taking the first image in the series, increasing the brightness, and marking cells that exhibited a baseline Pink Flamindo fluorescence. Occasional (<5%) cells exhibited a high baseline fluorescence relative to the general transfected cell population. These cells were excluded from analysis since they were close to saturation. This mask of identified cells (typically 15–25 per experiment) was then applied to the image series. Baseline fluorescence intensity was normalized to 100 based on the first two minutes of the time series.

2.4. Methods for Dorsal Root Ganglion Cell Culture

DRGs were harvested from P-0 through P-14 day old rat pups following strict IACUC guidelines for the ethical care and use of laboratory animals. Rats were euthanized using isoflurane inhalation and cervical dislocation. DRGs were harvested using the protocol described by Sleigh et al. [25]. Briefly, rats were sprayed with 70% ethanol and the dorsal side was opened along the longitudinal axis with surgical scissors. The spine was removed, cleaned of excess muscle, and cut longitudinally along the dorsal and ventral surfaces. It was then placed into cold Dissection Solution (Earl's Balanced Salt Solution (Gibco, 24010043), 10 mM $MgCl_2$, 1X GlutaMAX (Gibco, 35050061), Penicillin/ Streptomycin (500 μg/mL, Gibco, 15140122), and 10 mM HEPES (Thermo Fisher Scientific, BP310-1), and the spinal cord was carefully removed and discarded. DRGs were pulled from the vertebrate and placed into a 15 mL conical containing ice-cold Dissection Solution. DRGs were centrifuged at $100\times g$ at 4 °C and the media was replaced with Dissection Solution containing 10 mg/mL collagenase type II (Gibco, 17101015). DRGs were incubated at 37 °C for 20 min. Trypsin/EDTA (Gibco, 15090046) was then added to a final concentration of 0.05% and the tissue was incubated a further 3 min. Tissue was centrifuged at 4 °C as above and washed three times in ice-cold DMEM (Gibco, 11965126) containing 10% fetal calf serum. DRGs were then triturated 30 times in 4 mL of this medium, centrifuged, and resuspended in 3 mL of cold culture medium (Neurobasal A (Gibco, 10888022), 2.5 mg/mL insulin, 5 mg/mL transferrin, 5 mg/mL nerve growth factor-b (Sigma, SRP4304), 1X B27 (Gibco, 17504044), 1X GlutaMAX, 10% Fetal calf serum). Cells were then counted and plated at a concentration of ~5000 cells/cm^2 on coverslips coated with poly-D lysine (Sigma-Aldrich, St. Louis, MO, USA, P-7886), and laminin (Millipore, Burlington, VT, USA, SCR127). Cells were cultured at 37 °C and 5% CO_2 with one half of the media changed every 3–4 days.

2.5. Methods for Calcium Imaging

DRGs were treated with Fluo4-AM (5 µM) for 30 min at 37 °C after which the cells were washed in extracellular solution (see electrophysiology) for 20 min to allow for de-esterification of Fluo4-AM. Fluorescence was then monitored on a Nikon TE200 inverted microscope (Nikon Instruments, Melville, NY, USA) with a 10× objective, a Hamamatsu Photonics (Hamamatsu City, Japan, Flash 4.0 camera and Nikon Elements AR software, (version 4.50, Nikon Instruments, Melville, NY, USA) which controlled a Spectra X light engine (Lumencor Inc., Beaverton, OR, USA) for stimulation of fluorescence. Target DRGs were chosen based on neuronal morphology using a Brightfield image acquired before the experiment. This mask was then applied to the image series. Images were acquired every 30 s for 15 min and then analyzed using FIJI software with the 1-click ROI manager plugin [24], to measure the change in fluorescence intensity over 15 min. Baseline fluorescence intensity was normalized to zero based on the two minutes preceding drug application.

2.6. Statistics

For electrophysiology experiments, the data were fitted with a nonlinear regression (Sigmoidal dose response; GraphPad Prism 6, La Jolla, CA, USA), allowing calculation of an ED50, the effective dose or duration of depolarization at which a 50% inhibition is achieved. A statistically significant difference between these curves is defined as non-overlapping 95% confidence intervals of the ED50s. Values on graphs are presented as mean ± S.E.M. Comparisons of the effects of various THCV concentrations were made using a one-way ANOVA with Dunnett's post hoc vs. control. Statistical comparisons of single drug effects (e.g., baclofen alone vs. baclofen with CBDV) were conducted using an unpaired *t*-test.

For the cAMP assay, we used an area under the curve (AUC) analysis for time points from 0 to 15 min. Administration of a drug concentration series allowed the calculation of an IC50 for THCV in this system using GraphPad Prism 6. For a given experimental treatment, a same-day control forskolin-only experimental control was included. Experimental results were compared to their respective same-day controls.

2.7. Drugs

The drugs CBC, CBDA, CBDV, and 2-AG were purchased from Cayman Chemical (Ann Arbor, MI, USA), and CBDVA and THCV were purchased from Cerilliant Corporation (Round Rock, TX, USA).

3. Results

3.1. CBC Modestly Inhibits CB1 Signaling in Autaptic Hippocampal Neurons While CBDA, and CBDVA Are without Effect

Cannabichromene (CBC) is frequently cited as a phytocannabinoid with attractive properties [10]. Some interactions and similarities between CBC and Δ^9-THC were described in the early eighties [26,27]. CBC produced mild hypothermia in mice and affected motility in electroshock-induced model of seizures, but only at very high doses (75 mg/kg). Marketing for CBC-containing products often cite studies reporting anti-inflammatory [28] and analgesic [27] properties that can be mediated through CB2 receptors at which CBC has been described as more potent agonist than Δ^9-THC [29]. CBC has been shown to activate the transient receptor potential ankyrin type-1 (TRPA1) receptor at a relatively low concentration (EC50 = 90 nM) [30] and produced antinociceptive effects in rats following brainstem injection of low nanomole doses [31]. Other potential mechanisms of action include direct interaction with CB1 receptors, either at orthosteric or allosteric sites, and altered synthesis/metabolism of endocannabinoids. Although CBC has an impact on CB1-related behavior in mice, the effect is only prominent at high doses (100 mg/kg) and is not reversed by the CB1 inverse agonist SR141716 [32]. We tested CBC at 1 µM, a concentration chosen for CBC and other phytocannabinoids because it represents a likely physiological ceiling concentration that consumers might encounter (discussed in [13]).

It is also a concentration that has been shown to affect the activity of extracellular signal-regulated kinases 1 and 2 (ERK1/2) and viability of neuronal stem cells [33]. At 1 µM, CBC did not alter excitatory post-synaptic current (EPSC) amplitudes (Figure 2A, EPSC charge relative to baseline (1.0 = no effect) CBC: 1.04 ± 0.02, $n = 4$; $p = 0.23$ by one-sample t-test vs. baseline 1.0), indicating that CBC does not directly alter excitatory neurotransmission in this system.

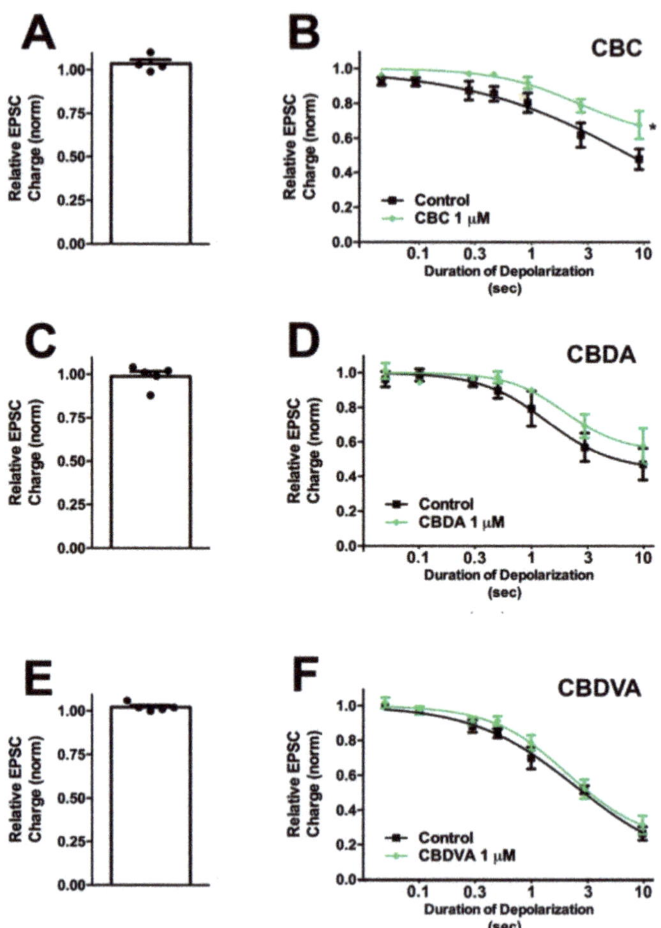

Figure 2. CBC modestly inhibits CB1 signaling in autaptic hippocampal neurons, while CBDA and CBDVA are without effect. (**A**) CBC, (**C**) CBDA, and (**E**) CBDVA have no direct effect on EPSCs. (**B**) CBC modestly inhibits maximal DSE. (**D,F**) CBDA and CBDV do not have a significant effect on DSE-mediated inhibition of EPSCs. *, $p < 0.05$, paired t-test for 10 s inhibition, drug vs. baseline.

To test whether CBC modulated cannabinoid signaling, we tested for its effects on depolarization-induced suppression of excitation (DSE), a form of endogenous 2-AG- and CB1-mediated retrograde signaling present in autaptic hippocampal neurons. As described in the methods section, successively longer depolarizations (100 ms, 300 ms, 500 ms, 1 s, 3 s, and 10 s) result in greater inhibition of EPSCs, yielding a 'depolarization dose-response' curve. A potentiator of cannabinoid signaling would be expected to shift this curve to the left, as is the case with positive allosteric modulators [34]. Conversely, an inhibitor of cannabinoid signaling would be expected to shift this curve to the right and depress maximal DSE, as seen with negative allosteric modulation [35]. We found that CBC did not alter the EC50 for DSE at 1 μM (Figure 2B, Table 1).

Table 1. Phytocannabinoid responses in autaptic hippocampal neurons. Values for EPSC inhibition in response to longest depolarization (10 s) for baseline and given phytocannabinoid. Paired *t*-test was used to compare maximal inhibition at 10 s depolarization vs. baseline in a given cell. Effective dose 50 (ED50, with 95% confidence interval) for depolarization-response curves indicating duration of depolarization that gave a 50% maximal response for the phytocannabinoids tested. None of the phytocannabinoids significantly altered the ED50.

	Inhibition at 10 s Depolarization					ED50 (95% CI)	
	Concentration	Control	Drug	Significant	*p* Value	Control	Drug
CBC	1 μM	0.48 ± 0.06	0.68 ± 0.08	Yes	0.01	1.58 s (0.78–3.18)	3.90 s (1.40–10.87)
CBDA	1 μM	0.47 ± 0.09	0.58 ± 0.10	No	0.11	1.72 s (0.84–3.54)	3.82 s (1.29–11.32)
CBDVA	1 μM	0.27 ± 0.04	0.32 ± 0.05	No	0.24	2.23 s (1.54–3.22)	3.10 s (1.93–4.99)
CBDV	100 nM	0.37 ± 0.05	0.44 ± 0.08	No	0.17	1.78 s (1.19–2.66)	6.05 s (2.32–15.72)
	1 μM	0.49 ± 0.07	0.85 ± 0.05	Yes	0.0013	1.78 s (1.19–2.66)	0.97 s (0.18–5.19)
THCV	100 nM	0.39 ± 0.04	0.88 ± 0.04	Yes	0.0018	1.84 s (1.20–2.82)	ambiguous

Cannabidiolic acid (CBDA) is the acidic precursor of CBD [36]. Although a U.S. patent was written, citing CBDA as a possible treatment for autism [37], and it was reported to have anti-nausea properties [38], little is known about its pharmacology. In our model at 1 μM, we found that CBDA did not alter EPSC amplitudes (Figure 2C, 0.99 ± 0.03, *n* = 5; *p* = 0.69 by one-sample *t*-test vs. baseline 1.0). We also did not see a significant change in DSE responses (Figure 2D, Table 1).

Cannabidivarinic acid (CBDVA) is the acidic precursor to CBDV and has received little attention until recently. CBDVA has a high oral bioavailability [39]; however, it seems to have poor brain penetration [40]. CBDVA was reported to inhibit DAGLα, however to a lesser extent than CBDV and CBDA [30]. We found that CBDVA did not alter excitatory neurotransmission at 1 μM (Figure 2E, 1 μM CBDVA: 1.02 ± 0.01, *n* = 5; *p* = 0.10 by one-sample *t*-test vs. baseline 1.0) and did not significantly alter DSE (Figure 2F, Table 1).

3.2. THCV Potently Inhibits CB1 Signaling

Tetrahydrocannabivarin (THCV) is a homolog of Δ^9-THC where, like CBDV, the lipophilic side chain is shortened by two methylene bridges [41]. THCV was reported to act as a competitive antagonist at CB1 receptors [42] and as an agonist at higher concentrations [43]. THCV was shown to have both anti-convulsant [44] and anti-inflammatory [38] properties, consistent with the findings by Thomas et al. [42]. While THCV did not alter neurotransmission on its own (Figure 3A, EPSC charge relative to baseline after THCV (1 μM): 1.01 ± 0.02, *n* = 5, *p* = 0.70 by one-sample *t*-test vs. baseline 1.0), THCV inhibited DSE in a concentration-dependent manner. The effect of THCV was surprisingly potent: 100 nM THCV was sufficient to fully block DSE in response to a 10 s depolarization and even 100 pM THCV significantly reduced DSE (Figure 3B,C; Table 2). The calculated IC50 for THCV in this system was 708 pM. The range of concentrations over which THCV acted was unusually long and a Schild plot yielded a slope of 0.52 (Figure 3D), potentially an indication of a non-competitive antagonism, negative cooperativity, or a second target in this system.

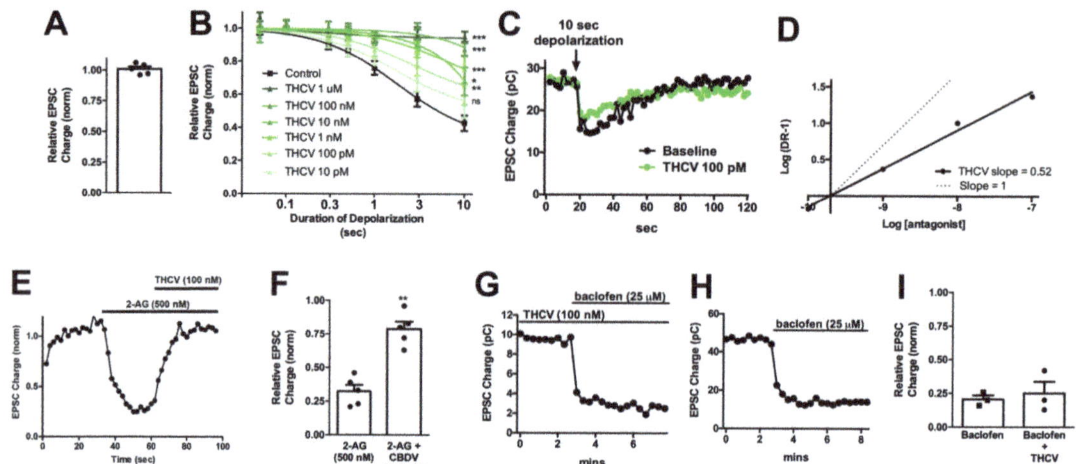

Figure 3. THCV potently inhibits presynaptic CB1 responses in autaptic neurons. (**A**) THCV (1 μM) has no direct effect on EPSCs. (**B**) THCV concentration-dependently reduces DSE inhibition of EPSCs, with significant effects even at 100 pM. *, $p < 0.05$, ***, $p < 0.005$ one-way ANOVA with Dunnett's post hoc vs. control. (**C**) Sample DSE responses before and after treatment with 1 μM THCV. (**D**) A Schild analysis shows that the Schild slope is less than 1. (**E**) Sample time course showing reversal of 2-AG inhibition by THCV (100 nM). (**F**) Summarized data showing that THCV reverses 2-AG action. **, $p < 0.01$ by paired *t*-test. (**G,H**) Sample time courses showing that baclofen (25 μM) responses are similar with and without pre-treatment with THCV (100 nM). (**I**) Summarized data for baclofen/THCV vs. baclofen alone. $p > 0.05$ unpaired *t*-test.

Table 2. THCV potently inhibits DSE in autaptic hippocampal neurons. Values for EPSC inhibition in response to longest depolarization (10 s) for baseline and THCV at various concentrations (10 pM–1 μM). One-way ANOVA with Dunnett's post hoc test was used to compare maximal inhibition at 10 s depolarization vs. controls. Effective dose 50 (ED50, with 95% confidence interval) for depolarization-response curves indicating duration of depolarization that gave a 50% maximal response for various concentrations of THCV.

	Concetration	Inhibition at 10 s	Significant	p Value	ED50 (95%CI)
Control	-	0.39 ± 0.04			1.84 s (1.20–2.82)
THCV	10 pM	0.56 ± 0.10	No	0.285	1.83 s (0.75–4.47)
	100 pM	0.66 ± 0.10	Yes	0.011	2.68 s (0.76–9.43)
	1 nM	0.75 ± 0.07	Yes	0.0008	7.35 s (1.15–47)
	10 nM	0.68 ± 0.08	Yes	0.014	ambiguous
	100 nM	0.87 ± 0.04	Yes	<0.0001	ambiguous
	1 μM	0.95 ± 0.03	Yes	<0.0001	ambiguous

Though Thomas et al. [42] reported that THCV is a competitive antagonist at CB1, DSE signaling occurs because 2-AG is synthesized postsynaptically, crosses the synaptic cleft, and acts at CB1 presynaptically. In principle, the effect of THCV on DSE might occur either presynaptically (i.e., at CB1 signaling) or post-synaptically (at some aspect of 2-AG production or transport). If the effect is pre-synaptic, then THCV should also inhibit bath-applied 2-AG. We tested this by applying 500 nM 2-AG and attempting to reverse 2-AG inhibition of EPSCs by switching into 2-AG + THCV (100 nM). We found that 100 nM THCV readily reversed inhibition by 500 nM 2-AG (Figure 3E,F, relative EPSC charge after 2-AG (500 nM): 0.32 ± 0.05; after 2-AG + THCV (100 nM): 0.79 ± 0.06; $n = 5$, $p < 0.01$ by paired *t*-test). Given then that THCV is acting presynaptically, it may be acting at CB1, but it might also be interfering more generally with presynaptic $G_{i/o}$ signaling. To test this possibility, we attempted to block the inhibition of EPSCs by GABA$_B$ agonist baclofen (25 μM), since activation of the GABA$_B$ receptor also inhibits EPSCs via the $G_{i/o}$ pathway in these neurons [45]. We found that baclofen responses were unimpeded by the presence of 100 nM THCV (Figure 3G,I; Relative EPSC charge after baclofen (25 μM) applied in presence of THCV (100 nM): 0.25 ± 0.09, $n = 3$; baclofen only: 0.21 ± 0.03, $n = 3$).

Given the potency of THCV to antagonize CB1 signaling during DSE, we tested for the effect of THCV on cAMP signaling to learn whether THCV would be similarly potent in other signaling pathways. CB1 activation is well known to inhibit the activity of adenylyl cyclase [46,47]. Using HEK293 cells transfected with CB1 and the cAMP indicator Pink Flamindo [23], we tested forskolin-induced changes in cAMP levels in response to 2-AG (2.5 μM) alone or co-treatment with various concentrations of THCV. The 2-AG reduces cAMP accumulation, an effect that is concentration-dependently inhibited by THCV with an IC50 of 7.3 nM (Figure 4A,B). THCV inhibition of 2-AG suppression of neurotransmitter release is therefore ~10× more potent than inhibition of cAMP accumulation in HEK293-CB1 cells.

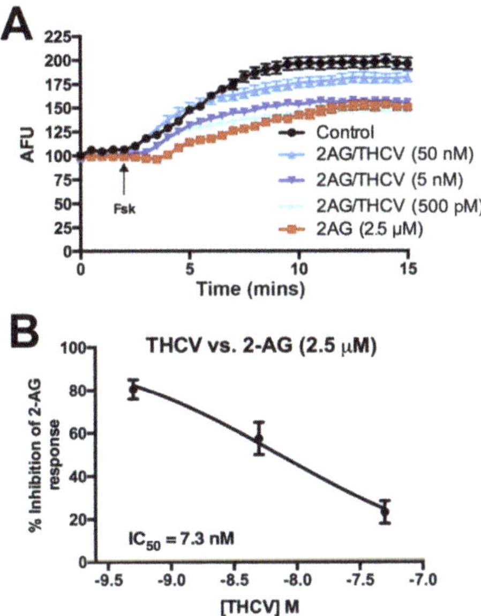

Figure 4. THCV inhibits 2-AG-mediated inhibition of adenylyl cyclase, but less potently than inhibition of neurotransmission. (**A**) Sample time courses from one set of experiments showing effects of drug combinations on forskolin-induced increases in cAMP in HEK293 cells transfected with mCB1 and the pink Flamindo cAMP indicator. (**B**) Summary cAMP responses show a concentration-dependent inhibition of 2-AG, with an IC50 of 7.3 nM.

3.3. CBDV Inhibits Endocannabinoid Signaling Postsynaptically

Cannabidivarin (CBDV) is a homolog of CBD where the lipophilic side chain is shortened by two methylene bridges [48]. CBDV has been shown to have anti-convulsant properties [49] though perhaps independently of CB1 [50]. CBDV activates and desensitizes TRPV1 transient receptor potential cation channel subfamily V member 1 (TRPV1) with promising anti-epileptic implications [51]. Moreover CBDV suppresses the expression of epilepsy-related genes following chemical convulsant treatment [52], suggesting it may be useful in preventing the development of epilepsy. CBDV has been the subject of preclinical studies for the treatment of epilepsy [53] and has been shown to rescue cognitive deficits and motor defects in a mouse model of Rett syndrome [54].

CBDV did not directly alter neurotransmission when applied (Figure 5A, CBDV 1 μM: 1.01 ± 0.02, $n = 5$, $p = 0.40$ by paired *t*-test). However, CBDV inhibited DSE at 1 μM but not at 100 nM (Figure 5B, Table 1). A sample trace is presented in Figure 5C.

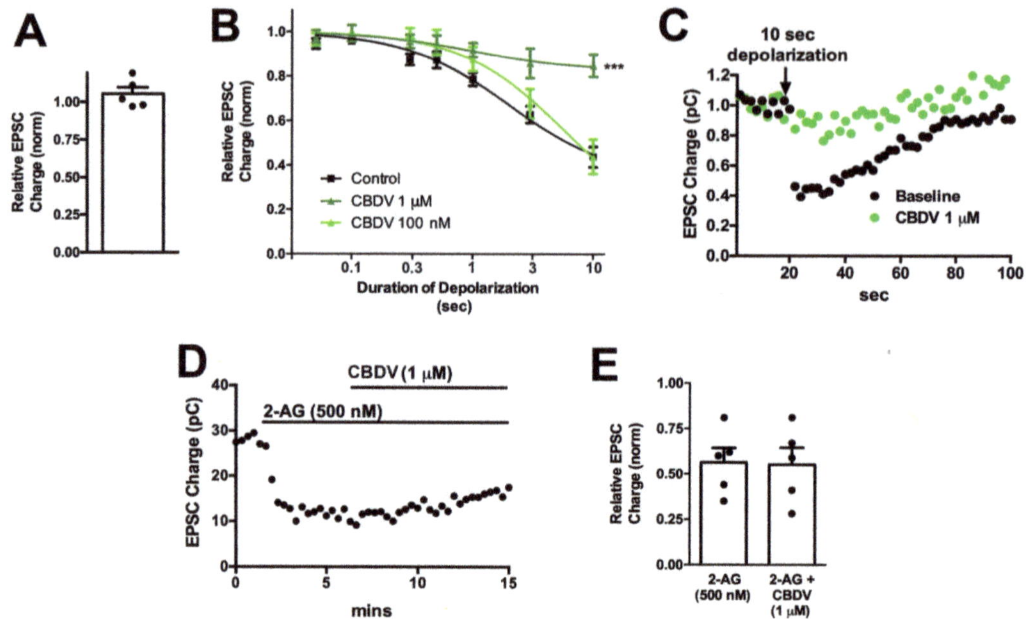

Figure 5. CBDV inhibits DSE post-synaptically in autaptic neurons.; (**A**) CBDV (1 µM) has no direct effect on EPSC; (**B**) CBDV blocks DSE at 1 µM but not at 100 nM; (**C**) sample DSE responses before and after treatment with 1 µM CBDV; (**D**) sample time course showing non-reversal of 2-AG inhibition by CBDV (1 µM). (**E**) Summarized data showing that CBDV (1 µM) does not reverse the effect of 2-AG (500 nM). ***, $p < 0.001$, one-way ANOVA with Dunnett's post-hoc test.

To explore the mechanism of CBDV action further, we tested the effect of CBDV on responses to bath-applied 2-AG. As noted for THCV, inhibition of DSE may occur due to altered CB1 signaling but also as a consequence of altered 2-AG production. If the effect of CBDV was due to inhibition of CB1 signaling, then CBDV should similarly inhibit the effects of bath-applied 2-AG. However, we did not see an inhibition of 2-AG responses, indicating that CBDV may act post-synaptically to impact 2-AG availability (Figure 5D,E: relative EPSC charge after 2-AG (500 nM): 0.56 ± 0.08; after 2-AG + CBDV (1 µM); 0.55 ± 0.09, $n = 5$, NS by paired t-test).

3.4. CBC, CBDA, CBDVA Do Not Alter Calcium Responses in Dorsal Root Ganglion Neurons

Several groups have reported that phytocannabinoids activate transient receptor potential (TRP) receptors (reviewed in [15]), though the effects often require concentrations in excess of 10 µM (e.g., [51]). Some however report responses at low-micromolar concentrations [30]. TRP receptors are ion channels that are opened by different stimuli that include chemicals but also temperature. We tested the activity of these phytocannabinoids in a second neuronal model, dorsal root ganglion neurons (DRGs) cultured from a rat. These neurons are known to natively express a variety of TRP channels, including TRPV1, TRPV3, TRPV4, and TRPA1, which have been reported to be activated by phytocannabinoids [30]. DRG subtypes differentially express these channels, underscoring one challenge in working with DRGs: they are not a uniform neuronal population. Several efforts have been made to classify DRG subpopulations [55,56]. We used calcium imaging to permit visualization of calcium influx of multiple neurons in response to TRP channel activation. We tested the TRPV1 agonist capsaicin (1 µM), finding that it activated ~40% of DRGs, consistent with reported literature (e.g., Figures 6A and 7A, [15]). According to De Petrocellis et al., THCV was one of the most potent agonists of TRPV1, with an EC50 of 1.5 µM, and one of the highest efficacies reported out of a dozen phytocannabinoids tested [30]. THCV proved to be the most likely to elicit a response, but most (94%) cells did not respond to THCV (Figure 6A). Those cells that did (6%), saw desensitizing (Figure 6B) or sustained (Figure 6C) responses, but these were infrequent (eight for each type of response (3%) of 249 cells). Significantly, most cells that responded to capsaicin failed to respond to THCV (e.g., Figure 6C). This suggests that the THCV-induced calcium responses that were observed were not due to TRPV1 activation by THCV. According to De Petrocellis, CBDV was also reported to activate TRP receptors [30], with an EC50 of 3.5 μM (for TRPV1), but their follow-on study using electrophysiological measurements only found effects at concentrations at 10 μM or higher [51]. We found that CBDV seldom (7%) activated calcium responses even in cells that were strongly activated by capsaicin (Figure 6D). On rare occasions, some cells appeared to see an increase in the frequency of spontaneous calcium transients (5 out of 158 cells Figure 6E) or activation of a steady calcium current (6 out of 158 cells Figure 6F). However, these responses were infrequent (3–4%).

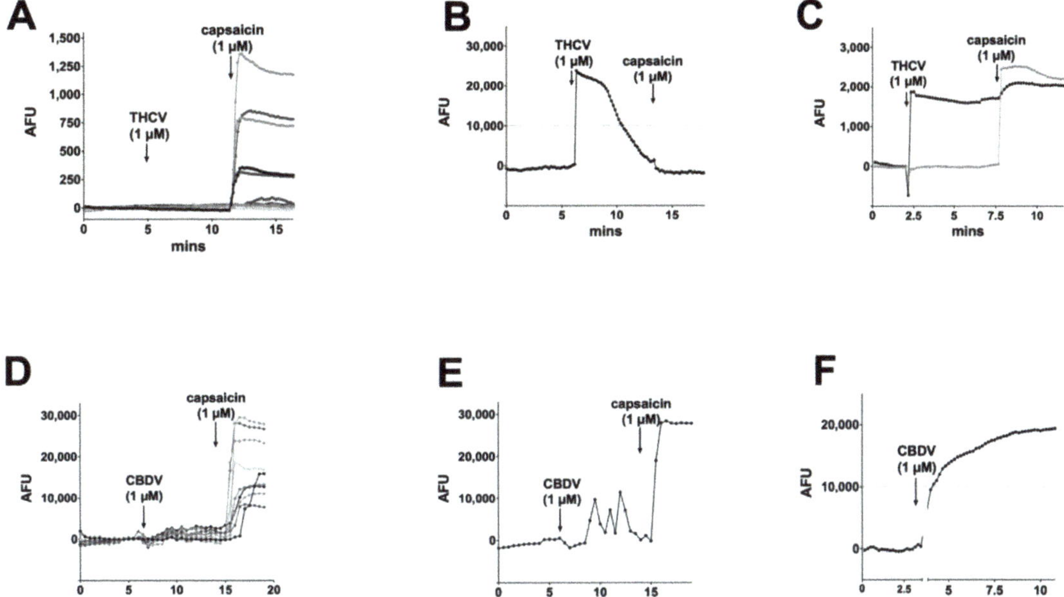

Figure 6. THCV and CBDV calcium responses in DRGs. (**A**) THCV (1 μM) rarely induced a calcium response in DRGs, while the TRPV1 agonist capsaicin (1 μM) induced responses in a large subset of DRGs. (**B**) In a small number of cells (~3%), THCV induced a desensitizing current. (**C**) A few cells (~3%) showed sustained responses to THCV. (**D**) CBDV (1 μM) did not typically induce a calcium response in DRGs. (**E**) In a few cells (~3%), CBDV appeared to increase spontaneous Ca transients. (**F**) A few cells (~4%) had a sustained calcium response to CBDV. AFU, arbitrary fluorescence units.

Of the remaining phytocannabinoids, CBDA (1 μM), on rare occasions (2%), elicited a sustained calcium response (2 out of 126 cells, Figure 7A,B). Similarly, CBC elicited a brief calcium response in a single cell out of 99 tested (Figure 7C,D). CBDVA never elicited significant calcium responses (Figure 7E).

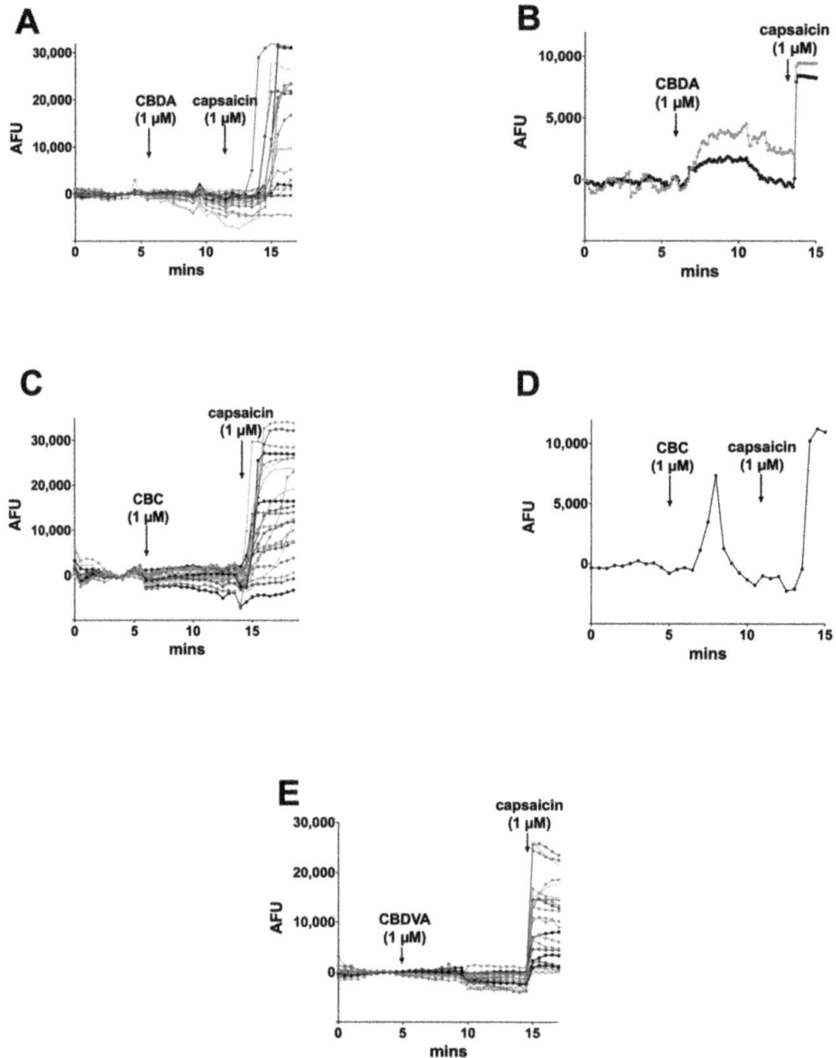

Figure 7. CBDA, CBC, and CBDVA rarely increased intracellular calcium in DRGs. (**A**) Sample time course shows calcium response in DRGs after treatment with CBDA (1 µM) followed by TRPV1 agonist capsaicin (1 µM), (**B**) on rare occasions (2%) DRG neurons responded to CBDA, (**C**) Sample time course shows calcium response in DRGs after treatment with CBC (1 µM) followed by capsaicin (1 µM), (**D**) on rare occasions (~1%) brief responses were seen after CBC treatment. (**E**) CBDVA failed to induce calcium responses in any cells tested.

4. Discussion

There has been increasing interest in 'minor' phytocannabinoids due to the changing legal landscape and advances in their synthesis and extraction. These compounds are now being introduced in consumer products and are marketed as having health benefits; however, they remain largely uncharacterized. We evaluated a panel of phytocannabinoids—CBC, CBDA, CBDV, CBDVA, and THCV—using two neuronal models. Autaptic hippocampal neurons express an endogenous CB1/2AG-based retrograde form of synaptic plasticity, while DRGs natively express a variety of TRP channels. Our chief findings show that three of the five compounds tested—CBC, CBDA, and CBDVA—had a slight or no effect in either model. However, CBDV and THCV each inhibited cannabinoid signaling, albeit via distinct mechanisms. THCV antagonized CB1 signaling as reported by others [42], but with unusually high potency, inhibiting endogenous 2-AG/CB1 signaling at concentrations as low as 100 pM. In our most striking finding, CBDV did not directly inhibit the CB1 receptor, but instead acted postsynaptically, perhaps by interfering with 2-AG production. In DRGs, despite several reports that phytocannabinoids activate calcium-permeable TRP channels, such as TRPV1, most compounds induced only infrequent, if any, calcium responses at 1 µM. These rare responses were therefore inconsistent with minor cannabinoid activation of TRPV1, for example, which is expressed in a large fraction of DRG neurons.

We chose 1 µM concentrations to test these compounds because this rests at the high end of the concentration range in which an individual is likely to encounter. We previously discussed this for the examples of THC [17] and CBD [13]. For example, Dravet Syndrome patients achieve ~1 µM blood plasma concentrations after 20 mg/kg CBD/Epidiolex treatments [57]. The pharmacokinetic properties of these compounds may nonetheless vary substantially and impact the final concentration and effect of a given phytocannabinoid [58,59]. Though minor cannabinoids are found at low concentrations in the cannabis plant, purification and synthesis of these compounds allow their incorporation into products to deliver doses comparable to THC and CBD. Nonetheless, it is unlikely that the minor cannabinoids that were inactive in these two model systems have activity at CB1 receptors or TRP channels in human neurons.

The THCV findings are interesting in several respects. THCV was reported to produce hypophagia in both non-fasted and fasted mice at doses of 3 mg/kg [60], possibly mediated by CB1 antagonism. A study in healthy volunteers showed that THCV increases neural response to rewards and aversive stimuli connected with food [61], while another report indicated that antagonism by THCV of CB1 signaling seems to be free of adverse events associated with CB1 inverse agonists, such as rimonabant [62]. As much of the interest in phytocannabinoids has to do with their CNS effects, the high potency of THCV in a neuronal model is significant.

The minor cannabinoids have been dismissed by some because their concentrations in the plant are relatively low. However, if a compound that is present in cannabis at 1% of the concentration of THC is 100-fold more potent, then the contribution of this compound to the net effect of cannabis may be significant and must be considered. This point also ties into the proposed 'entourage effect' [63]. The entourage effect generally refers to synergistic action by compounds present with THC in cannabis. In the case of THCV, the presumed net effect would be to oppose the action of THC at CB1 and may contribute to the net effect of cannabis in cultivars that have higher levels of THCV. Notable also was the broad distribution of THCV concentrations that affected neuronal cannabinoid signaling. A Schild analysis of these data is consistent with negative cooperativity between THCV and CB1 in this system or the possibility that THCV acts on a second target.

Our most novel finding is that CBDV interferes with cannabinoid signaling, not by inhibiting CB1 receptors, but postsynaptically, perhaps by hindering the production of the endocannabinoid 2-AG. However, the underlying mechanism for this remains to be elucidated and can be investigated in future studies.

Our negative findings for CBDA and CBDVA in autaptic neurons do not rule out activity of these compounds at other components of the cannabinoid signaling system. This includes other receptors [64], enzymes such as ABHD6 and ABHD12 [65], and members of the TRP family of ion channels not expressed in DRG neurons, several of which are modulated by endocannabinoids [16,66]. Several of the phytocannabinoids tested are entering into clinical trials based on their proposed health effects.

Our study of TRP responses in DRGs yielded mostly negative results. This can likely be attributed to two factors, one being the higher concentrations employed by most studies that have reported effects, and the second due to differences in expression systems versus a natively expressing neuronal population. Our results for CBDV and TRPV1 are in agreement with [51], who saw no effects for CBDV at concentrations below 10 μM using electrophysiological measures in TRPV1-expressing HEK293 cells. However, based on the finding by De Petrocellis et al. [30] that THCV activated TRPV1 with an EC50 of 1.5 μM and an efficacy of >60% of ionomycin, one would have expected a TRPV1 response in ~40% of DRG neurons. The difference may lie in the assay employed, which relied on responses in a homogenized sample and was therefore an additional step removed from an intact expression system. While we did observe occasional responses to phytocannabinoids, they were prohibitively infrequent to permit further investigation.

In summary, we found that, in a sampling of five phytocannabinoids that have attracted general interest in the population, two exerted substantial effects on CB1 and 2-AG-mediated cannabinoid signaling in a neuronal model. THCV and CBDV both inhibited cannabinoid signaling. However, while THCV acted as a CB1 antagonist, CBDV acted postsynaptically to inhibit DAGLα-mediated 2-AG production. These findings highlight the importance of testing phytocannabinoid interaction with all components of the cannabinoid signaling system; moreover, the remaining 'minor' phytocannabinoids may offer more interesting surprises.

Author Contributions: Conceptualization, A.S. and K.M.; methodology, A.S., J.W.-M. and K.M.; formal analysis, A.S., M.D., S.W., J.W.-M. and T.B.; investigation, S.W., A.S., M.D., J.T., W.C., K.H., T.B., C.W.; writing—original draft preparation, A.S., W.C.; writing—review & editing, A.S., M.D. and K.M.; project administration, A.S. and K.M.; funding acquisition, K.M. All authors have read and agreed to the published version of the manuscript.

Funding: This work was supported by NIH grant AT011162 (KM).

Institutional Review Board Statement: The study was conducted according to the guidelines of the Declaration of Helsinki, and approved by the Animal Care Committee of Indiana University (protocol code 18-033, 29 April 2018).

Data Availability Statement: The data presented in this study are available on request from the corresponding author.

Conflicts of Interest: The authors declare no conflict of interest.

References

1. Elsohly, M.A.; Slade, D. Chemical constituents of marijuana: The complex mixture of natural cannabinoids. *Life Sci.* **2005**, *78*, 539–548. [CrossRef] [PubMed]
2. Gaoni, Y.; Mechoulam, R. Isolation, structure and partial synthesis of an active constituent of hashish. *J. Am. Chem. Soc.* **1964**, *86*, 1646–1647. [CrossRef]
3. Howlett, A.C.; Barth, F.; Bonner, T.I.; Cabral, G.; Casellas, P.; Devane, W.A.; Felder, C.C.; Herkenham, M.; Mackie, K.; Martin, B.R.; et al. International Union of Pharmacology. XXVII. Classification of cannabinoid receptors. *Pharmacol. Rev.* **2002**, *54*, 161–202. [CrossRef] [PubMed]
4. Matsuda, L.A.; Lolait, S.J.; Brownstein, M.J.; Young, A.C.; Bonner, T.I. Structure of a cannabinoid receptor and functional expression of the cloned cDNA. *Nature* **1990**, *346*, 561–564. [CrossRef] [PubMed]
5. Munro, S.; Thomas, K.L.; Abu-Shaar, M. Molecular characterization of a peripheral receptor for cannabinoids. *Nature* **1993**, *365*, 61–65. [CrossRef]
6. Piomelli, D. The molecular logic of endocannabinoid signalling. *Nat. Rev. Neurosci.* **2003**, *4*, 873–884. [CrossRef] [PubMed]
7. Adams, R.; Hunt, M.; Clark, J.H. Structure of cannabidiol, a product isolated from the marihuana extract of Minnesota wild hemp. *J. Am. Chem. Soc.* **1940**, *62*, 4. [CrossRef]
8. Billakota, S.; Devinsky, O.; Marsh, E. Cannabinoid therapy in epilepsy. *Curr. Opin. Neurol.* **2019**, *32*, 220–226. [CrossRef]
9. Luo, X.; Reiter, M.A.; d'Espaux, L.; Wong, J.; Denby, C.M.; Lechner, A.; Zhang, Y.; Grzybowski, A.T.; Harth, S.; Lin, W.; et al. Complete biosynthesis of cannabinoids and their unnatural analogues in yeast. *Nature* **2019**, *567*, 123–126. [CrossRef]
10. Russo, E.B. Taming THC: Potential cannabis synergy and phytocannabinoid-terpenoid entourage effects. *Br. J. Pharmacol.* **2011**, *163*, 1344–1364. [CrossRef]
11. Mechoulam, R.; Parker, L.A.; Gallily, R. Cannabidiol: An overview of some pharmacological aspects. *J. Clin. Pharmacol.* **2002**, *42*, 11S–19S. [CrossRef] [PubMed]
12. Laprairie, R.B.; Bagher, A.M.; Kelly, M.E.; Denovan-Wright, E.M. Cannabidiol is a negative allosteric modulator of the cannabinoid CB1 receptor. *Br. J. Pharmacol.* **2015**, *172*, 4790–4805. [CrossRef] [PubMed]

13. Straiker, A.; Dvorakova, M.; Zimmowitch, A.; Mackie, K. Cannabidiol Inhibits Endocannabinoid Signaling in Autaptic Hippocampal Neurons. *Mol. Pharmacol.* **2018**, *94*, 743–748. [CrossRef]
14. Murataeva, N.; Straiker, A.; Mackie, K. Parsing the players: 2-arachidonoylglycerol synthesis and degradation in the CNS. *Br. J. Pharmacol.* **2014**, *171*, 1379–1391. [CrossRef]
15. Muller, C.; Morales, P.; Reggio, P.H. Cannabinoid Ligands Targeting TRP Channels. *Front. Mol. Neurosci.* **2018**, *11*, 487. [CrossRef] [PubMed]
16. Smart, D.; Gunthorpe, M.J.; Jerman, J.C.; Nasir, S.; Gray, J.; Muir, A.I.; Chambers, J.K.; Randall, A.D.; Davis, J.B. The endogenous lipid anandamide is a full agonist at the human vanilloid receptor (hVR1). *Br. J. Pharmacol.* **2000**, *129*, 227–230. [CrossRef] [PubMed]
17. Straiker, A.; Mackie, K. Depolarization-induced suppression of excitation in murine autaptic hippocampal neurones. *J. Physiol.* **2005**, *569*, 501–517. [CrossRef]
18. Straiker, A.; Mackie, K. Metabotropic suppression of excitation in murine autaptic hippocampal neurons. *J. Physiol.* **2007**, *578*, 773–785. [CrossRef]
19. Kellogg, R.; Mackie, K.; Straiker, A. Cannabinoid CB1 receptor-dependent long-term depression in autaptic excitatory neurons. *J. Neurophysiol.* **2009**, *102*, 1160–1171. [CrossRef]
20. Bekkers, J.M.; Stevens, C.F. Excitatory and inhibitory autaptic currents in isolated hippocampal neurons maintained in cell culture. *Proc. Natl. Acad. Sci. USA* **1991**, *88*, 7834–7838. [CrossRef]
21. Furshpan, E.J.; MacLeish, P.R.; O'Lague, P.H.; Potter, D.D. Chemical transmission between rat sympathetic neurons and cardiac myocytes developing in microcultures: Evidence for cholinergic, adrenergic, and dual-function neurons. *Proc. Natl. Acad. Sci. USA* **1976**, *73*, 4225–4229. [CrossRef]
22. Levison, S.W.; McCarthy, K.D. Characterization and partial purification of AIM: A plasma protein that induces rat cerebral type 2 astroglia from bipotential glial progenitors. *J. Neurochem.* **1991**, *57*, 782–794. [CrossRef]
23. Harada, K.; Ito, M.; Wang, X.; Tanaka, M.; Wongso, D.; Konno, A.; Hirai, H.; Hirase, H.; Tsuboi, T.; Kitaguchi, T. Red fluorescent protein-based cAMP indicator applicable to optogenetics and in vivo imaging. *Sci. Rep.* **2017**, *7*, 7351. [CrossRef]
24. Thomas, L.S.; Gehrig, J. ImageJ/Fiji ROI 1-click tools for rapid manual image annotations and measurements. *MicroPubl. Biol.* 2020. [CrossRef]
25. Sleigh, J.N.; Weir, G.A.; Schiavo, G. A simple, step-by-step dissection protocol for the rapid isolation of mouse dorsal root ganglia. *BMC Res. Notes* **2016**, *9*, 82. [CrossRef]
26. Hatoum, N.S.; Davis, W.M.; Elsohly, M.A.; Turner, C.E. Cannabichromene and delta 9-tetrahydrocannabinol: Interactions relative to lethality, hypothermia and hexobarbital hypnosis. *Gen. Pharmacol.* **1981**, *12*, 357–362. [CrossRef]
27. Davis, W.M.; Hatoum, N.S. Neurobehavioral actions of cannabichromene and interactions with delta 9-tetrahydrocannabinol. *Gen. Pharmacol.* **1983**, *14*, 247–252. [CrossRef]
28. Wirth, P.W.; Watson, E.S.; ElSohly, M.; Turner, C.E.; Murphy, J.C. Anti-inflammatory properties of cannabichromene. *Life Sci.* **1980**, *26*, 1991–1995. [CrossRef]
29. Udoh, M.; Santiago, M.; Devenish, S.; McGregor, I.S.; Connor, M. Cannabichromene is a cannabinoid CB2 receptor agonist. *Br. J. Pharmacol.* **2019**, *176*, 4537–4547. [CrossRef]
30. De Petrocellis, L.; Ligresti, A.; Moriello, A.S.; Allara, M.; Bisogno, T.; Petrosino, S.; Stott, C.G.; Di Marzo, V. Effects of cannabinoids and cannabinoid-enriched Cannabis extracts on TRP channels and endocannabinoid metabolic enzymes. *Br. J. Pharmacol.* **2011**, *163*, 1479–1494. [CrossRef]
31. Maione, S.; Piscitelli, F.; Gatta, L.; Vita, D.; De Petrocellis, L.; Palazzo, E.; de Novellis, V.; Di Marzo, V. Non-psychoactive cannabinoids modulate the descending pathway of antinociception in anaesthetized rats through several mechanisms of action. *Br. J. Pharmacol.* **2011**, *162*, 584–596. [CrossRef] [PubMed]
32. DeLong, G.T.; Wolf, C.E.; Poklis, A.; Lichtman, A.H. Pharmacological evaluation of the natural constituent of Cannabis sativa, cannabichromene and its modulation by Delta(9)-tetrahydrocannabinol. *Drug Alcohol Depend.* **2010**, *112*, 126–133. [CrossRef] [PubMed]
33. Shinjyo, N.; Di Marzo, V. The effect of cannabichromene on adult neural stem/progenitor cells. *Neurochem. Int.* **2013**, *63*, 432–437. [CrossRef]
34. Mitjavila, J.; Yin, D.; Kulkarni, P.M.; Zanato, C.; Thakur, G.A.; Ross, R.; Greig, I.; Mackie, K.; Straiker, A. Enantiomer-specific positive allosteric modulation of CB1 signaling in autaptic hippocampal neurons. *Pharmacol. Res.* **2018**, *129*, 475–481. [CrossRef]
35. Straiker, A.; Mitjavila, J.; Yin, D.; Gibson, A.; Mackie, K. Aiming for allosterism: Evaluation of allosteric modulators of CB1 in a neuronal model. *Pharmacol. Res.* **2015**, *99*, 370–376. [CrossRef]
36. Taura, F.; Morimoto, S.; Shoyama, Y. Purification and characterization of cannabidiolic-acid synthase from Cannabis sativa L.. Biochemical analysis of a novel enzyme that catalyzes the oxidocyclization of cannabigerolic acid to cannabidiolic acid. *J. Biol. Chem.* **1996**, *271*, 17411–17416. [CrossRef]
37. Guy, G.; Wright, S.; Brodie, J.; Woolley-Roberts, M.; Maldonado, R.; Parolaro, D.; Luongo, L. Use of Cannabidiolic Acid in the Treatment of Autism Spectrum Disorder and Associated Disorders. 2019. Available online: https://patentscope.wipo.int/search/en/detail.jsf?docId=US241444657&docAn=16092560 (accessed on 15 July 2021).

38.	Bolognini, D.; Costa, B.; Maione, S.; Comelli, F.; Marini, P.; Di Marzo, V.; Parolaro, D.; Ross, R.A.; Gauson, L.A.; Cascio, M.G.; et al. The plant cannabinoid Delta9-tetrahydrocannabivarin can decrease signs of inflammation and inflammatory pain in mice. *Br. J. Pharmacol.* **2010**, *160*, 677–687. [CrossRef] [PubMed]

39.	Kleinhenz, M.D.; Magnin, G.; Lin, Z.; Griffin, J.; Kleinhenz, K.E.; Montgomery, S.; Curtis, A.; Martin, M.; Coetzee, J.F. Plasma concentrations of eleven cannabinoids in cattle following oral administration of industrial hemp (Cannabis sativa). *Sci. Rep.* **2020**, *10*, 12753–12763. [CrossRef]

40.	Anderson, L.L.; Low, I.K.; Banister, S.D.; McGregor, I.S.; Arnold, J.C. Pharmacokinetics of Phytocannabinoid Acids and Anticonvulsant Effect of Cannabidiolic Acid in a Mouse Model of Dravet Syndrome. *J. Nat. Prod.* **2019**, *82*, 3047–3055. [CrossRef]

41.	Merkus, F.W. Cannabivarin and tetrahydrocannabivarin, two new constituents of hashish. *Nature* **1971**, *232*, 579–580. [CrossRef]

42.	Thomas, A.; Stevenson, L.A.; Wease, K.N.; Price, M.R.; Baillie, G.; Ross, R.A.; Pertwee, R.G. Evidence that the plant cannabinoid Delta9-tetrahydrocannabivarin is a cannabinoid CB1 and CB2 receptor antagonist. *Br. J. Pharmacol.* **2005**, *146*, 917–926. [CrossRef]

43.	Pertwee, R.G. The diverse CB1 and CB2 receptor pharmacology of three plant cannabinoids: Delta9-tetrahydrocannabinol, cannabidiol and delta9-tetrahydrocannabivarin. *Br. J. Pharmacol.* **2008**, *153*, 199–215. [CrossRef]

44.	Hill, A.J.; Weston, S.E.; Jones, N.A.; Smith, I.; Bevan, S.A.; Williamson, E.M.; Stephens, G.J.; Williams, C.M.; Whalley, B.J. Δ9-Tetrahydrocannabivarin suppresses in vitro epileptiform and in vivo seizure activity in adult rats. *Epilepsia* **2010**, *51*, 1522–1532. [CrossRef] [PubMed]

45.	Straiker, A.J.; Borden, C.R.; Sullivan, J.M. G-Protein alpha Subunit Isoforms Couple Differentially to Receptors that Mediate Presynaptic Inhibition at Rat Hippocampal Synapses. *J. Neurosci.* **2002**, *22*, 2460–2468. [CrossRef] [PubMed]

46.	Howlett, A.C. Cannabinoid inhibition of adenylate cyclase: Relative activity of constituents and metabolites of marihuana. *Neuropharmacology* **1987**, *26*, 507–512. [CrossRef]

47.	Howlett, A.C. Pharmacology of cannabinoid receptors. *Annu Rev. Pharmacol. Toxicol.* **1995**, *35*, 607–634. [CrossRef]

48.	Vollner, L.; Bieniek, D.; Korte, F. Hashish. XX. Cannabidivarin, a new hashish constituent. *Tetrahedron Lett.* **1969**, *3*, 145–147. [CrossRef]

49.	Hill, A.J.; Mercier, M.S.; Hill, T.D.; Glyn, S.E.; Jones, N.A.; Yamasaki, Y.; Futamura, T.; Duncan, M.; Stott, C.G.; Stephens, G.J.; et al. Cannabidivarin is anticonvulsant in mouse and rat. *Br. J. Pharmacol.* **2012**, *167*, 1629–1642. [CrossRef]

50.	Hill, T.D.; Cascio, M.G.; Romano, B.; Duncan, M.; Pertwee, R.G.; Williams, C.M.; Whalley, B.J.; Hill, A.J. Cannabidivarin-rich cannabis extracts are anticonvulsant in mouse and rat via a CB1 receptor-independent mechanism. *Br. J. Pharmacol.* **2013**, *170*, 679–692. [CrossRef]

51.	Iannotti, F.A.; Hill, C.L.; Leo, A.; Alhusaini, A.; Soubrane, C.; Mazzarella, E.; Russo, E.; Whalley, B.J.; Di Marzo, V.; Stephens, G.J. Nonpsychotropic plant cannabinoids, cannabidivarin (CBDV) and cannabidiol (CBD), activate and desensitize transient receptor potential vanilloid 1 (TRPV1) channels in vitro: Potential for the treatment of neuronal hyperexcitability. *ACS Chem. Neurosci.* **2014**, *5*, 1131–1141. [CrossRef]

52.	Amada, N.; Yamasaki, Y.; Williams, C.M.; Whalley, B.J. Cannabidivarin (CBDV) suppresses pentylenetetrazole (PTZ)-induced increases in epilepsy-related gene expression. *PeerJ* **2013**, *1*, e214. [CrossRef]

53.	Huizenga, M.N.; Sepulveda-Rodriguez, A.; Forcelli, P.A. Preclinical safety and efficacy of cannabidivarin for early life seizures. *Neuropharmacology* **2019**, *148*, 189–198. [CrossRef]

54.	Zamberletti, E.; Gabaglio, M.; Piscitelli, F.; Brodie, J.S.; Woolley-Roberts, M.; Barbiero, I.; Tramarin, M.; Binelli, G.; Landsberger, N.; Kilstrup-Nielsen, C.; et al. Cannabidivarin completely rescues cognitive deficits and delays neurological and motor defects in male Mecp2 mutant mice. *J. Psychopharmacol.* **2019**, *33*, 894–907. [CrossRef] [PubMed]

55.	Schroeder, J.E.; McCleskey, E.W. Inhibition of Ca^{2+} currents by a mu-opioid in a defined subset of rat sensory neurons. *J. Neurosci.* **1993**, *13*, 867–873. [CrossRef] [PubMed]

56.	Wangzhou, A.; McIlvried, L.A.; Paige, C.; Barragan-Iglesias, P.; Shiers, S.; Ahmad, A.; Guzman, C.A.; Dussor, G.; Ray, P.R.; Gereau, R.W.; et al. Pharmacological target-focused transcriptomic analysis of native vs cultured human and mouse dorsal root ganglia. *Pain* **2020**, *161*, 1497–1517. [CrossRef]

57.	Devinsky, O.; Patel, A.D.; Thiele, E.A.; Wong, M.H.; Appleton, R.; Harden, C.L.; Greenwood, S.; Morrison, G.; Sommerville, K.; On behalf of the GWPCARE1 Part A Study Group. Randomized, dose-ranging safety trial of cannabidiol in Dravet syndrome. *Neurology* **2018**, *90*, 1204–1211. [CrossRef] [PubMed]

58.	Anderson, L.L.; Etchart, M.G.; Bahceci, D.; Golembiewski, T.A.; Arnold, J.C. Cannabis constituents interact at the drug efflux pump BCRP to markedly increase plasma cannabidiolic acid concentrations. *Sci. Rep.* **2021**, *11*, 14948. [CrossRef]

59.	Doohan, P.T.; Oldfield, L.D.; Arnold, J.C.; Anderson, L.L. Cannabinoid Interactions with Cytochrome P450 Drug Metabolism: A Full-Spectrum Characterization. *AAPS J.* **2021**, *23*, 91. [CrossRef]

60.	Riedel, G.; Fadda, P.; McKillop-Smith, S.; Pertwee, R.G.; Platt, B.; Robinson, L. Synthetic and plant-derived cannabinoid receptor antagonists show hypophagic properties in fasted and non-fasted mice. *Br. J. Pharmacol.* **2009**, *156*, 1154–1166. [CrossRef]

61.	Tudge, L.; Williams, C.; Cowen, P.J.; McCabe, C. Neural effects of cannabinoid CB1 neutral antagonist tetrahydrocannabivarin on food reward and aversion in healthy volunteers. *Int. J. Neuropsychopharmacol.* **2015**, *18*, 1–9. [CrossRef] [PubMed]

62.	McPartland, J.M.; Duncan, M.; Di Marzo, V.; Pertwee, R.G. Are cannabidiol and Delta(9) -tetrahydrocannabivarin negative modulators of the endocannabinoid system? A systematic review. *Br. J. Pharmacol.* **2015**, *172*, 737–753. [CrossRef] [PubMed]

63. Ben-Shabat, S.; Fride, E.; Sheskin, T.; Tamiri, T.; Rhee, M.H.; Vogel, Z.; Bisogno, T.; De Petrocellis, L.; Di Marzo, V.; Mechoulam, R. An entourage effect: Inactive endogenous fatty acid glycerol esters enhance 2-arachidonoyl-glycerol cannabinoid activity. *Eur. J. Pharmacol.* **1998**, *353*, 23–31. [CrossRef]
64. Mackie, K.; Stella, N. Cannabinoid receptors and endocannabinoids: Evidence for new players. *AAPS J.* **2006**, *8*, 298–306. [CrossRef]
65. Blankman, J.L.; Simon, G.M.; Cravatt, B.F. A comprehensive profile of brain enzymes that hydrolyze the endocannabinoid 2-arachidonoylglycerol. *Chem. Biol.* **2007**, *14*, 1347–1356. [CrossRef] [PubMed]
66. Morales, P.; Hurst, D.P.; Reggio, P.H. Molecular Targets of the Phytocannabinoids: A Complex Picture. *Prog. Chem. Org. Nat. Prod.* **2017**, *103*, 103–131. [PubMed]

Article

Contribution of the Adenosine 2A Receptor to Behavioral Effects of Tetrahydrocannabinol, Cannabidiol and PECS-101

Todd M. Stollenwerk, Samantha Pollock and Cecilia J. Hillard *,†

Department of Pharmacology and Toxicology and Neuroscience Research Center, Medical College of Wisconsin, Milwaukee, WI 53226, USA; tstollenwerk@mcw.edu (T.M.S.); smnthpllck@gmail.com (S.P.)
* Correspondence: chillard@mcw.edu
† Mechoulam Awardee in 2011.

Abstract: The cannabis-derived molecules, Δ^9 tetrahydrocannabinol (THC) and cannabidiol (CBD), are both of considerable therapeutic interest for a variety of purposes, including to reduce pain and anxiety and increase sleep. In addition to their other pharmacological targets, both THC and CBD are competitive inhibitors of the equilibrative nucleoside transporter-1 (ENT-1), a primary inactivation mechanism for adenosine, and thereby increase adenosine signaling. The goal of this study was to examine the role of adenosine A2A receptor activation in the effects of intraperitoneally administered THC alone and in combination with CBD or PECS-101, a 4′-fluorinated derivative of CBD, in the cannabinoid tetrad, elevated plus maze (EPM) and marble bury assays. Comparisons between wild-type (WT) and A2AR knock out (A2AR-KO) mice were made. The cataleptic effects of THC were diminished in A2AR-KO; no other THC behaviors were affected by A2AR deletion. CBD (5 mg/kg) potentiated the cataleptic response to THC (5 mg/kg) in WT but not A2AR-KO. Neither CBD nor THC alone affected EPM behavior; their combination produced a significant increase in open/closed arm time in WT but not A2AR-KO. Both THC and CBD reduced the number of marbles buried in A2AR-KO but not WT mice. Like CBD, PECS-101 potentiated the cataleptic response to THC in WT but not A2AR-KO mice. PECS-101 also reduced exploratory behavior in the EPM in both genotypes. These results support the hypothesis that CBD and PECS-101 can potentiate the cataleptic effects of THC in a manner consistent with increased endogenous adenosine signaling.

Keywords: 4′-fluoro-cannabidiol; cannabinoid tetrad; elevated plus maze; catalepsy; marble bury; HUF-101; equilibrative nucleoside transporter

Citation: Stollenwerk, T.M.; Pollock, S.; Hillard, C.J. Contribution of the Adenosine 2A Receptor to Behavioral Effects of Tetrahydrocannabinol, Cannabidiol and PECS-101. *Molecules* **2021**, *26*, 5354. https://doi.org/10.3390/molecules26175354

Academic Editor: Mauro Maccarrone

Received: 31 July 2021
Accepted: 1 September 2021
Published: 2 September 2021

Publisher's Note: MDPI stays neutral with regard to jurisdictional claims in published maps and institutional affiliations.

1. Introduction

Δ^9-Tetrahydrocannabinol (THC) and cannabidiol (CBD) are terpene phenols synthesized by the cannabis plant that can produce therapeutically important effects in humans. For example, evidence is accumulating that THC is an effective analgesic in humans, particularly in the treatment of chronic pain [1], while CBD is currently FDA-approved to treat severe childhood seizures [2]. Human studies also indicate that both THC [3] and CBD [4] can promote sleep and reduce anxiety. Although THC and CBD are structurally very similar and share some therapeutic benefits, there is little overlap in the mammalian proteins with which they interact. THC is a partial agonist with moderate affinity (K_{DS} between 40 and 100 nM) at both CB1 and CB2 cannabinoid receptors, and the majority of its pharmacological effects at moderate doses are the result of interactions with these G protein-coupled receptors (GPCRs) [5]. CBD, having greater molecular flexibility, can interact with multiple receptors, including the GPCRs serotonin 1A receptor (5HT1A) and GPR55, as well as several members of the transient receptor potential family of inotropic receptors [6]. The differences in protein targets contribute to the significant differences in adverse effects of THC and CBD. THC, by virtue of its activity as a CB1R agonist, interferes with complex tasks, such as driving, and has dependence liability [7]. On the other hand,

107

the adverse effects of CBD are relatively mild, including somnolence and gastrointestinal disturbances, although incidences of liver toxicity have also been seen [1].

Cellular studies have demonstrated that the equilibrative nucleoside transporter type 1 (ENT-1) is a molecular target that is shared by both THC and CBD. CBD and THC are both competitive inhibitors of ENT-1 nucleoside binding sites with IC_{50} values less than 200 nM [8]. ENT-1 is a major regulator of extracellular and signaling concentrations of adenosine, and ENT-1 inhibitors, including CBD, act as indirect agonists of adenosine receptor signaling [9]. Indeed, multiple studies have demonstrated that some effects of CBD are blocked by adenosine receptor antagonists, including its anti-inflammatory effects [10–12].

Interestingly, preclinical studies indicate significant interactions between type 1 cannabinoid receptors (CB1R) and adenosine A2 receptors (A2AR). Both CB1R and A2AR are highly expressed in the striatum, and multiple studies have demonstrated that CB1R and A2AR can form heterodimers [13]. There is functional evidence for interactions between the two systems; for example, the hypolocomotor and rewarding effects of CB1R agonists are diminished by A2AR antagonism [14]. One goal of the studies reported here was to expand this understanding to examine the requirement for A2AR in the effects of THC in the cannabinoid tetrad (locomotor activity, catalepsy, body temperature and spinal pain reflexes) and anxiety assays.

Considerable preclinical and clinical studies have been carried out to describe the interactions between THC and CBD [15]. Published studies suggest that co-treatment with CBD can modulate the effects of THC; however, the mechanisms that underlie these effects are not clear. Important factors include the dose ratios used; the absolute dose of each drug; and the behavioral or physiological response to THC that is measured. Given the data discussed above that some effects of CB1R agonists are modulated by changes in A2AR activity and that CBD can act as an indirect agonist of A2AR, the second goal of these studies was to explore the requirement of A2AR in CBD-induced modulation of THC effects in the cannabinoid tetrad and anxiety assays.

A significant difficulty with the use of CBD as an oral therapeutic is its low and variable oral bioavailability [16]. A series of fluorinated derivatives of CBD have been synthesized in an attempt to increase potency and reduce pharmacokinetic variability [17]. Among these derivatives, 4′-fluoro-CBD (now called PECS-101, formerly called HU-474 and HUF-101) shares many of the pharmacological effects of CBD, including reduced anxiety-like behaviors [17] and reduced responses to painful stimuli [18]. The third goal of these studies was to compare the effects of CBD and PECS-101 as modulators of THC effects in the cannabinoid tetrad and anxiety assays.

2. Results

2.1. THC Dose Response Studies

2.1.1. Cannabinoid Tetrad

WT and A2AR-KO mice were injected with vehicle or THC (1, 10 and 100 mg/kg) and then were assayed in the cannabinoid tetrad in the order and with the timing described in the Methods and shown in Supplementary Materials Figure S4.

- Locomotor activity: Vehicle-treated, A2AR-KO mice exhibited a greater mean distance traveled in the open field than vehicle-treated, WT mice (Figure 1A). THC produced a dose-dependent reduction in locomotor activity in both WT and A2AR-KO mice. Two-way ANOVA indicated significant effects of both THC ($F_{3,40} = 10.3$, $p < 0.0001$) and genotype ($F_{1,40} = 7.7$, $p < 0.01$) without a significant interaction ($F_{3,40} = 1.5$, $p = 0.22$).
- Catalepsy: THC produced a dose-dependent increase in the time with front paws on a ring stand, a commonly employed assay for cannabinoid-induced catalepsy (Figure 1B). Two-way ANOVA indicated a significant effect of THC ($F_{3,40} = 47.4$, $p < 0.0001$) and a significant interaction between THC and genotype ($F_{3,40} = 3.4$, $p < 0.05$) without a significant effect of genotype alone ($F_{1,40} = 2.4$, $p = 0.13$). Sidak's

multiple comparison post hoc test revealed that, following treatment with 100 mg/kg THC, A2AR-KO mice exhibited significantly less catalepsy than WT.

- Body temperature: Rectal temperature was measured as an index of body temperature (Supplementary Materials Figure S1A). Two-way ANOVA indicated a significant effect of THC treatment ($F_{3,40}$ = 35.0, $p < 0.0001$); genotype did not significantly affect rectal temperature ($F_{1,40}$ = 0.9, $p = 0.33$), and the interaction was not significant ($F_{3,40}$ = 0.3, $p = 0.82$).
- Nociceptive reflex: Latency to move the tail in response to a heat stimulus was used to assess the antinociceptive effects of THC in both genotypes (Supplementary Materials Figure S1B). Two-way ANOVA indicated a significant effect of THC ($F_{3,40}$ = 17.0, $p < 0.0001$); genotype did not significantly affect the tail-flick latency ($F_{1,40}$ = 0.9, $p = 0.35$) and the interaction was not significant ($F_{3,40}$ = 1.3, $p = 0.30$).

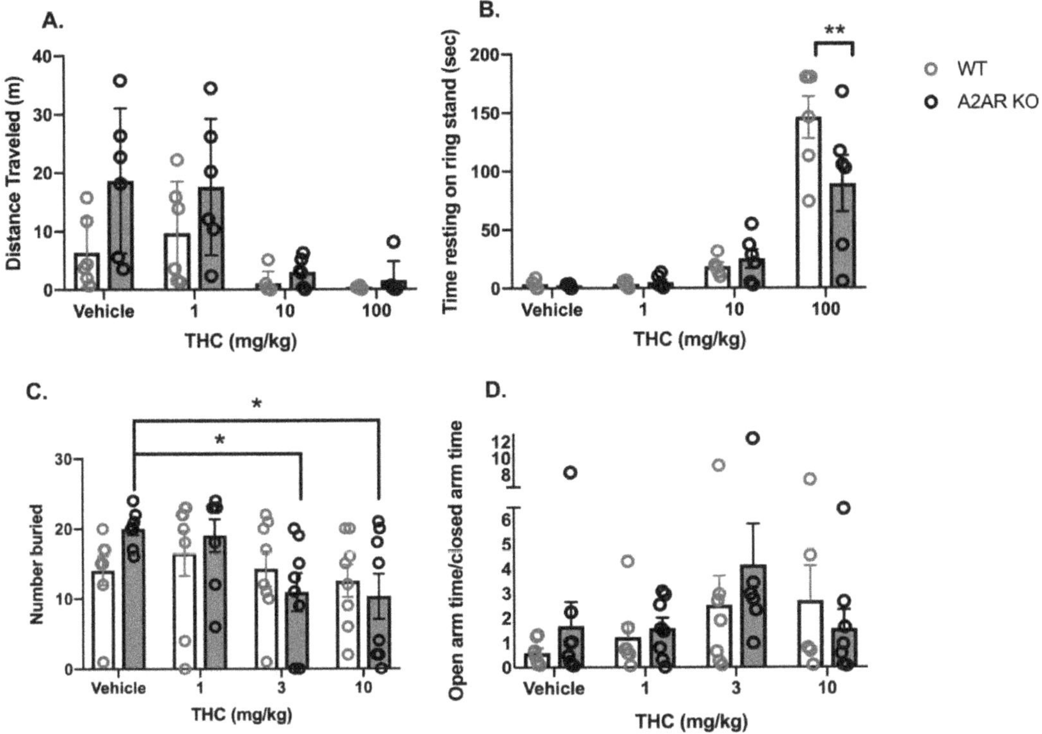

Figure 1. Comparison of some behavioral effects of THC in wild type (WT, open bars and gray symbols) and A2AR null (A2AR-KO, closed bars and black symbols) mice. (**A**) Effects of THC on locomotor activity in an open field. Mice were placed into a circular field for 15 min, and the distance moved was determined. (**B**) Effects of THC on cataleptic behavior in the ring stand assay. Sidak's multiple comparison test was used to compare all groups to each other. (**C**) Effects of THC on the number of marbles buried. Dunnett's multiple comparison test was used to compare each drug group to the vehicle control. (**D**) Effects of THC on the ratio of time spent in the open and closed arms of the EPM. A total of 8 mice were removed from this analysis because they entered the open arm and became immobile (1 mouse each from the WT/vehicle, WT/1 mg/kg, WT/3 mg/kg groups; 3 from the WT/10 mg/kg group and 2 from the KO/3 mg/kg group). Bars represent the mean, and vertical lines are the standard error of the mean. * $p < 0.05$ and ** $p < 0.01$.

2.1.2. Anxiety Assays

WT and A2AR-KO mice were injected with vehicle or THC (1, 3 and 10 mg/kg). Mice were assessed in the marble bury assay, followed by the elevated plus maze (EPM).

- Marble Bury Assay: Two-way ANOVA indicated a significant effect of THC ($F_{3,56}$ = 3.2, $p < 0.05$) without a significant effect of genotype ($F_{1,56}$ = 0.16, $p = 0.88$) or a significant interaction ($F_{3,56}$ = 1.5, $p = 0.41$) (Figure 1C). Dunnett's *t*-tests indicate that treatment with 3 and 10 mg/kg THC significantly reduces the number of marbles buried compared to vehicle treated in the A2AR-KO mice; there were no significant differences in the WT mice.
- EPM: Two-way ANOVA indicated no significant effects of either THC ($F_{3,48}$ = 2.15, $p = 0.1$) or genotype ($F_{1,48}$ = 0.52, $p = 0.47$) and no significant interaction ($F_{3,48}$ = 0.74, $p = 0.53$) on the ratio of time spent in the open and closed arms (OAT/CAT) (Figure 1D). There were no significant effects of THC ($F_{3,48}$ = 0.30, $p = 0.73$) or genotype ($F_{3,48}$ = 0.05, $p = 0.26$) on the total number of arm entries (Supplementary Materials Figure S1D).

2.2. THC/CBD Combination Studies

WT and A2AR-KO mice were treated with 5 mg/kg of CBD or THC or their combination. All mice received two injections, with vehicle substituting for the drug when required.

- Locomotor Activity: The distance traveled in the open field was measured in the eight treatment groups (Figure 2A). Three-way ANOVA indicated significant effects of both THC ($F_{1,55}$ = 13.0, $p < 0.001$) and genotype ($F_{1,55}$ = 6.3, $p < 0.05$) but not CBD ($F_{1,55}$ = 0.025, $p = 0.87$). There was a significant interaction between THC and genotype ($F_{1,55}$ = 6.6, $p < 0.05$). Post hoc tests revealed that THC produced a significant reduction in locomotor activity in A2AR-KO mice also treated with CBD.
- Catalepsy: Three-way ANOVA indicated significant effects of THC ($F_{1,52}$ = 20.0, $p < 0.0001$) and CBD ($F_{1,52}$ = 5.0, $p < 0.05$) and a trend toward a significant effect of genotype ($F_{1,52}$ = 2.4, $p = 0.13$) (Figure 2B). There was a significant interaction between CBD and genotype ($F_{1,52}$ = 5.3, $p < 0.05$). Post hoc tests demonstrate that WT mice treated with a combination of THC and CBD exhibited significantly greater time resting on the ring stand than WT mice treated with either drug alone. Neither THC nor CBD or their combination produced catalepsy in A2AR-KO mice. Post hoc tests revealed a significant difference between the THC/CBD cotreatment groups in WT and A2AR-KO mice.
- Body Temperature: Three-way ANOVA indicates a significant effect of THC ($F_{1,56}$ = 4.26, $p < 0.05$) without significant effects of either CBD ($F_{1,56}$ = 0.9, $p = 0.35$) or genotype ($F_{1,56}$ = 2.6, $p = 0.11$) (Supplementary Materials Figure S2A).
- Nociceptive Reflex: Three-way ANOVA indicates a significant effect of THC ($F_{1,56}$ = 19, $p < 0.001$) without significant effects of either CBD ($F_{1,56}$ = 0.8, $p = 0.79$) or genotype ($F_{1,56}$ = 0.3, $p = 0.58$) (Supplementary Materials Figure S2B).
- Marble Bury Assay. Three-way ANOVA indicates significant effects of both THC ($F_{1,56}$ = 10.4, $p < 0.01$) and CBD ($F_{1,56}$ = 6.2, $p < 0.05$) but not genotype ($F_{1,56}$ = 0.55, $p = 0.46$) (Figure 2C). The interaction between CBD and genotype is significant ($F_{1,56}$ = 7.1, $p < 0.01$); post hoc tests demonstrated a significant reduction in the number of marbles buried between the vehicle treated and THC/CBD treated A2AR-KO mice.
- EPM: Three-way ANOVA indicated that both THC ($F_{1,48}$ = 4.9, $p < 0.05$) and CBD ($F_{1,48}$ = 5.4, $p < 0.05$) had significant effects on the OAT/CAT ratio, while genotype trended to a significant effect ($F_{1,48}$ = 2.9, $p = 0.09$) (Figure 2D). The interaction between genotype and CBD was also significant ($F_{1,48}$ = 4.1, $p < 0.05$), while the interaction between CBD and THC trended toward significance ($F_{1,48}$ = 2.6, $p = 0.1$). Post hoc tests revealed a significant increase in OAT/CAT in WT mice treated with a combination of THC/CBD compared to those treated with vehicle and those treated with THC alone. Three-way ANOVA of the total arm entries indicated no significant effects of either THC ($F_{1,48}$ = 0.5, $p = 0.49$) or CBD ($F_{1,48}$ = 1.1, $p = 0.31$) (Supplementary Materials Figure S2C). While there was a significant effect of genotype ($F_{1,48}$ = 5.0, $p < 0.05$) and a significant interaction between THC and genotype ($F_{1,48}$ = 5.9, $p < 0.05$), post hoc tests did not elucidate any significant group differences.

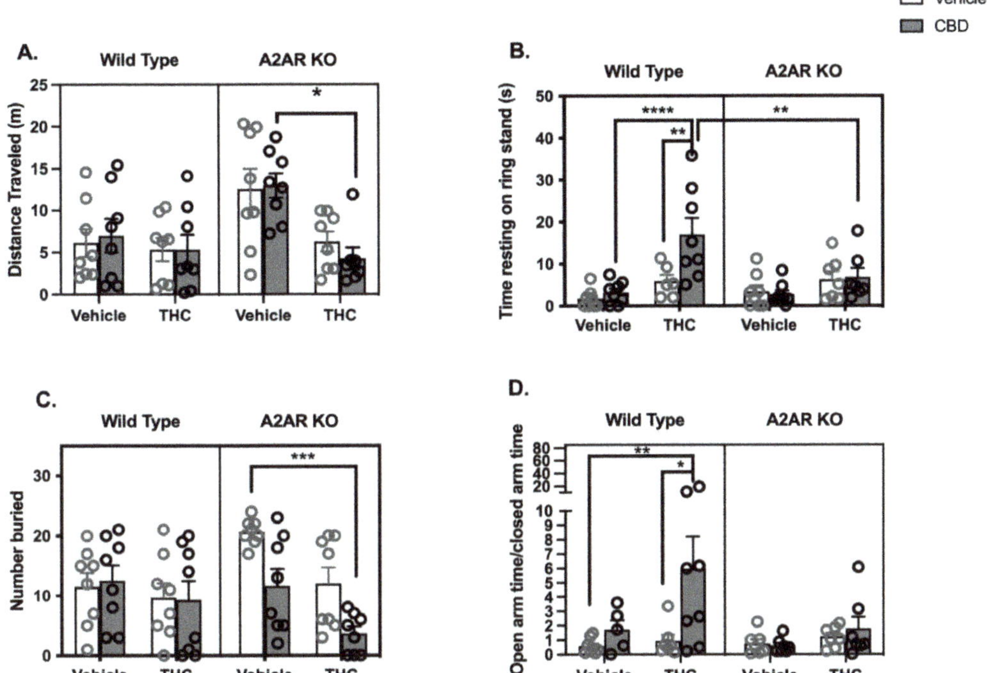

Figure 2. Behavioral effects of THC (5 mg/kg), CBD (5 mg/kg) and their combination in wild type (WT, open bars and gray symbols) and A2AR null (A2AR-KO, closed bars and black symbols) mice. (**A**) Drug effects on distance moved in 15 min while in a cylindrical open field arena. One outlier was identified in the CBD/THC/A2AR-KO group. (**B**) Drug effects on cataleptic behavior in the ring stand assay. Outlier analysis indicated 4 outliers in this data set (2 in the WT/VEH/THC group, 1 in the KO/VEH/THC group and 1 in the KO/THC/CBD group). (**C**) Drug effects on the number of marbles buried. (**D**) Drug effects on the ratio of time spent in the open and closed arms of the EPM. A total of 8 mice were removed from the analysis; 3 mice became immobile on the open arm (2 in the WT/veh/CBD group and 1 in the KO/THC/CBD group); and 5 additional mice were identified as statistical outliers (1 each in the WT/veh/CBD, WT/veh/THC, KO/veh/veh, KO/veh/THC and KO/veh/CBD groups). Bars represent the mean, and vertical lines are the standard error of the mean. * $p < 0.05$, ** $p < 0.01$, *** $p < 0.001$, **** $p < 0.0001$.

2.3. THC/PECS-101 Combination Studies

WT and A2AR-KO mice were treated with 5 mg/kg of PECS-101 or THC or their combination. All mice received two injections, with vehicle substituting for the drug when required.

- Locomotor Activity: Three-way ANOVA indicated a significant effect of genotype ($F_{1,53} = 4.2$, $p < 0.05$), while the effects of both THC ($F_{1,53} = 4.0$, $p = 0.051$) and PECS-101 trended to significance ($F_{1,53} = 3.4$, $p = 0.07$) (Figure 3A). There was a significant interaction between THC and PECS-101 ($F_{1,53} = 6.0$, $p < 0.05$) and a trend to an interaction among THC and PECS-101 and genotype ($F_{1,53} = 3.4$, $p = 0.07$).
- Catalepsy: Three-way ANOVA indicated significant effects of THC ($F_{1,56} = 12.8$, $p < 0.001$) and genotype ($F_{1,56} = 4.3$, $p < 0.05$) but not PECS-101 ($F_{1,56} = 0.5$, $p = 0.48$) (Figure 3B). There was a significant interaction between THC and genotype ($F_{1,56} = 5.3$, $p < 0.05$). Post hoc tests demonstrate that WT mice treated with a combination of THC and PECS-101 exhibited significantly greater time resting on the ring stand than WT mice treated with PECS-101 alone. However, this did not occur in

A2AR-KO mice, and there was a significant difference between the WT and A2AR-KO THC/PECS-101 combined treatment groups.

- Body Temperature: Three-way ANOVA indicates a significant effect of THC ($F_{1,56}$ = 4.1, $p < 0.05$) without significant effects of either PECS-101 ($F_{1,56}$ = 0.5, $p = 0.97$) or genotype ($F_{1,56}$ = 1.9, $p = 0.17$) (Supplementary Materials Figure S3A).
- Nociceptive Reflex: Three-way ANOVA indicates significant effects of THC ($F_{1,56}$ = 8.3, $p < 0.01$) and PECS-101 ($F_{1,56}$ = 10.8, $p < 0.01$) but not genotype ($F_{1,56}$ = 0.3, $p = 0.87$) (Supplementary Materials Figure S3B). The examination of the data suggest that THC inhibits the nociceptive reflex in the absence but not in the presence of PECS-101, although the interaction of THC and PECS-101 was not significant ($p = 0.15$).
- Marble Bury Assay: Three-way ANOVA indicates significant effects of THC ($F_{1,56}$ = 10.8, $p < 0.01$) and PECS-101 ($F_{1,56}$ = 6.6, $p < 0.05$) but not genotype ($F_{1,56}$ = 1.8, $p = 0.19$) (Figure 3C). There was a nearly significant interaction between THC and genotype ($F_{1,56}$ = 3.9, $p = 0.052$). Post hoc tests demonstrated that, in the A2AR-KO mice only, THC reduced the number of marbles buried compared to both vehicle and PECS-101 treated mice.
- EPM: Three-way ANOVA indicates that none of the factors had a significant effect on the OAT/CAT ratio (THC: $F_{1,48}$ = 1.2, $p = 0.28$; PECS-101: $F_{1,48}$ = 0.7, $p = 0.42$; and genotype: $F_{1,48}$ = 1.0, $p = 0.34$) (Figure 3D). There was, however, a significant interaction between THC and PECS-101 ($F_{1,48}$ = 4.8, $p < 0.05$) and a trending interaction among THC, PECS-101 and genotype ($F_{1,48}$ = 3.0, $p = 0.09$). Post hoc tests indicated that the combination of THC and PECS-101 significantly reduced the OAT/CAT ratio compared to the effect of PECS-101 alone in the A2AR-KO mice. Surprisingly, PECS-101 had a very significant effect on the total arm entries; the three-way ANOVA results for PECS-101 were $F_{1,48}$ = 21, $p < 0.0001$; neither THC ($F_{1,48}$ = 2.1, $p = 0.16$) nor genotype ($F_{1,48}$ = 0.5, $p = 0.46$) significantly affected the total arm entries (Supplementary Materials Figure S3C).

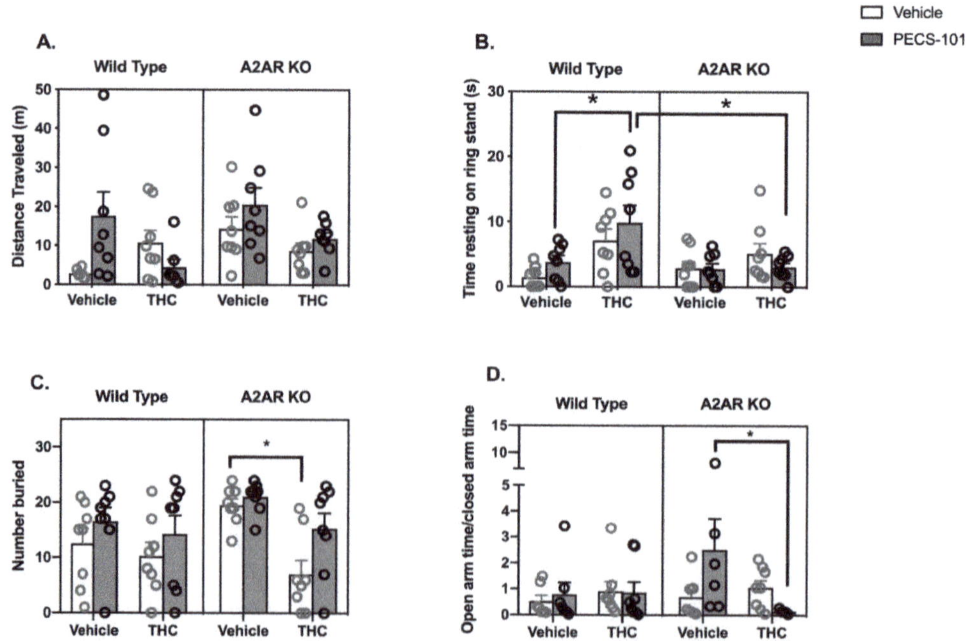

Figure 3. Behavioral effects of THC (5 mg/kg), PECS-101 (5 mg/kg) and their combination in wild type (WT, open bars

and gray symbols) and A2AR null (A2AR-KO, closed bars and black symbols) mice. (**A**) Drug effects on distance moved in 15 min while in a cylindrical open field arena. Three outliers were identified (2 in the WT/veh/veh group and 1 in the WT/THC/PECS-101 group). (**B**) Drug effects on cataleptic behavior in the ring stand assay. (**C**) Drug effects on the number of marbles buried. (**D**) Drug effects on the ratio of time spent in the open and closed arms of the EPM. A total of 8 mice were eliminated from this data set; three mice froze on the open arm of the maze (2 in the KO/veh/PECS-101 group and 1 in the KO/THC/PECS-101 group) and five mice were statistical outliers (one each in the WT/veh/veh, WT/veh/PECS-101 and KO/veh/veh groups; and 2 in the KO/THC/PECS-101 group). Bars represent the mean, and vertical lines are the standard error of the mean. * $p < 0.05$.

3. Discussion

The first goal of the study was to determine whether six behavioral effects commonly seen with CB1R agonist treatment (reduced spontaneous movement, catalepsy, hypothermia, antinociceptive reflex inhibition, marble burying and behavior in the elevated plus maze) were altered in mice with genetic deletion of the A2A subtype of adenosine receptor. We examined a range of THC doses in each of the assays, and the only assay in which a significant interaction between THC treatment and genotype occurred was catalepsy. THC produced significantly less catalepsy in the A2AR-KO mice than WT, suggesting that signaling through the A2AR is required for the full cataleptic effect of THC.

Catalepsy has long been recognized as a cardinal behavioral sign of THC intoxication in rodents [19]. The cataleptic effect of cannabinoid agonists is characterized by immobility when placed in a position that would normally evoke immediate movement. It is not that animals are unable to move, but rather that they are in a trance-like state. It has been suggested that THC-induced catalepsy is responsible for motor vehicle accidents in cannabis-intoxicated individuals [20]. The cataleptic effects of THC and other CB1R agonists are completely dependent on the expression of the CB1 subtype of the cannabinoid receptor [21]. Cell type-specific CB1R deletion strategies [21] and rescue studies in which CB1R are added back to specific neuronal subtypes in otherwise CB1R-null mice [22] both indicate that CB1R expression in D1 dopamine receptor-expressing medium spiny neurons (MSN) of the striatum is sufficient for THC-induced catalepsy. The results of a recent study support this conclusion and further suggest that the CB1R pool responsible is present on mitochondria (mtCB1R) in the axon terminals of D1-expressing striatonigral neurons [23].

Previous studies have demonstrated that CB1R and A2AR can form functionally relevant heterodimers in the striatum [24] and hippocampus [25], and these heterodimers have been suggested to mediate the motor-depressant and addictive effects of the cannabinoid agonists [14]. However, while evidence suggests that the CB1Rs involved in the cataleptic response to THC are in D1R-expressing MSN, A2AR are expressed most abundantly on dendrites of D2 dopamine receptor-expressing MSNs [26,27]. Additionally, A2AR is not found in regions of the striatum that are enriched in D1R-expressing neurons [28]. While studies of the interactions between CB1R and A2AR in striatal slices suggest that A2AR activation positively regulates the synaptic effects of CB1R agonists, the results suggest that the interaction is indirect and not mediated by receptors that are expressed in the same cell [29]. Thus, available evidence suggests that A2AR is critical in the neuronal circuit through which CB1R agonists act to produce catalepsy, but CB1R/A2AR heterodimers are not likely involved. Earlier work from our laboratory demonstrated that THC is an effective inhibitor of the ENT-1, with an IC_{50} value of 170 nM [8]. Therefore, THC-induced catalepsy could be the result of both its CB1R agonist and A2AR indirect agonist effects. Support for this hypothesis comes from data that A2AR agonists are cataleptic per se [30] and can enhance the cataleptic response to haloperidol [31]. This mechanism is consistent with the current finding that the cataleptic effect of THC is reduced in the absence of A2AR.

We have compared the effects of a single combination of THC and CBD to the effects of the same dose of each drug alone in the classic cannabinoid tetrad and on several anxiety-like behaviors. We chose to use a 1:1 dose ratio, as there is evidence that high doses of CBD can affect THC metabolism [19]. As our goal was to explore CBD-mediated enhancement of behavioral responses to THC, we chose to use a threshold dose of THC

(5 mg/kg). Indeed, in WT mice, the only consistent effect of 5 mg/kg THC was an increase in tail-flick latency; THC exhibited inconsistent effects in the catalepsy response and did not produce significant effects in the locomotor, temperature, marble bury and EPM assays.

Consistent with previous studies [18,32], 5 mg/kg CBD had no effect on the tetrad behaviors. We also did not see significant effects of CBD at this dose in either the marble bury or EPM assays. Previous findings are mixed in this regard; some earlier studies have shown an anti-anxiety effect of CBD in mice using the EPM [33,34], while others have not [35].

The vehicle-treated A2AR-KO mice travelled significantly greater distances in the open field compared to their WT littermates and, while 5 mg/kg THC did not suppress locomotor activity in the WT, the same dose suppressed locomotor activity in the A2AR-KO mice, particularly in the presence of CBD. These findings suggest that A2AR activity opposes the hypolocomotor effects of THC in WT mice. This finding is at odds with previously published data that striatal A2AR are required for the hypolocomotor effects of cannabinoid agonists [24,29,36].

In WT mice, the combination of THC and CBD produced significantly greater cataleptic behavior than either drug alone. This appears to be a synergistic rather than additive effect, although more extensive dose-response studies are needed to confirm this conclusion. The CBD enhancement of THC-induced catalepsy did not occur in the A2AR-KO mice, suggesting that intact A2AR signaling is required for the effect of CBD, the effect of THC, or the effect of both. Given previous studies that CBD can act as an indirect agonist of adenosine signaling at A2AR in this dose range [8,12], it is possible that CBD enhances THC-mediated catalepsy because it functions as an A2AR activator. This mechanism is analogous to that described above for THC, and we hypothesize that the synergistic effect of CBD on THC-induced catalepsy occurs because CBD can increase adenosine signaling in the circuit.

We also found evidence for a synergistic interaction between THC and CBD to produce an anxiolytic-like response in the EPM that did not occur in the A2AR-KO mice. Previous studies have shown that both the adenosine receptor antagonist, caffeine [37], and the genetic deletion of A2AR [38] are associated with increased anxiety-like behaviors in mice. Importantly, nitrobenzylthioinosine, an ENT-1 inhibitor, produces anxiolytic effects in the EPM when injected into the amygdala [39]. These findings, together with our current results, suggest that the hypothesis that CBD and THC synergize to reduce anxiety and that the ability of one or both of the cannabinoids to inhibit the ENT-1 contributes to this effect under the dose conditions studied.

The third goal of the studies in this project was to compare the effects of 4'-fluoro-CBD (PECS-101) to those of CBD. Earlier studies found that PECS-101 increased open arm time in the EPM assay, reduced immobility in the forced swim assay and enhanced prepulse inhibition at a dose of 3 mg/kg [17]. Higher doses (30 mg/kg and greater) were active in various assays of nociception [18] and exhibited neuroprotective characteristics in rats [40]. In all of these studies, PECS-101 produced effects similar to those of CBD, although was more potent.

We have found similarities between the effects of PECS-101 and CBD in our studies. Like CBD, PECS-101 did not exhibit consistent effects in the tetrad behaviors. Additionally, like CBD, PECS-101 potentiated the cataleptic effects of THC in an A2AR-dependent manner. However, mice treated with PECS-101 exhibited a tendency to bury more marbles regardless of co-treatment with THC or mouse genotype. In addition, mice treated with PECS-101 showed a significant reduction in total arm entries in the EPM in all treatment conditions, suggesting a reduction in exploratory behavior. On the other hand, WT mice treated with PECS-101 trended to increased distance traveled in the open field, suggesting that PECS-101 has a complex effect on locomotor behavior. Analysis of variance indicated that PECS-101 treatment had a significant effect on the tail-flick latency, and examination of the data indicates that it tended to reduce the effect of THC to increase latency. However, post hoc tests did not support a significant difference between THC alone and THC/PECS-101 in either genotype.

Although future studies are required to examine the interaction of PECS-101 with the ENT-1 directly, we hypothesize that it shares the ability of CBD to act as an indirect agonist of the A2AR receptor, particularly in the striatal circuit involved in cataleptic behavior.

There are several limitations of this study. First, it was conducted only in male mice; whether similar interactions occur in female animals is an open and important question. Second, we only examined a single dose combination of THC and CBD/PECS-101. Given the large number of potential targets for CBD and its congeners, it is highly likely that their interactions with the effects of THC will differ at different doses and dose ratios. Finally, because the dose of THC used was low, there is considerable variability in its behavioral effects. In addition, a number of mice in the EPM froze on the open arms and were therefore eliminated from the analysis. This is likely due to the use of a protocol in which the open arms were brightly lit and thus very aversive. This was deliberate in order to potentiate observation of anxiolytic effects but reduced the number of mice per group.

In summary, these findings add to our understanding of the mechanisms of action of THC and potential interactions between CBD and THC. They indicate that, while CB1R agonism is essential to the effects of THC, other mechanisms, including inhibition of adenosine reuptake, could synergize with this mechanism. Importantly, while other studies indicate that CBD and THC could have opposing effects, our data suggest that they produce synergistic effects on catalepsy. Assuming that catalepsy is an undesirable effect of cannabis, these data indicate that combined THC/CBD preparations could be more harmful than either drug alone. In addition, some cannabinoid users combine cannabis and CBD with coffee and other caffeinated beverages to modulate the psychological effects. It is possible that caffeine moderates the THC and/or CBD experience by inhibiting their effects on A2AR-mediated signaling. On the other hand, recent data indicate that activation of A2AR signaling can have beneficial effects in the context of substance use disorders [41], which together with human studies demonstrating that CBD can reduce anxiety and craving in opiate-dependent and abstinent individuals [42], suggests that the ability of CBD to elevate A2AR signaling could be an important therapeutic mechanism.

4. Materials and Methods

4.1. Animals

All of the animal experimentation reported herein were carried out in accord with ARRIVE guidelines and was approved by the Institutional Animal Care and Use Committee at the Medical College of Wisconsin.

The animal subjects for this study were male adult mice between 8 and 12 weeks of age. All mice were obtained from in-house breeding of 129S-*Adora*2a^{tm1Jfc}/J mice; breeders were originally obtained from The Jackson Laboratory, Bar Harbor, ME, USA (Stock Number 010685). The mice were originally developed by Chen and colleagues [43] and are on a mixed background of 129S and C57-Bl6/J. Tissue from the ear pinnae was used as a source of DNA for genotyping; tissue was added to 0.3 mL of 10 mM NaOH containing 1 mM EDTA and heated at 95 °C for 13 min. All mice were genotyped at least once, and genotypes were re-assessed if ambiguous bands were seen. The primers used were common forward primer: GGG CTC CTC GGT GTA CAT; reverse WT primer: CCC ACA GAT CTA GCC TTA; and reverse knock out primer: CAT TTG TCA CGT CCT GCA CGA C. For the WT reaction, samples were held at 94 °C for 2 min; then cycled 35 times (94 °C for 45 s, 56 °C for 30 s, and 72 °C for 2 min) followed by holding at 72 °C for 10 min. For the KO reaction, the cycle temperature and times were 94 °C for 45 s, 60 °C for 1 min and 72 °C for 1 min, followed by 2 min at 72 °C. Male WT and A2AR-KO offspring of het by het breeders were used as experimental subjects in this study. Because of the mixed strain background, the mice exhibited a variety of coat colors, from white to black. We did not use mice with white fur in these studies because it was more difficult for the tracking software to identify the mice, and they tended to exhibit greater sensitivity to the light source in the tail-flick assay.

4.2. Drugs

All drugs were delivered by intraperitoneal injection in a volume of 0.1 mL/25 g body weight. Drugs were administered individually, and multiple injections were given as close in time as possible using opposite abdominal sides. The order of drugs and the side given were randomized. Drug emulsions were prepared using an emulphor:ethanol:saline (1:1:18) vehicle as described previously [44]. Briefly, drugs were dissolved in 100% ethanol at a concentration 20 times greater than the final desired concentration. An equal volume of Kolliphor EL (Sigma Chemical, St. Louis, MO, USA, C5135) was added, and the mixture was vortexed well. Sterile saline was added in a dropwise fashion with continuous vortexing. THC and CBD were obtained from the NIDA Drug Supply Program. PECS-101 was obtained from Gary Hiller (Phytecs, Inc., Los Angeles, CA, USA). The studies in which the effect of THC was examined in the behavioral tetrad were carried out using doses of 1, 10 and 100 mg/kg based upon previous dose-response studies [45] and with the goal of dose range-finding. Because 100 mg/kg THC resulted in a complete loss of locomotor activity, we chose to use lower doses (1, 3 and 10 mg/kg) in the anxiety assays as both of these assays require animal movement to be useful. Our choice of 5 mg/kg THC for the combination studies was driven by our goal of investigating additive and synergistic effects of THC and CBD, so a low THC dose was utilized for these studies.

4.3. Behavioral Assays

Two sets of mice were used; the cannabinoid tetrad was conducted in one set, and the EPM/MB assays in the second set.

For the tetrad, mice were acclimated to the experimental room for at least 30 min prior to baseline measurements (Supplementary Materials Figure S4). Baseline measurements included body weight, rectal temperature, and pre-treatment latency in the tail-flick assay. Mice were injected at $t = 0$ and allowed to remain undisturbed in their home cage until $t = 25$ min. The mice were placed into the open field at $t = 25$ min and behavior recorded for 15 min, followed immediately by rectal temperature measurement and placement of the mice into the home cage. At $t = 50$ min, the tail-flick assay was conducted, followed at $t = 60$ min by the ring stand (catalepsy) assay.

For the open field assay component of the tetrad, mice were placed into a round plexiglass arena (diameter 19 inches and height 13 inches) that was cleaned with 70% isopropyl alcohol between mice. Behavior was recorded with a ceiling-mounted Sony Handycam (HDR-CX405) and was analyzed by AnyMaze Behavior Tracking Software (Stoelting, Wood Dale, IL, USA). Rectal temperature measurements were determined at baseline and after drug treatment; a thermistor probe (Physitemp RET-3 probe and BAT-12 thermometer, Clifton, NJ, USA) was lubricated and inserted to a depth of 25 mm and held in place until a stable reading was obtained. For the tail-flick reflex, latency to move the tail away from a heat-generating light (IITC Tail Flick Analgesia Meter Series 8, Woodland Hills, CA, USA) was recorded, and a cut-off time of 10 s was used to prevent injury to the tail. For the ring stand assay, mice were placed with their front paws on a metal ring 4.5 cm above the bench top, and the time to remove both paws was recorded with a stopwatch.

For the anxiety assays (which were carried out in a separate set of mice), mice were habituated to the testing room for at least 30 min, followed by injections and return to the home cage for 30 min (Supplementary Materials Figure S5). The marble bury assay was completed first. Mice were placed individually in clean cages containing 5 cm of bedding that was smoothed and slightly compacted; twenty-four 1.5 cm blue marbles were arranged in a 4×6 array on top of the bedding. After 30 min with full room lighting, mice were removed, and the number of marbles covered to a depth of at least 2/3 with bedding was recorded. After 25 min in the home cage, the mice were placed in the center space of an elevated plus maze and allowed to explore for 5 min. The EPM apparatus consisted of two open arms (30 cm long × 5 cm wide) and two enclosed arms (30 cm × 5 cm × 15 cm walls) elevated 40 cm from the floor. Room lighting was turned off, and the open arms of the EPM were lighted. Behavior on the EPM was recorded using the ceiling-mounted camera described above and analyzed using AnyMaze Behavior Tracking Software.

4.4. Statistical Analyses

Statistical analyses were carried out using Prism version 9 (GraphPad). All data sets were analyzed using ANOVA followed by post hoc tests if appropriate. For the THC dose response studies, 2-way ANOVA was used with drug dose and genotype as the factors; for the drug combination studies, 3-way ANOVA was used with THC, CBD (or PECS-101) and genotype as the factors. Sidak's Multiple Comparison's post hoc tests were used when significant interaction terms occurred. In one study (the effects of THC on marble burying response), Dunnett's *t*-tests were used to compare the effects of THC doses to the vehicle group. Statistical information for the interaction terms in the 3-way ANOVAs is only provided if significant. In all studies, the original number of replicates was 8; however, outlier analysis (ROUT method with $Q = 1.0$) was applied to each set of replicates which occasionally resulted in a reduction in n. For the EPM studies, there were multiple instances of mice that entered and remained immobile in the open arm. These mice were removed from the EPM analysis, although their marble bury response was still analyzed.

Supplementary Materials: The following are available online, Figure S1: Effects of THC in WT and A2AR-KO mice on rectal temperature, tail-flick, and total arm entries in the EPM. Figure S2: Effects of THC in combination with CBD in WT and A2AR-KO mice on rectal temperature, tail-flick, and total arm entries in the EPM. Figure S3: Effects of THC in combination with PECS-101 in WT and A2AR-KO mice on rectal temperature, tail-flick, and total arm entries in the EPM. Figure S4: Timeline for the tetrad studies. Figure S5: Timeline for the anxiety assays.

Author Contributions: The following are the author contributions to this study: Conceptualization, T.M.S. and C.J.H.; methodology, S.P. and C.J.H.; writing—original draft preparation, C.J.H.; writing—review and editing, T.M.S., S.P. and C.J.H.; funding acquisition, C.J.H. All authors have read and agreed to the published version of the manuscript.

Funding: This research was funded by Phytecs, Inc. and by the National Heart, Lung and Blood Institute grant R01 HL139008. T.M.S. is a member of the Medical College of Wisconsin-Medical Scientist Training Program (MCW-MSTP), which is partially supported by a T32 grant from the National Institute of General Medical Sciences (NIGMS), GM080202.

Institutional Review Board Statement: All animals studies were approved by the Medical College of Wisconsin's Institutional Animal Care and Use Committee (Approval number 141).

Informed Consent Statement: Not applicable.

Data Availability Statement: Authors will provide raw data upon request.

Acknowledgments: We acknowledge the Drug Supply Program of the National Institute on Drug Abuse for the generous gift of THC and CBD used in this study.

Conflicts of Interest: This study was partially funded by Phytecs, Inc through a grant to the Medical College of Wisconsin. Phytecs, Inc. played no role in the design, collection, analysis or interpretation of the data, the writing of the manuscript, or in the decision to publish the results. C.J.H. serves as a member of the scientific advisory board of Phytecs, Inc.

Sample Availability: Please contact Phytecs, Inc. for samples of PECS-101.

References

1. Dos Santos, R.G.; Hallak, J.E.C.; Crippa, J.A.S. Neuropharmacological Effects of the Main Phytocannabinoids: A Narrative Review. *Adv. Exp. Med. Biol.* **2021**, *1264*, 29–45. [PubMed]
2. Britch, S.C.; Babalonis, S.; Walsh, S.L. Cannabidiol: Pharmacology and therapeutic targets. *Psychopharmacology* **2021**, *238*, 9–28. [CrossRef] [PubMed]
3. Fernandez-Ruiz, J.; Galve-Roperh, I.; Sagredo, O.; Guzman, M. Possible therapeutic applications of cannabis in the neuropsychopharmacology field. *Eur. Neuropsychopharmacol.* **2020**, *36*, 217–234. [CrossRef] [PubMed]
4. O'Sullivan, S.E.; Stevenson, C.W.; Laviolette, S.R. Could Cannabidiol Be a Treatment for Coronavirus Disease-19-Related Anxiety Disorders? *Cannabis Cannabinoid Res.* **2021**, *6*, 7–18. [CrossRef]

5. Mechoulam, R.; Hanus, L.O.; Pertwee, R.; Howlett, A.C. Early phytocannabinoid chemistry to endocannabinoids and beyond. *Nat. Rev. Neurosci.* **2014**, *15*, 757–764. [CrossRef]

6. Lee, J.L.C.; Bertoglio, L.J.; Guimaraes, F.S.; Stevenson, C.W. Cannabidiol regulation of emotion and emotional memory processing: Relevance for treating anxiety-related and substance abuse disorders. *Br. J. Pharmacol.* **2017**, *174*, 3242–3256. [CrossRef]

7. Ford, T.C.; Hayley, A.C.; Downey, L.A.; Parrott, A.C. Cannabis: An Overview of its Adverse Acute and Chronic Effects and its Implications. *Curr. Drug Abuse Rev.* **2017**, *10*, 6–18. [CrossRef]

8. Carrier, E.J.; Auchampach, J.A.; Hillard, C.J. Inhibition of an equilibrative nucleoside transporter by cannabidiol: A mechanism of cannabinoid immunosuppression. *Proc. Natl. Acad. Sci. USA* **2006**, *103*, 7895–7900. [CrossRef]

9. Pastor-Anglada, M.; Perez-Torras, S. Who Is Who in Adenosine Transport. *Front. Pharmacol.* **2018**, *9*, 627. [CrossRef]

10. Liou, G.I.; Auchampach, J.A.; Hillard, C.J.; Zhu, G.; Yousufzai, B.; Mian, S.; Khan, S.; Khalifa, Y.M. Mediation of Cannabidiol Anti-inflammation in the Retina by Equilibrative Nucleoside Transporter and A2A Adenosine Receptor. *Investig. Ophthalmol. Vis. Sci.* **2008**, *49*, 5526–5531. [CrossRef]

11. Castillo, A.; Tolon, M.R.; Fernandez-Ruiz, J.; Romero, J.; Martinez-Orgado, J. The neuroprotective effect of cannabidiol in an in vitro model of newborn hypoxic-ischemic brain damage in mice is mediated by CB(2) and adenosine receptors. *Neurobiol. Dis.* **2010**, *37*, 434–440. [CrossRef]

12. Mecha, M.; Feliu, A.; Inigo, P.M.; Mestre, L.; Carrillo-Salinas, F.J.; Guaza, C. Cannabidiol provides long-lasting protection against the deleterious effects of inflammation in a viral model of multiple sclerosis: A role for A2A receptors. *Neurobiol. Dis.* **2013**, *59*, 141–150. [CrossRef]

13. Borroto-Escuela, D.O.; Wydra, K.; Fores-Pons, R.; Vasudevan, L.; Romero-Fernandez, W.; Frankowska, M.; Ferraro, L.; Beggiato, S.; Crespo-Ramirez, M.; Rivera, A.; et al. The Balance of MU-Opioid, Dopamine D2 and Adenosine A2A Heteroreceptor Complexes in the Ventral Striatal-Pallidal GABA Antireward Neurons May Have a Significant Role in Morphine and Cocaine Use Disorders. *Front. Pharmacol.* **2021**, *12*, 627032. [CrossRef]

14. Ferre, S.; Lluis, C.; Justinova, Z.; Quiroz, C.; Orru, M.; Navarro, G.; Canela, E.I.; Franco, R.; Goldberg, S.R. Adenosine-cannabinoid receptor interactions. Implications for striatal function. *Br. J. Pharmacol.* **2010**, *160*, 443–453. [CrossRef]

15. Boggs, D.L.; Nguyen, J.D.; Morgenson, D.; Taffe, M.A.; Ranganathan, M. Clinical and Preclinical Evidence for Functional Interactions of Cannabidiol and Delta(9)-Tetrahydrocannabinol. *Neuropsychopharmacology* **2018**, *43*, 142–154. [CrossRef] [PubMed]

16. Perucca, E.; Bialer, M. Critical Aspects Affecting Cannabidiol Oral Bioavailability and Metabolic Elimination, and Related Clinical Implications. *CNS Drugs* **2020**, *34*, 795–800. [CrossRef]

17. Breuer, A.; Haj, C.G.; Fogaca, M.V.; Gomes, F.V.; Silva, N.R.; Pedrazzi, J.F.; Del Bel, E.A.; Hallak, J.C.; Crippa, J.A.; Zuardi, A.W.; et al. Fluorinated Cannabidiol Derivatives: Enhancement of Activity in Mice Models Predictive of Anxiolytic, Antidepressant and Antipsychotic Effects. *PLoS ONE* **2016**, *11*, e0158779.

18. Silva, N.R.; Gomes, F.V.; Fonseca, M.D.; Mechoulam, R.; Breuer, A.; Cunha, T.M.; Guimaraes, F.S. Antinociceptive effects of HUF-101, a fluorinated cannabidiol derivative. *Prog. Neuropsychopharmacol. Biol. Psychiatry* **2017**, *79*, 369–377. [CrossRef] [PubMed]

19. Jones, G.; Pertwee, R.G. A metabolic interaction in vivo between cannabidiol and 1 -tetrahydrocannabinol. *Br. J. Pharmacol.* **1972**, *45*, 375–377. [CrossRef] [PubMed]

20. Martin, J.L.; Gadegbeku, B.; Wu, D.; Viallon, V.; Laumon, B. Cannabis, alcohol and fatal road accidents. *PLoS ONE* **2017**, *12*, e0187320. [CrossRef]

21. Monory, K.; Blaudzun, H.; Massa, F.; Kaiser, N.; Lemberger, T.; Schutz, G.; Wotjak, C.T.; Lutz, B.; Marsicano, G. Genetic dissection of behavioural and autonomic effects of Delta(9)-tetrahydrocannabinol in mice. *PLoS Biol.* **2007**, *5*, e269. [CrossRef]

22. De Giacomo, V.; Ruehle, S.; Lutz, B.; Haring, M.; Remmers, F. Differential glutamatergic and GABAergic contributions to the tetrad effects of Delta(9)-tetrahydrocannabinol revealed by cell-type-specific reconstition of the CB1 receptor. *Neuropharmacology* **2020**, *179*, 108287. [CrossRef]

23. Soria-Gomez, E.; Pagano Zottola, A.C.; Mariani, Y.; Desprez, T.; Barresi, M.; Bonilla-Del Rio, I.; Muguruza, C.; Le Bon-Jego, M.; Julio-Kalajzic, F.; Flynn, R.; et al. Subcellular specificity of cannabinoid effects in striatonigral circuits. *Neuron* **2021**, *109*, 1513–1526.e11. [CrossRef]

24. Carriba, P.; Ortiz, O.; Patkar, K.; Justinova, Z.; Stroik, J.; Themann, A.; Muller, C.; Woods, A.S.; Hope, B.T.; Ciruela, F.; et al. Striatal Adenosine A(2A) and Cannabinoid CB(1) Receptors Form Functional Heteromeric Complexes that Mediate the Motor Effects of Cannabinoids. *Neuropsychopharmacology* **2007**, *32*, 2249–2259. [CrossRef]

25. Aso, E.; Fernandez-Duenas, V.; Lopez-Cano, M.; Taura, J.; Watanabe, M.; Ferrer, I.; Lujan, R.; Ciruela, F. Adenosine A2A-Cannabinoid CB1 Receptor Heteromers in the Hippocampus: Cannabidiol Blunts Delta(9)-Tetrahydrocannabinol-Induced Cognitive Impairment. *Mol. Neurobiol.* **2019**, *56*, 5382–5391. [CrossRef]

26. DeMet, E.M.; Chicz-DeMet, A. Localization of adenosine A2A-receptors in rat brain with [3H]ZM-241385. *Naunyn Schmiedebergs Arch. Pharmacol.* **2002**, *366*, 478–481. [CrossRef] [PubMed]

27. Rosin, D.L.; Hettinger, B.D.; Lee, A.; Linden, J. Anatomy of adenosine A2A receptors in brain: Morphological substrates for integration of striatal function. *Neurology* **2003**, *61*, S12–S18. [CrossRef] [PubMed]

28. Gangarossa, G.; Espallergues, J.; Mailly, P.; De Bundel, D.; de Kerchove d'Exaerde, A.; Herve, D.; Girault, J.A.; Valjent, E.; Krieger, P. Spatial distribution of D1R- and D2R-expressing medium-sized spiny neurons differs along the rostro-caudal axis of the mouse dorsal striatum. *Front. Neural Circuits* **2013**, *7*, 124. [CrossRef]

29. Tebano, M.T.; Martire, A.; Chiodi, V.; Pepponi, R.; Ferrante, A.; Domenici, M.R.; Frank, C.; Chen, J.F.; Ledent, C.; Popoli, P. Adenosine A2A receptors enable the synaptic effects of cannabinoid CB1 receptors in the rodent striatum. *J. Neurochem.* **2009**, *110*, 1921–1930. [CrossRef] [PubMed]

30. Ferre, S.; Rubio, A.; Fuxe, K. Stimulation of adenosine A2 receptors induces catalepsy. *Neurosci. Lett.* **1991**, *130*, 162–164. [CrossRef]

31. Mandhane, S.N.; Chopde, C.T.; Ghosh, A.K. Adenosine A2 receptors modulate haloperidol-induced catalepsy in rats. *Eur. J. Pharmacol.* **1997**, *328*, 135–141. [CrossRef]

32. Long, L.E.; Chesworth, R.; Huang, X.F.; McGregor, I.S.; Arnold, J.C.; Karl, T. A behavioural comparison of acute and chronic Delta9-tetrahydrocannabinol and cannabidiol in C57BL/6JArc mice. *Int. J. Neuropsychopharmacol.* **2009**, *13*, 861–876. [CrossRef]

33. Schiavon, A.P.; Bonato, J.M.; Milani, H.; Guimaraes, F.S.; Weffort de Oliveira, R.M. Influence of single and repeated cannabidiol administration on emotional behavior and markers of cell proliferation and neurogenesis in non-stressed mice. *Prog. Neuropsychopharmacol. Biol. Psychiatry* **2016**, *64*, 27–34. [CrossRef]

34. Zieba, J.; Sinclair, D.; Sebree, T.; Bonn-Miller, M.; Gutterman, D.; Siegel, S.; Karl, T. Cannabidiol (CBD) reduces anxiety-related behavior in mice via an FMRP-independent mechanism. *Pharmacol. Biochem. Behav.* **2019**, *181*, 93–100. [CrossRef] [PubMed]

35. Liu, J.; Scott, B.W.; Burnham, W.M. Effects of cannabidiol and Delta9-tetrahydrocannabinol in the elevated plus maze in mice. *Behav. Pharmacol.* **2021**. [CrossRef] [PubMed]

36. Borgkvist, A.; Marcellino, D.; Fuxe, K.; Greengard, P.; Fisone, G. Regulation of DARPP-32 phosphorylation by Delta(9)-tetrahydrocannabinol. *Neuropharmacology* **2008**, *54*, 31–35. [CrossRef] [PubMed]

37. Marangos, P.J.; Boulenger, J.P. Basic and clinical aspects of adenosinergic neuromodulation. *Neurosci. Biobehav. Rev.* **1985**, *9*, 421–430. [CrossRef]

38. Johansson, B.; Halldner, L.; Dunwiddie, T.V.; Masino, S.A.; Poelchen, W.; Gimenez-Llort, L.; Escorihuela, R.M.; Fernandez-Teruel, A.; Wiesenfeld-Hallin, Z.; Xu, X.J.; et al. Hyperalgesia, anxiety, and decreased hypoxic neuroprotection in mice lacking the adenosine A1 receptor. *Proc. Natl. Acad. Sci. USA* **2001**, *98*, 9407–9412. [CrossRef]

39. Chen, J.; Rinaldo, L.; Lim, S.J.; Young, H.; Messing, R.O.; Choi, D.S. The type 1 equilibrative nucleoside transporter regulates anxiety-like behavior in mice. *Genes Brain Behav.* **2007**, *6*, 776–783. [CrossRef]

40. Perez, M.; Cartarozzi, L.P.; Chiarotto, G.B.; Oliveira, S.A.; Guimaraes, F.S.; Oliveira, A.L.R. Neuronal preservation and reactive gliosis attenuation following neonatal sciatic nerve axotomy by a fluorinated cannabidiol derivative. *Neuropharmacology* **2018**, *140*, 201–208. [CrossRef]

41. Wydra, K.; Gawlinski, D.; Gawlinska, K.; Frankowska, M.; Borroto-Escuela, D.O.; Fuxe, K.; Filip, M. Adenosine A2AReceptors in Substance Use Disorders: A Focus on Cocaine. *Cells* **2020**, *9*, 1372. [CrossRef] [PubMed]

42. Hurd, Y.L.; Spriggs, S.; Alishayev, J.; Winkel, G.; Gurgov, K.; Kudrich, C.; Oprescu, A.M.; Salsitz, E. Cannabidiol for the Reduction of Cue-Induced Craving and Anxiety in Drug-Abstinent Individuals with Heroin Use Disorder: A Double-Blind Randomized Placebo-Controlled Trial. *Am. J. Psychiatry* **2019**, *176*, 911–922. [CrossRef] [PubMed]

43. Chen, J.F.; Huang, Z.; Ma, J.; Zhu, J.; Moratalla, R.; Standaert, D.; Moskowitz, M.A.; Fink, J.S.; Schwarzschild, M.A. A(2A) adenosine receptor deficiency attenuates brain injury induced by transient focal ischemia in mice. *J. Neurosci.* **1999**, *19*, 9192–9200. [CrossRef] [PubMed]

44. Cradock, J.C.; Davignon, J.P.; Litterst, C.L.; Guarino, A.M. An intravenous formulation of 9-tetrahydrocannabinol using a non-ionic surfactant. *J. Pharm. Pharmacol.* **1973**, *25*, 345. [CrossRef] [PubMed]

45. Marshell, R.; Kearney-Ramos, T.; Brents, L.K.; Hyatt, W.S.; Tai, S.; Prather, P.L.; Fantegrossi, W.E. In vivo effects of synthetic cannabinoids JWH-018 and JWH-073 and phytocannabinoid Delta-THC in mice: Inhalation versus intraperitoneal injection. *Pharmacol. Biochem. Behav.* **2014**, *124C*, 40–47. [CrossRef] [PubMed]

Article

Subsynaptic Distribution, Lipid Raft Targeting and G Protein-Dependent Signalling of the Type 1 Cannabinoid Receptor in Synaptosomes from the Mouse Hippocampus and Frontal Cortex

Miquel Saumell-Esnaola [1,2,3], Sergio Barrondo [1,2,4], Gontzal García del Caño [2,5], María Aranzazu Goicolea [6], Joan Sallés [1,2,4], Beat Lutz [3] and Krisztina Monory [3,*]

1   Department of Pharmacology, Faculty of Pharmacy, University of the Basque Country UPV/EHU, 01006 Vitoria-Gasteiz, Spain; miquel.saumell@ehu.eus (M.S.-E.); sergio.barrondo@ehu.eus (S.B.); joan.salles@ehu.eus (J.S.)
2   Bioaraba, Neurofarmacología Celular y Molecular, 01009 Vitoria-Gasteiz, Spain; gontzal.garcia@ehu.eus
3   Institute of Physiological Chemistry, University Medical Center of the Johannes Gutenberg University Mainz, 55128 Mainz, Germany; beat.lutz@uni-mainz.de
4   Centro de Investigación Biomédica en Red de Salud Mental (CIBERSAM), 28029 Madrid, Spain
5   Department of Neurosciences, Faculty of Pharmacy, University of the Basque Country UPV/EHU, 01006 Vitoria-Gasteiz, Spain
6   Department of Analytical Chemistry, Faculty of Pharmacy, University of the Basque Country UPV/EHU, 01006 Vitoria-Gasteiz, Spain; mariaaranzazu.goicolea@ehu.eus
*   Correspondence: monory@uni-mainz.de; Tel.: +49-6131-39-24551

Citation: Saumell-Esnaola, M.; Barrondo, S.; García del Caño, G.; Goicolea, M.A.; Sallés, J.; Lutz, B.; Monory, K. Subsynaptic Distribution, Lipid Raft Targeting and G Protein-Dependent Signalling of the Type 1 Cannabinoid Receptor in Synaptosomes from the Mouse Hippocampus and Frontal Cortex. *Molecules* **2021**, *26*, 6897. https://doi.org/10.3390/molecules26226897

Academic Editors: Mauro Maccarrone and Eva de Lago Femia

Received: 14 October 2021
Accepted: 11 November 2021
Published: 16 November 2021

Publisher's Note: MDPI stays neutral with regard to jurisdictional claims in published maps and institutional affiliations.

Abstract: Numerous studies have investigated the roles of the type 1 cannabinoid receptor (CB1) in glutamatergic and GABAergic neurons. Here, we used the cell-type-specific CB1 rescue model in mice to gain insight into the organizational principles of plasma membrane targeting and $G\alpha i/o$ protein signalling of the CB1 receptor at excitatory and inhibitory terminals of the frontal cortex and hippocampus. By applying biochemical fractionation techniques and Western blot analyses to synaptosomal membranes, we explored the subsynaptic distribution (pre-, post-, and extra-synaptic) and CB1 receptor compartmentalization into lipid and non-lipid raft plasma membrane microdomains and the signalling properties. These data infer that the plasma membrane partitioning of the CB1 receptor and its functional coupling to $G\alpha i/o$ proteins are not biased towards the cell type of CB1 receptor rescue. The extent of the canonical $G\alpha i/o$ protein-dependent CB1 receptor signalling correlated with the abundance of CB1 receptor in the respective cell type (glutamatergic versus GABAergic neurons) both in frontal cortical and hippocampal synaptosomes. In summary, our results provide an updated view of the functional coupling of the CB1 receptor to $G\alpha i/o$ proteins at excitatory and inhibitory terminals and substantiate the utility of the CB1 rescue model in studying endocannabinoid physiology at the subcellular level.

Keywords: type 1 cannabinoid receptor CB1; cholesterol; hippocampus; frontal cortex; synaptosomes; rescue model; anti-CB1 antibody

1. Introduction

The physiological role of the activation of the presynaptically located CB1 receptor in balancing excitatory and inhibitory neurotransmission is essential for various behaviours [1–4]. Various mouse lines have been developed that rescue CB1 receptor expression specifically in dorsal telencephalic glutamatergic neurons (Glu-CB1-RS) and in forebrain GABAergic neurons (GABA-CB1-RS) [5,6]. Importantly, the results demonstrated that rescue strategies re-establish existing levels of CB1 receptors expressed in glutamatergic and GABAergic cell types accurately, without the interference of additional cells expressing CB1 receptors [5–7]. Moreover, this cell-type selective CB1 receptor expression has made

121

it possible to define the contributions of both glutamatergic and GABAergic CB1 receptor to the tetrad effects of Δ^9-tetrahydrocannabinol [8]. Interestingly, this genetic rescue approach has revealed functions of CB1 receptor subpopulations that remain undetected when relying solely on a conditional knockout approach [8,9]. Previously, we have addressed cell-type specificity of the functional CB1 receptor coupling to Gαi/o proteins in hippocampal homogenates of conditional knockout mice [10]. Data showed that the CB1 receptor was more efficiently coupled to Gαi/o protein signalling in glutamatergic neurons than in GABAergic neurons [10]. The cell type-specific effects on agonist efficacy at the CB1 receptor observed in conditional mutant mouse lines prompted us to focus on CB1 receptors located specifically at nerve terminals, a physiologically relevant location, and to the proximal components of the signalling machinery in this subsynaptic compartment, the Gαi/o protein family. In this context, biochemical constraints related to the activation of G protein coupled receptors (GPCRs), such as membrane lipid composition, and specifically the effects of membrane cholesterol abundance, are highly relevant as well. We decided to explore these concepts in synaptosomal fractions purified from frontal cortex and hippocampus derived from Glu-CB1-RS and GABA-CB1-RS mouse lines [5,6]. Moreover, we took advantage of biochemical fractionation techniques to separate either the subsynaptic domains (pre-, post-, and extra-synaptic fractions) [11] that have been successfully applied to the study of the subsynaptic location of CB1 receptor in rat striatal synaptosomes [12] or the biochemically defined lipid and non-lipid raft plasma membrane microdomains [13]. We have also explored the effects of membrane cholesterol on CB1 signalling by the cholesterol depletion agent methyl-β-cyclodextrin (MβCD). Although previous results using electron microscopy, a method that preserves the structure of the synapse, demonstrated that rescue strategies re-establish existing CB1 receptor levels expressed in glutamatergic and GABAergic cells [7], the accessibility of the CB1 receptor epitopes by large molecules such as antibodies could hamper their detection. Here, an alternative approach was used, which takes advantage of biochemical fractionation techniques allowing the immunological detection of solubilized plasma membrane proteins by a subsequent Western blot analysis.

The results obtained provide an updated view of the functional coupling of the CB1 receptor to the canonical Gαi/o protein signalling at excitatory and inhibitory terminals of the mouse frontal cortex and hippocampus, and demonstrate the potential of our approach to gain insight into the organizational principles of the CB1 receptor plasma membrane location and Gαi/o protein signalling.

2. Results

2.1. Validation of the Enriched Synaptosomal Fraction from Mouse Frontal Cortex

Synaptosomes were purified using a fractionation protocol based on sucrose gradients and differential centrifugation, which allowed the separation of synaptic terminals from other particles of different subcellular origin according to their density. To assess the suitability of the synaptosome enriched fraction, Western blot and epifluorescence microscope techniques were applied. To determine the purity of the synaptosomal fraction (SYN), Western blot assays were carried out using antibodies raised against several proteins that have been used as markers of specific sub-cellular compartments. As shown in the Supplementary Figure S1, the immunoreactivity for different synaptic proteins (synaptophysin, syntaxin 1a, and the NMDA receptor subunit NR1) was enriched in the synaptosomal fraction. The immunoreactivity of Ras-related protein (Rab11b), which is found in synaptic endosomes among other cellular compartments, was also preferentially enriched in the synaptosome enriched fraction. These markers were also detected in the nuclear fraction (P1), and in the crude plasmatic membrane fraction (P2), although their signals were significantly lower than in synaptosomes. On the other hand, the signals for non-synaptic markers were faint or undetectable in synaptosomes, indicating the low contamination of this fraction with non-synaptic membranes. Specifically, the immunoreactivity for the nuclear marker histone H3 and the glial fibrillary acid protein (GFAP) was highest in the nucleus-enriched fraction (P1), whereas the cytosolic marker glyceraldehyde-3-phosphate dehydrogenase (GAPDH) was enriched in the cytoplasm fraction (S1) (Supplementary Figure S1). We also examined the synaptosomes by double immunofluorescence and high-resolution microscopy, combining MAP2/GFAP or SNAP25/GFAP double immunofluorescence labelling with the membrane staining dye DiIC16 and high-resolution fluorescence microscopy analysis (Supplementary Figure S2). The DiIC16 dye allowed us to quantify the size and the origin of all particles found in the preparation. Immunofluorescence assays showed that about 80% of particles displayed a size between 0.25–1.5 μm, which is consistent with that described for synaptosomes. On the other hand, 15% and 5% of the particles showed a size less than 0.25 μm and greater than 1.5 μm, respectively. Half of the DiIC16 positive particles within 0.25–1.5 μm range size were identified as of neuronal origin by MAP2 and SNAP25 staining, whereas a very low glial contamination was observed by GFAP-immunostaining. SNAP25 and MAP2 labelling also revealed that about half of the particles in the synaptosome-enriched fraction were composed of presynaptic or postsynaptic elements. These values are in good agreement with other published data showing that isolated nerve terminals made up approximately 50% of the structures revealed by electron microscopy [14]. These results demonstrated the suitability of the efficiency protocol used to purify mouse brain synaptosomes.

2.2. Characteristics of the Immunoreactive Signals Provided by Anti-CB1 Antibodies in Frontal Cortical Synaptosomes Derived from Wild Type Mice

To study the CB1 receptor protein located in the synaptosomal fraction by Western blot assays, we used three commercially available antibodies (CB1-Immunogenes, CB1-Go-Af450 and CB1-Rb-Af380) that were raised against the 31 amino acids of the extreme carboxy-terminus of the mouse CB1 receptor. These have been recently shown to be the most reliable ones for Western blot [15]. As negative control, we used brain cortical tissue of CB1-deficient mice (CB1-KO) to test the specificity of these antibodies for the CB1 receptor. All three antibodies recognized a specific band at ~50 kDa consistent with the 52 kDa theoretical molecular mass of the mouse CB1 receptor, which was absent in synaptosomes derived from the cortical tissue of CB1-KO (Figure 1A). Furthermore, the CB1-Immunogenes and the CB1-Go-Af450 antibodies clearly recognized a specific extra band at ~35 kDa, which was also absent in synaptosomes obtained from CB1-KO (Figure 1A). Strikingly, the lower molecular weight band was hardly detectable with the CB1-Rb-Af380 antibody in most experiments (Figure 1A). To analyse whether the migration of the observed ~50 kDa and ~35 kDa bands could be modified by proteolytic degradation of the CB1 receptor, we performed Western blot assays of synaptosome samples subjected to a potentially proteolytic condition by their incubation at 37 °C in the absence and the presence of protease inhibitors. No changes were observed in the immunoreactivity of the ~50 kDa and ~35 kDa bands after incubation of synaptosomes at 37 °C for 1 or 2 h in the absence or in the presence of protease inhibitors (Supplementary Figure S3A). Moreover, no changes were observed when the synaptosomal enriched fraction was obtained in the absence or the presence of protease inhibitors during the fractionation procedure (see Supplementary Table S2). These results suggest that under our experimental conditions, the appearance of the lower molecular mass band of ~35 kDa is not the product of the proteolytic degradation of the ~50 kDa band protein, at least during the fractionation procedure or handling and processing synaptosomes.

As the CB1 receptor has two consensus sequences for N-linked glycosylation at the N-terminal tail, we examined whether the two immunoreactive bands detected in Western blot assay could be glycosylated and non-glycosylated forms of the receptor. To answer this question, the frontal cortical synaptosomes were treated with the peptide N-glycosylase enzyme (PNGase F). PNGase F is the most effective enzymatic method for removing almost all N-linked oligosaccharides from glycoproteins, because it cleaves between the innermost GlcNAc and asparagine residues of high mannose, hybrid, and complex oligosaccharides. As recommended by the manufacturer, we performed Western blot assays to analyse the migration profile of the CB1 receptor bands obtained with each one of the three antibodies after incubating synaptosomes for one hour at 37 °C with the PNGase F enzyme (25 UI/μg total synaptosomal protein). N-glycosidase treatment of synaptosomes resulted in a clear shift in the migration profile of the ~50 kDa band, which was not present anymore for any of the three antibodies used. Instead, two new specific bands migrating at ~40 kDa and ~37 kDa were detected with all three antibodies (Figure 1B; Supplementary Figure S3B,C). However, no changes in the intensity of the ~35 kDa band were observed when the CB1-Go-450 or the CB1-Immunogenes antibody was used. Strikingly, the CB1-Rb-Af380 antibody, which hardly detected the ~35 kDa band in untreated samples, recognized a stronger ~35 kDa band in the PNGase F treated synaptosomes (Figure 1B). However, this band was unspecific because it was also detected in cortical synaptosomes of the CB1-KO mice (Supplementary Figure S3C). The possibility that the ~40 kDa and ~37 kDa immunoreactive band could be products of a partial deglycosylation of the CB1 receptor was tested by doubling both the PNGase amount and incubation time, but no changes were observed (Supplementary Figure S3B).

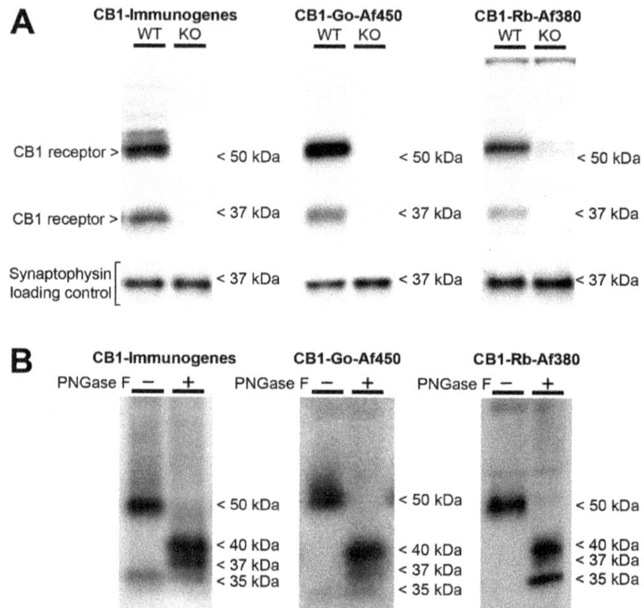

Figure 1. Immunoblot against CB1 receptor protein using CB1-Immunogenes, CB1-Go-Af450 and CB1-Rb-Af380 antibodies. (**A**) Representative Western blots carried out loading the same amount of total protein (20 µg/lane) from synaptosomes of brain cortical tissue of wild-type (WT) and CB1-KO mice. The molecular weights depicted correspond to the signal of the standard markers. (**B**) Representative Western blots carried out loading the same amount of control and PNGase F treated frontal cortical synaptosomes (20 µg total protein/lane). The approximate molecular masses of the immunoreactive species detected on the blot are indicated.

2.3. Subsynaptic Compartmentalization of the CB1 Receptor and Other Proteins of the Endocannabinoid System in Frontal Cortex Synaptosomes Derived from Wild-Type and CB1-RS Mice

To investigate the synaptic distribution of CB1 receptor and other proteins of the endocannabinoid system, cortical synaptosomes were fractionated in three major subsynaptic domains: the presynaptic active zone (PAZ), the postsynaptic density (PSD), and the extra-synaptic zone (EXTRA). The extra-synaptic region consists of plasma membrane not specialized in synapses and of cytoplasm of synaptic terminal, whereas the presynaptic active zone and the postsynaptic density consist of "particle web" components and protein dense specialization attached to the presynaptic and postsynaptic membrane, respectively [11]. We recovered 67%, 12%, and 5% of the total amount of synaptosomal membrane protein in the EXTRA, PSD, and PAZ fractions, respectively. Thus, proteins from the extra-synaptic region contribute in the largest proportion to the synaptosomal fraction. The efficiency of the protocol was validated by Western blot assays. Equal amounts of total protein of the three isolated subsynaptic fractions (PAZ, PSD, and EXTRA), and increasing amounts of total protein of the initial synaptosomal fraction were loaded on the same gel. We used antibodies raised against PSD-95, Shank3, and gephyrin as markers of PSD, and Munc-18 and SNAP-25 as markers of PAZ and EXTRA subsynaptic domains. As expected, the immunoreactivity of PSD-95, Shank3, and gephyrin was only detected in the PSD fraction, and the intensity of the signal was significantly higher than in the synaptosome fraction, which is in line with the fact that the PSD fraction was purified approximately eight times compared to synaptosomes, considering the protein yield of each subsynaptic fraction (Supplementary Figure S4). On the other hand, the presynaptic proteins Munc-18 and SNAP-25 were detected in PAZ and EXTRA fractions, although they showed higher enrichment in the EXTRA fraction than in the PAZ (Supplementary Figure S4). The functional profile of these two proteins is consistent with what might be expected because they reflect synaptic and non-synaptic populations of proteins found in the synaptic terminal. These results showed that our protocol is adequate for obtaining subsynaptic domains from synaptosomes.

Once the efficiency of the protocol was established, we analysed the subsynaptic compartmentalization of the CB1 receptor using the three antibodies described above: CB1-Immunogenes, CB1-Go-Af450, and CB1-Rb-Af380. With respect to the ~50 kDa band, the immunoreactivity was highest in the EXTRA fraction, although a clearly detectable but considerably less intense signal was detected in the PSD fraction (Figure 2A). Furthermore, a weak band was detected in the PAZ fraction. Relative to the total receptor signal detected in EXTRA and PSD compartments, 68% and 32% of the immunoreactivity was present in each of the domains, respectively, with no differences between the results obtained with the three antibodies (Figure 2C). In summary, the density of the CB1 receptor (~50 kDa band) in the extra-synaptic membrane was considerably higher than in the postsynaptic domain. Densitometric analysis of the ~35 kDa immunoreactive bands showed similar values in the EXTRA and PSD fractions, indicating that the ~50 kDa and ~35 kDa proteins partition differently (Figure 2C). Although the same amount of total protein from these fractions was loaded for Western blot analysis, the yield of total synaptosome protein in the EXTRA fraction was almost 5.8-fold higher than in the PSD fraction, revealing that most CB1 receptor is located in the extra-synaptic membrane (about 90% of the total amount of synaptic immunoreactivity). Therefore, the immunoreactive signal detected in synaptosomes is mainly derived from the EXTRA fraction and the contribution of the CB1 receptor signal of the PSD (about 8%) and the PAZ (about 1%) is very low. Altogether, the distribution of CB1 receptor in the frontal cortex is similar to rat striatal CB1 receptor, which was found in all subsynaptic fractions [12]. Regarding the Gαi/o proteins, the canonical transducers coupled to CB1 receptor at the plasma membrane, three of the four αi/o subunits studied (Gαo, Gαi1, and Gαi3) were found exclusively in the EXTRA fraction (Figure 2A). Interestingly, Gαi2 was mostly detected in the PSD fraction, although a weak signal could be observed in the EXTRA fraction. Like the Gαo, Gαi1, and Gαi3 proteins, the CB1 receptor interacting protein-1a (CRIP1a) was only detected in the EXTRA fraction. The proteins involved in the synthesis and degradation of the major endocannabinoid 2-arachidonylglycerol (2-AG), Gαq/11 subunit, phospholipase C-β1 (PLC-β1) and monoacylglycerol lipase (MAGL) were found in the EXTRA fraction, whereas diacylglycerol lipase-α (DAGL-α) was mostly enriched in the PSD fraction (Figure 2B). Finally, the Gβ subunit signal was highest in the EXTRA fraction, but also clearly detectable in the PAZ and, to a lesser extent, in the PSD fraction (Figure 2A). These results are consistent with the synaptic retrograde signalling function assigned to 2-AG.

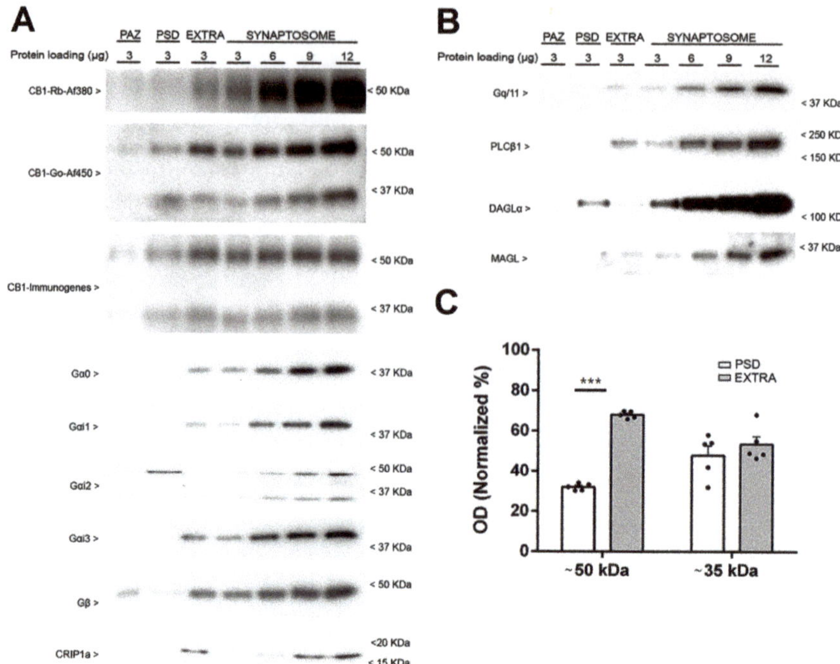

Figure 2. Subsynaptic compartmentalization of the CB1 receptor, the canonical transducers coupled to CB1 receptors and other proteins of the endocannabinoid system in PAZ, PSD, and EXTRA fractions isolated from cortical synaptosomes derived from wild-type mice. Representative Western blots carried out by immunoblotting increasing amounts of cortical synaptosomes (3, 6, 9, and 12 µg/lane) and different subsynaptic fractions of wild-type mice (3 µg/lane) using antibodies against CB1 receptor, Gαi/o subtypes, Gβ and Crip1a (**A**) and Gαq/11, PLC-β1, DAGL-α, and MAGL (**B**). Presynaptic fraction PAZ, postsynaptic fraction PSD, and extrasynaptic fraction EXTRA. Protein migration was consistent with their expected molecular mass. For the CB1 receptor and the Gαi2 protein, extra bands migrating at ~35 kDa and ~36 kDa were detected, respectively (CB1, 52.8 kDa; Gαo 40.1 kDa; Gαi1, 40.5 kDa; Gαi2, 40.4 kDa; Gαi3, 40.5 kDa; Gβ (common), 37.3 kDa and 36.3 kDa 1 and 2 isoforms; CRIP1a, 18.6 kDa; Gαq/11, 42.0 kDa; PLC-β1, 138.3 kDa and 133.3 kDa, the β1a and β1b isoforms, respectively; DAGL-α, 115.3 kDa; MAGL, 33.3 kDa). The molecular weights depicted correspond to the signal of the standard markers. (**C**) The bar graphs show the subsynaptic distribution of the CB1 receptor immunoreactive signals of ~50 kDa and ~35 kDa bands obtained with the CB1-Immunogenes, CB1-Go-Af450 and CB1-Rb-Af380 antibodies. The quantification was performed using data from the three antibodies together. The immunoreactive signals of PSD and EXTRA fractions are shown normalized to the total signal detected in both compartments. ~50 kDa: EXTRA 67.9 ± 0.7 vs. PSD 32.0 ± 0.7; ~35 kDa: EXTRA 52.8 ± 3.9 vs. PSD 47.9 ± 4.8. Values correspond to the means ± SEM of five independent assays, using subsynaptic fraction preparations obtained from a pool of cerebral cortices of eight adult mice. Unpaired two tailed t test. *** = $p < 0.001$.

Then, we examined the subsynaptic distribution of the CB1 receptor in synaptosomes derived from frontal cortex of CB1-RS mice. The subsynaptic marker distribution was qualitatively indistinguishable between the wild-type and CB1-RS (Supplementary Figures S4 and S5). Semiquantitative analysis of the CB1-immunoreactive bands showed no statistically significant differences between wild-type and CB1-RS mice (Supplementary Table S3). In other words, the ~50 kDa and ~35 kDa bands detected by the three anti-CB1 antibodies were equally distributed in wild-type and in CB1-RS mice (Figure 2A,C; Supplementary Figure S6A,C). We also studied the subsynaptic distribution of proteins of the endocannabinoid system in CB1-RS mice. The subsynaptic profile of different elements of the endocannabinoid system and the signalling proteins coupled to CB1 receptor was qualitatively similar between wild-type and CB1-RS mice (Figure 2B and Supplementary Figure S6B).

2.4. Localization of CB1 Receptors in Lipid Raft and Non-Lipid Raft Microdomains of Synaptosomal Plasma Membranes Obtained from Frontal Cortical Brain Tissue of Wild-Type and CB1-RS Mice

We further examined the localization of the CB1 receptor in "raft" and "non-raft" microdomains derived from the synaptosomal plasma membranes of the frontal cortex of wild-type and CB1-RS mice. Typically, a total of 12 fractions of increasing sucrose density were obtained and were biochemically characterized by quantitative analysis of alkaline phosphatase enzymatic activity, determination of the total protein amount, and the use of raft and non-raft markers (Figure 3; Supplementary Figure S7). Low protein content and high alkaline phosphatase activity are characteristic of lipid raft fractions. In the Western blot assays, we used antibodies raised against thymocyte-1 (Thy-1) and flotillin proteins, and Na^+/K^+-ATPase protein as markers of raft and non-raft microdomains, respectively. In the wild type mice, alkaline phosphatase activity was highest in fractions four and five along with a low protein content (Figure 3B,C). We also detected increased immunoreactivity for raft markers and decreased or absent immunoreactivity for non-raft markers in these two fractions, suggesting that they were enriched in raft microdomains (Figure 3A). Specifically, the immunoreactivity of Thy-1 was only detected in fractions four and five and the highest intensity signal of flotillin was also detected in these two fractions, with a tendency to weaken in higher density fractions. Furthermore, it should be noted that the amount of total protein loading of the fractions four and five was lower compared to the others because the same volumes of fractions were loaded in these Western blot assays. On the other hand, the fractions between 8 and 12 displayed no alkaline phosphatase activity, high protein concentration, and high and low immunoreactivity for Na^+/K^+-ATPase and flotillin, respectively (Figure 3A–C). With these results, we concluded that fractions four and five were enriched in raft microdomains and fractions 6 to 12, on the other hand, were non-raft fractions. Subsequently, the expression of CB1 receptor and Gαi/o protein subtypes was analysed (Figure 3A). Unexpectedly, CB1 immunoreactivity distribution profile varied depending on the antibody used. Whereas CB1-Rb-Af380 antibody recognized a single specific band at ~50 kDa exclusively in the raft fraction, the CB1-Immunogenes antibody recognized ~50 kDa and ~35 kDa specific bands in both raft and non-raft fractions. On the other hand, the CB1-Go-Af450 antibody did not detect any CB1 receptor signal in any of the raft and non-raft fractions. Different levels of Gαi/o protein subtypes were detected in both raft and non-raft fractions, suggesting that the CB1 receptor can interact with different Gαi/o subtypes in both compartments.

In the CB1-RS mice, alkaline phosphatase activity was highest in fractions five and six along with a stronger immunoreactivity for lipid raft markers, and lower or absent for non-raft markers, suggesting that these fractions were enriched in raft microdomains (Supplementary Figure S7A–C). On the other hand, the fractions between 8 and 12 displayed no alkaline phosphatase activity, high protein concentration, and high and low immunoreactivity for non-raft and raft markers, respectively (Supplementary Figure S7A–C). The raft vs non-raft partitioning of the CB1 receptor did not differ qualitatively between wild-type and CB1-RS mice with both CB1-immunogenes and CB1-Rb-Af380 anti-CB1 antibodies. The raft vs. non-raft partitioning profile of the α subunits of the Gαi/o protein family analysed was also similar in wild-type and CB1-RS mice (Figure 3A; Supplementary Figure S7A).

2.5. Analysis of the Coupling of the CB1 Receptor to Gαi/o Proteins in Frontal Cortical and Hippocampal Synaptosomes Obtained from CB1-RS and Wild-Type Brain Mice

The relative expression of the CB1 receptor in wild-type and CB1-RS mice was analysed in frontal cortex and hippocampal synaptosomes using CB1-Immunogenes, given that this antibody recognizes the CB1 receptor located in both raft and non-raft compartments derived from synaptosomal membranes. The expression of CB1 receptor was higher in hippocampal synaptosomes than in frontal cortical synaptosomes in both wild-type and CB1-RS mice. With respect to the immunoreactivity of the ~50 kDa band, no statistical differences were observed either in hippocampal or in frontal cortical synaptosomes (Supplementary Figure S8 and Table S4). We did not observe statistically significant difference in the immunoreactive signals of the ~35 kDa band in frontal cortical synaptosomes. However, in hippocampal synaptosomes, the immunoreactivity of the ~35 kDa band was 25% lower in CB1-RS mice than in wild-type mice (Supplementary Figure S8 and Table S4). Finally, synaptosomes from CB1-RS mice were characterized for canonical functionality of the CB1 receptor, and results were compared with synaptosomes obtained from wild-type mice. For this purpose, we performed [^{35}S]GTPγS binding assays stimulated by the cannabinoid agonist CP 55,940 in synaptosomes purified from frontal cortex and hippocampus (Supplementary Figure S9). The analysis of the CP 55,940 concentration–response curves for stimulation of the specific [^{35}S]GTPγS binding provided the same maximal percent stimulation (%E$_{max}$) and pEC$_{50}$ values both in frontal cortical and hippocampal synaptosomes from wild-type and CB1-RS mice (Supplementary Table S5).

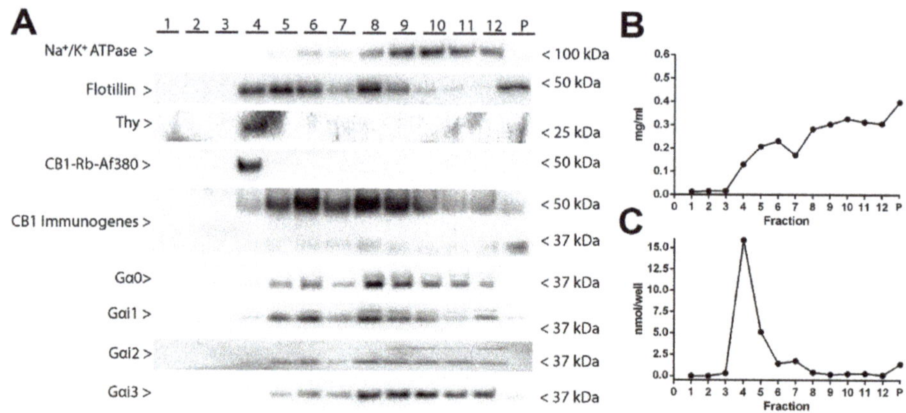

Figure 3. Alkaline phosphatase enzymatic activity, total protein amount and distribution of raft and non-raft markers in lipid raft and non-lipid raft fractions isolated from frontal cortical synaptosomes derived from wild-type mice. (**A**) Representative Western blots running in parallel same volume (20 µL/lane) of the collected 12 fractions and of the pellet (P). Immunoblot against Na$^+$/K$^+$-ATPase, Flotillin, Thymocyte (Thy-1), CB1 receptor, and Gαi/o subtypes. Protein migration was consistent with their expected molecular mass. For the CB1 receptor and the Gαi2 protein, an extra band migrating at ~35 kDa and ~36 kDa was detected, respectively. Na$^+$/K$^+$-ATPase, 112.3 kDa; Flotillin, 47.5 kDa; thymocyte 1 (Thy-1), 18.1 kDa; CB1 receptor, 52.8 kDa; Gαo 40.1 kDa; Gαi1, 40.5 kDa; Gαi2, 40.4 kDa; Gαi3, 40.5 kDa. The molecular weights depicted correspond to the signal of the standard markers. (**B**) Total protein content of the collected 12 fractions and of the pellet (P). (**C**) Alkaline phosphatase activity of the collected 12 fractions and of the pellet (P).

2.6. Analysis of the CB1 Receptor Protein Expression and Gαi/o Protein Coupling in Synaptosomes Obtained from Frontal Cortical and Hippocampal Tissue of Glu-CB1-RS, GABA-CB1-RS and CB1-RS Mice

Once the CB1-RS mouse model was validated, the expression and functional coupling of the CB1 receptor was analysed in brain synaptosomal membranes from Glu-CB1-RS and GABA-CB1-RS mice. Increasing amount of total protein of CB1-RS, Glu-CB1-RS, and GABA-CB1-RS frontal cortical synaptosomes were resolved by SDS-PAGE and CB1 receptor expression was analysed by immunoblot using the CB1-Immunogenes antibody (Figure 4A). Anti-syntaxin antibody was used as a protein loading control. A semiquantitative analysis of immunoreactive signals was performed comparing slopes values, which were obtained by regression analysis of curves that were generated plotting OD values for each protein loading (Figure 4B). Regression analysis of standard curves revealed a linear relationship (r^2 = 0.98) between the amount of protein and the relative optical density for each sample (see legend to Figure 4). The immunoreactivity for the CB1 receptor ~50 kDa band was similar in synaptosomal fractions from Glu-CB1-RS and GABA-CB1-RS, reaching in both partial rescue mice about 45% of the signal found in CB1-RS, with no statistical differences between Glu-CB1-RS and GABA-CB1-RS (Supplementary Table S6). The same relative pattern was observed for the ~35 kDa band. Furthermore, the immunoreactivity level in these two types of neurons reached around 85% of the CB1 signal seen in CB1-RS samples, indicating that in the frontal cortical synaptic terminals the CB1 receptor is expressed dominantly in these two types of neurons. As in the frontal cortex, a semiquantitative analysis of immunoreactive signals in hippocampal synaptosomes was performed comparing slopes values (Figure 4C,D). Levels of 28% and 70% of the immunoreactivity of the ~50 kDa band found in CB1-RS were present in the Glu-CB1-RS and GABA-CB1-RS mice, respectively (Supplementary Table S6). As expected, the slope values in synaptosomal samples from either partial rescue mice were significantly lower than in CB1-RS samples. The same relative pattern was observed for the ~35 kDa band. Furthermore, the synaptosomal immunoreactivity level in these two neuronal types reached around 100% of the total signal of the CB1 receptor, indicating that in the hippocampal synaptic terminals the CB1 receptor is expressed almost exclusively in GABAergic and glutamatergic neurons.

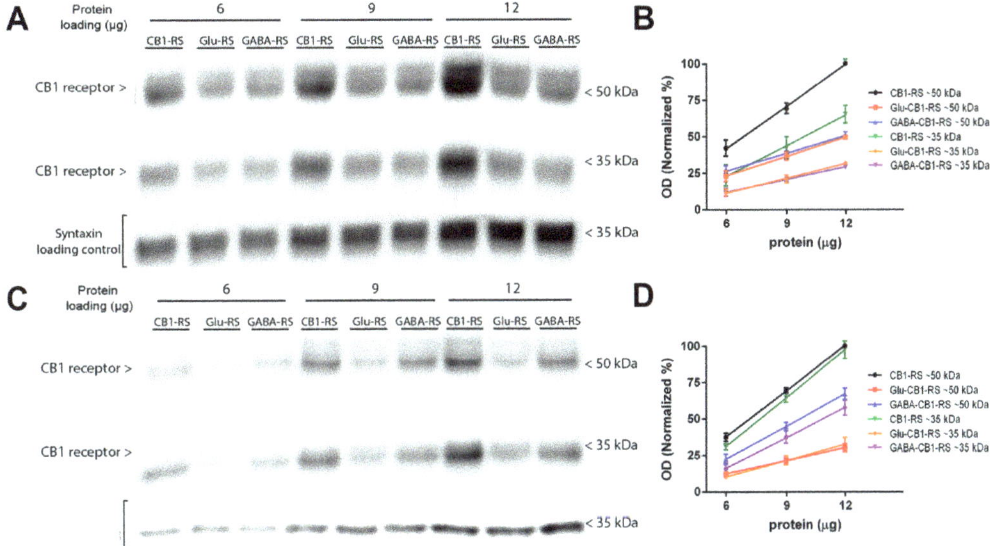

Figure 4. CB1 receptor protein levels in synaptosomes obtained from frontal cortical and hippocampal tissue of Glu-CB1-RS, GABA-CB1-RS, and CB1-RS mice. Representative Western blots carried out by immunoblotting increasing amounts of frontal cortical (**A**) and hippocampal (**C**) synaptosomes (6, 9 or 12 μg/line). CB1-Immunogenes antibody was used for detecting CB1 receptor protein, and anti-syntaxin antibody was used as a loading control. The molecular weights depicted correspond to the signal of the standard markers. (**B**). Regression analysis of curves generated by optical density (OD) values of the immunoreactive signals of CB1 receptor from frontal cortical synaptosome membranes. ~50 kDa: CB1-RS: $y = 9.64x - 16.34$, $r^2 = 0.99$. Glu-CB1-RS: $y = 4.39x - 3.08$, $r^2 = 0.99$. GABA-CB1-RS: $y = 4.41x - 1.81$, $r^2 = 0.99$. ~35 kDa: CB1-RS: $y = 6.79x - 22.23$, $r^2 = 0.98$. Glu-CB1-RS: $y = 3.33x - 8.41$, $r^2 = 0.99$. GABA-CB1-RS: $y = 2.88x - 5.20$, $r^2 = 0.99$. (**D**). Regression analysis of curves generated by optical density (OD) values of the immunoreactive signals of CB1 receptor from hippocampal synaptosome membranes. ~50 kDa: CB1-RS: $y = 10.52x - 26.04$; $r^2 = 0.99$; Glu-CB1-RS: $y = 2.97x - 5.87$; $r^2 = 0.99$. GABA-CB1-RS: $y = 7.48x - 22.97$, $r^2 = 0.99$. ~35 kDa: CB1-RS: $y = 11.13x - 36.19$; $r^2 = 0.99$. Glu-CB1-RS: $y = 3.72x - 12.54$; $r^2 = 0.99$. GABA-CB1-RS: $y = 6.97x - 26.17$; $r^2 = 0.99$. Analysis of the CB1 receptor protein expression by the slope comparison method in frontal cortical an in hippocampal synaptosomes is shown in the Supplementary Table S6.

The functional coupling of the CB1 receptor was then assessed in synaptosomal membranes obtained from frontal cortex of CB1-RS, Glu-CB1-RS, and GABA-CB1-RS mice by CP 55,940- and WIN 55,212-2-stimulated specific [^{35}S]GTPγS binding. Similar values of %E_{max} and pEC_{50} parameters were obtained in Glu-CB1-RS and GABA-CB1-RS mice, with no significant differences between them (Figure 5A,B; Table 1). The %E_{max} values in synaptosomal samples from either partial rescue mice were significantly lower than in CB1-RS samples, whereas no differences were observed in the pEC_{50} values (Table 1). As expected, no cannabinoid agonist-stimulated [^{35}S]GTPγS binding was observed in Stop-CB1 mice (Figure 5A,B). Next, we assessed the functional coupling of the CB1 receptor in synaptosomes obtained from hippocampus of Glu-CB1-RS and GABA-CB1-RS mice by CP 55,940 and WIN 55,212-2-stimulated specific [^{35}S]GTPγS binding (Figure 5C,D). The $\%E_{max}$ value in synaptosomal samples from Glu-CB1-RS rescue mice was significantly lower than in CB1-RS synaptosomes, whereas no differences were observed between GABA-CB1-RS and CB1-RS synaptosomes (Table 1). The $\%E_{max}$ values differed between synaptosomal fractions from partial rescue mice, reaching a statistical significance when CP 55,940 agonist was used in the assay. In contrast, no significant difference (Figure 5C,D; Table 1) was obtained between partial rescue mice $\%E_{max}$ with WIN 55,212-2, although the value of the Glu-CB1-RS mouse was 40% lower than of GABA-CB1-RS. Similar values of pEC_{50} parameters were obtained in all three genotypes, with no significant differences (Table 1). Again, no cannabinoid agonist-stimulated [^{35}S]GTPγS binding was observed in synaptosomes of Stop-CB1 mice (Figure 5C,D).

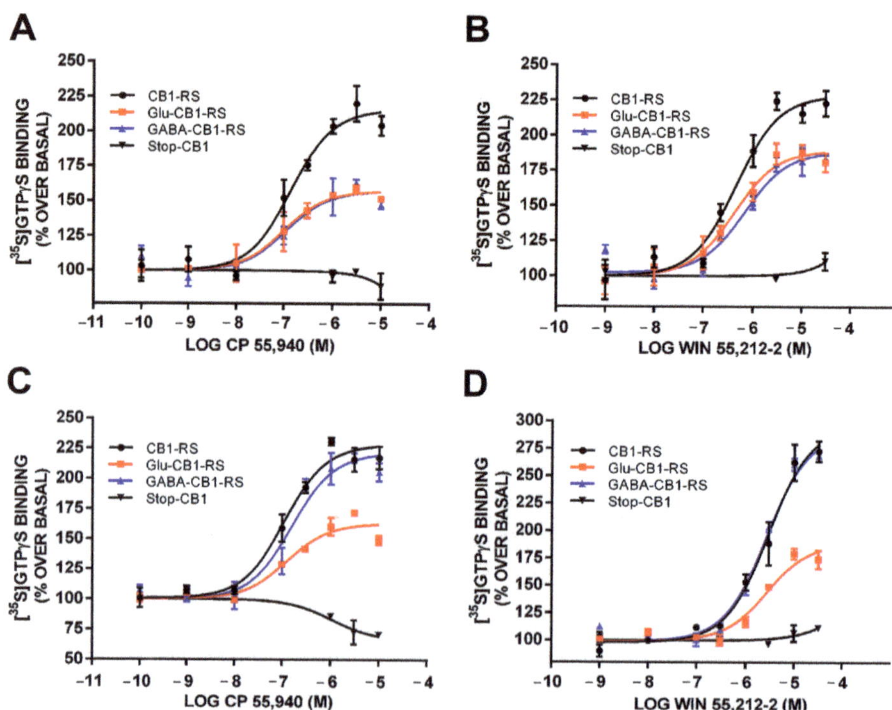

Figure 5. CB1 receptor coupling to Gαi/o proteins in synaptosomes obtained from frontal cortical and hippocampal tissue of Glu-CB1-RS, GABA-CB1-RS, CB1-RS, and Stop-CB1 mice. CP 55,940- (**A**) and WIN 55,212-2- (**B**) stimulated [^{35}S]GTPγS binding in frontal cortical synaptosomes. CP 55,940- (**C**) and WIN 55,212-2- (**D**) stimulated [^{35}S]GTPγS binding in hippocampal synaptosomes. Concentration–response curves were constructed using mean values $\pm$ SEM from triplicate data points of three independent experiments. Emax values are expressed as % specific [^{35}S]GTPγS bound of basal.

Table 1. Concentration–response curves for agonist-stimulated specific [^{35}S]GTPγS binding in frontal cortical and hippocampal synaptosomes derived from CB1-RS, Glu-CB1-RS, and GABA-CB1-RS mice. Values correspond to the means ± SEM of three independent experiments. Both in frontal cortex and hippocampus the experiments were carried out using two preparations enriched in synaptosomes, each of them obtained from pools of the frontal cortices and hippocampi of eight adult mice. Unpaired (%E_{max}) or paired (pEC_{50}, Basal) one-way ANOVA followed by sidak test.

	CB1-RS	Glu-CB1-RS	GABA-CB1-RS
Frontal Cortex			
CP 55,940			
%E_{max}	211.7 ± 2.37	157.11 ± 1.79 *	159.00 ± 1.10 *
pEC_{50}	6.85 ± 0.07	6.77 ± 0.12	6.72 ± 0.12
WIN 55,221-2			
%E_{max}	248.0 ± 18.74	189.55 ± 12.17 *	184.8 ± 6.39 *
pEC_{50}	6.75 ± 0.11	6.14 ± 0.12	6.13 ± 0.04
Hippocampus			
CP 55,940			
%E_{max}	248.60 ± 16.36	183.43 ± 14.4 *	240.60 ± 10.67 #
pEC_{50}	6.61 ± 0.09	6.59 ± 0.16	6.61 ± 0.09
WIN 55,221-2			
%E_{max}	270.90 ± 9.68	195.85 ± 5.44 *	258.1 ± 12.70
pEC_{50}	6.11 ± 0.12	5.93 ± 0.13	6.02 ± 0.12
Basal (cpm)	28,040 ± 2102	23,344 ± 1767 *	23,559 ± 1725 *

* = significantly different from CB1-RS, $p < 0.05$; # = significantly different from Glu-CB1-RS, $p < 0.05$.

2.7. Analysis of the CB1 Receptor Coupling to Gαi/o Proteins in Control and MβCD Pretreated Synaptosomes Obtained from Frontal Cortical Tissue of Glu-CB1-RS, GABA-CB1-RS, and CB1-RS Mice

We also assessed whether cholesterol exerted its negative regulation on agonist efficacy differently on CB1 receptor in glutamatergic or GABAergic terminals. To this end, we first determined the concentration of methyl-β-cyclodextrin (MβCD) necessary to observe an increase in the maximal responses to full efficacy cannabinoid agonists in [^{35}S]GTPγS binding assays. Thus, first, we analysed the effect of the pretreatment of synaptosomal membranes with MβCD (5 mM, 10 mM, and 20 mM) on CP 55,940-stimulated [^{35}S]GTPγS binding at a maximal concentration of the agonist (10 µM). The results showed an increase in the efficacy in comparison to the control (vehicle pretreated synaptosomal membranes) at 10 mM and 20 mM MβCD (Supplementary Figure S10; Supplementary Table S7). Because the maximal increase in efficacy with respect to control was achieved with 10 mM MβCD, this concentration was used for subsequent experiments. CP 55,940- and WIN 55,212-2-stimulated [^{35}S]GTPγS binding assays were performed in control and MβCD-treated frontal cortical synaptosomes from Glu-CB1-RS and GABA-CB1-RS mice. The cholesterol depletion (30% decrease in plasma membrane levels) by MβCD (10 mM) increased the maximal CP 55,940- and WIN 55,212-2-stimulated [^{35}S]GTPγS specific binding, and the magnitude of this effect was not affected by the genotype (Figure 6A,B; Supplementary Table S8). Next, we generated concentration–response curves for CP 55,940-stimulated [^{35}S]GTPγS binding to determine whether cholesterol depletion also impacted the agonist potency (pEC_{50} parameter). No statistically significant changes were observed for this parameter between MβCD treated and control synaptosomes in Glu-CB1-RS and GABA-CB1-RS mice (Figure 6C,D; Table 2). Again, the increase in the efficacy of CP 55,940 agonist induced by MβCD treatment did not differ statistically between CB1 receptor in excitatory or inhibitory terminals (Figure 6C,D; Table 2).

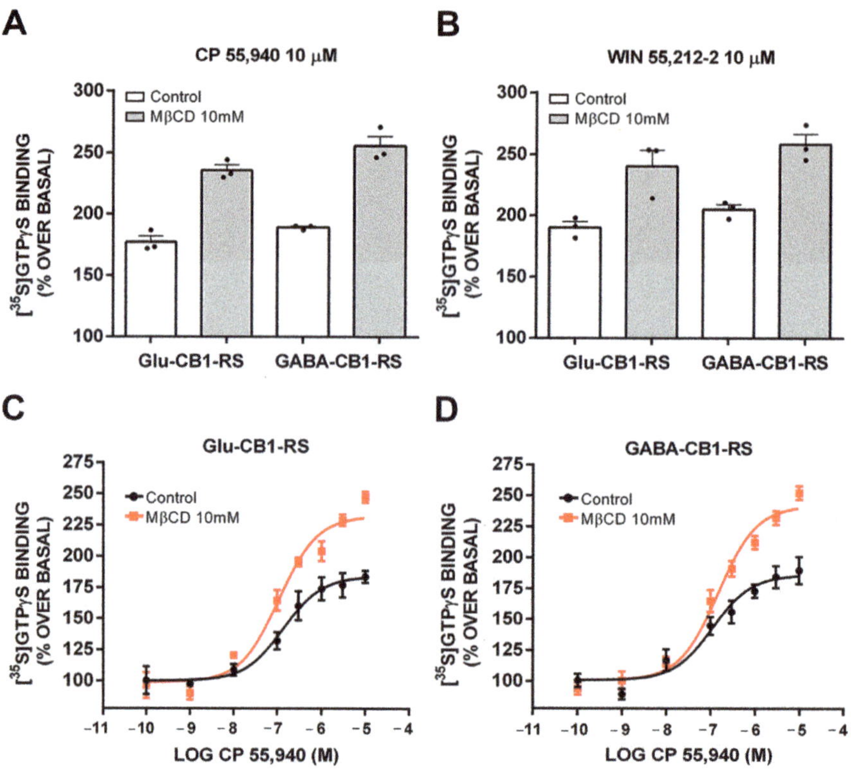

Figure 6. CB1 receptor coupling to Gαi/o proteins in control and 10 mM MβCD pretreated synaptosomes from frontal cortical tissue of Glu-CB1-RS and GABA-CB1-RS mice. (**A,B**) Bar graph of 10 μM CP 55,940 and 10 μM WIN 55,212-2-stimulated maximal [^{35}S]GTPγS binding. (**C,D**) Concentration–response curves for the CP 55,940-stimulated [^{35}S]GTPγS binding. Concentration–response curves were constructed using mean values ± SEM from triplicate data points of three independent experiments. Emax values are expressed as % specific [^{35}S]GTPγS bound of basal.

Table 2. Concentration–response curves for the CP 55,940-stimulated specific [^{35}S]GTPγS binding in vehicle (control) or MβCD pretreated frontal cortical synaptosomes derived from Glu-CB1-RS and GABA-CB1-RS mice. Values correspond to the means ± SEM of three independent experiments performed in triplicate, using synaptosomes enriched preparations obtained from a pool of the frontal cortices of eight adult mice. Unpaired (%E$_{max}$) or paired (pEC$_{50}$, Basal) two-tailed t-test.

	Glu-CB1-RS		GABA-CB1-RS	
	Control	MβCD	Control	MβCD
%E$_{max}$	170.95 ± 3.23	225.87 ± 8.75 *	167.43 ± 9.17	217 ± 14.20 *
pEC$_{50}$	6.61 ± 0.12	6.78 ± 0.06	7.02 ± 0.08	6.96 ± 0.13
Basal (cpm)	11,175 ± 264	7443 ± 267 *	10,324 ± 457	8696 ± 95

* = significantly different from control, $p < 0.05$.

3. Discussion

To gain insight into the organizational principles of plasma membrane location and Gαi/o protein signalling of the CB1 receptor at glutamatergic and GABAergic terminals of the mouse frontal cortex and hippocampus, the use of a highly specific anti-CB1 receptor antibody is mandatory. Although many antibodies designed against distinct antigenic sequences of the CB1 receptor have been developed, the interpretation of results has been controversial, at times providing poorly reproducible data. Therefore, proper antibody testing and validation must be considered when studies using anti-CB1 antibodies are conducted. In this sense, we have recently provided robust data on the suitability for different applications of several anti-CB1 antibodies [15], highlighting the need for the fit-for-purpose (F4P) approach for validation of antibodies and the importance of choosing the platform that best fits their end-use. In this previous work, the CB1-Rb-Af380 and CB1-Go-Af450 antibodies, both raised against the carboxy-terminal 31 amino acids of the mouse CB1 receptor, provided excellent results for the recognition of the denatured CB1 receptor from brain tissue in Western blot assay. Hence, in the present study, we used the commercial CB1-Rb-Af380, CB1-Go-Af450, and CB1-Immunogenes antibodies (all of them raised against an identical antigenic sequence) for the immunodetection of the CB1 receptor at synaptic terminals of the frontal cortical and hippocampal mouse brain tissue. All three antibodies recognized a specific band at ~50 kDa consistent with the 52 kDa predicted molecular mass of mouse CB1 receptor. Additionally, a specific extra band at ~35 kDa was clearly recognized with CB1-Immunogenes and CB1-Go-Af450 antibodies, whereas it was hardly detectable with the CB1-Rb-Af380 antibody in most experiments. The specificity of the detected signals was validated using cortical synaptosomes from CB1-KO animals (see Figure 1A). The molecular weight of the ~50 kDa-specific band detected by Western blot agrees with previous results where CB1-Rb-Af380 and CB1-Go-Af450 antibodies have been used [14,16,17], including some from our laboratory [5,10,15,18,19]. However, the second less intense but clearly positive ~35 kDa band detected here was not previously reported in mice. The discrepancy could be partly explained by the different subcellular fractions and/or experimental conditions used between the studies, which could impact the sensitivity of the antibodies for the detection of this CB1 receptor species. Although detection of unexpected bands at low molecular weight can be indicative of proteolytic degradation, we did not observe changes in the immunoreactivity of the ~50 kDa and ~35 kDa bands by the preincubation of synaptosomal membranes at 37 °C or the inclusion of protease inhibitors. Previous results from our laboratory and other authors have reported that the gel migration of the CB1 receptor can be altered by modifying its N-glycosylation status [20–23]. Given that the extracellular N-terminus of the mouse CB1 receptor has two consensus sequences for N-linked glycosylation [24], we examined the effects of the pretreatment of cortical synaptosomes with PNGase F. This enzymatic pretreatment resulted in a clear shift in the migration profile of the ~50 kDa immunoreactivity band, rendering it virtually undetectable with any of the three antibodies used. This observation agrees with some previous reports [22,23] and it indicates that the ~50 kDa band represents an N-glycosylated species of the CB1 receptor. At the same time, other studies reported a detection of a major band of about 60 kDa using different anti-CB1 antibodies [20,21,25–28], and although it has been explained as a result of glycosylation of the CB1 receptor, the discrepancies between these reports and the one presented here must be due to other factors. In the PNGase F pretreated samples, we detected new CB1 receptor-specific bands migrating at ~40 kDa and ~37 kDa, although we were not able to observe an increase in the ~35 kDa signal. In addition, the CB1-Rb-Af380 antibody recognized a new unspecific strong signal at ~35 kDa (see Figure 1B). Probably, this signal corresponds to the cross-reactivity of the CB1-Rb-Af380 antibody with the 35 kDa PNGase F from *Flavobacterium meningosepticum* [29,30], which was present in abundance in the pretreated synaptosomal sample. The emergence of CB1-specific immunoreactive proteins with different apparent molecular masses on SDS-PAGE after its deglycosylation could be explained, at least in part, by the formation of a tandem electrophoretic mobility shift (EMS-shift) motif within the sequence of the N-terminus of the CB1 receptor as a consequence of PNGase F-mediated deamination of asparagine residues. Recently, it has been reported that the mobility shift often observed in post-translationally phosphorylated proteins (phosphorylation-dependent electrophoretic mobility shift; PDMES), rather than by the molecular mass of covalently linked phosphate groups, is caused by the presence of negatively charged amino acids around the phosphorylation site that generate an electrophoretic mobility shift (EMS)-related motif $\theta X_{1-3}\theta X_{1-3}\theta$, where θ corresponds to an acidic or phosphorylated amino acid and X represents any amino acid [31]. As these authors proposed, EMS-motifs inhibit the binding of SDS to the peptide bond of proteins by charge–charge repulsion (see Supplementary Figure S11A,B), which results in a decreased ratio of SDS/peptide stoichiometry causing a mobility shift. It is likely that generation of a tandem EMS-motif in the sequence of the canonical mouse CB1 receptor following PNGase F-catalysed deamination of Asn-78 and Asn-84 (see Supplementary Figure S11C) could be sufficient to cause a mobility shift of about 5 kDa, which could also account for greater apparent molecular mass of deglycosylated species than the non-glycosylated one (~35 kDa).

To fully investigate the synaptic distribution of the CB1 receptor, mouse cortical synaptosomes from wild-type and CB1-RS mice were subjected to a fractionation protocol based on the differential pH and detergent sensitivity of three major subsynaptic domains [11]: the presynaptic fraction PAZ, the postsynaptic fraction PSD and the extrasynaptic fraction EXTRA. In agreement with previously reported electron microscopy data [32], our data revealed that the CB1 receptors are primarily located in the extrasynaptic membrane of the terminals together with the $G\alpha i/o$ subunits involved in its canonical downstream signalling and with CRIP1a, a protein that interacts with the CB1 receptor to modulate its functional state [33,34]. A smaller but clearly detectable pool of CB1 receptor was located in the PSD fraction, which is consistent with some previous reports [12,35]. In some experiments, a very weak signal close to the detection limit was observed in the PAZ fraction. This is also consistent with immunogold electron microscopy because the CB1 receptor can hardly be found inside the presynaptic active zone [32]. Based on the protein yield for each subsynaptic fraction, it can be concluded that about 90% of the total CB1 receptor expressed in cortical synaptosomes is found in the extrasynaptic fraction. The CB1 receptor immunoreactivity, found extrasynaptically, which may indicate recycling and/or newly synthesized pools of the CB1 receptors, is concordant with previous electron microscopy findings in the hippocampus, where presynaptic CB1 receptor was found primarily in extrasynaptic membranes of GABAergic boutons [36,37]. Of course, in the present study, receptors in the extrasynaptic fraction may comprise postsynaptic receptors outside the postsynaptic density as well, and we found CB1 receptor also in the postsynaptic density. With respect to other proteins involved in the synthesis and degradation of the 2-AG, the $G\alpha q/11$ subunits, PLC-β1, and MAGL were found in the EXTRA fraction, whereas DAGL-α was mostly enriched in the postsynaptic density fraction. Although both PLC-β1 and DAGL-α are located around the postsynaptic dense zone at the edge of glutamatergic synapses [38–42], DAGL-α contains binding motifs that allow it to interact with the postsynaptic scaffold protein Homer [43], which could explain the immunoreactivity in the PSD fraction.

In agreement with previous studies, our results indicate that several $G\alpha i/o$ protein subtypes coexist in the extrasynaptic region (EXTRA), while only the $G\alpha i2$ subtype was detected in the postsynaptic density (PSD). One of the earliest discoveries in the cannabinoid field has been the dependence of cannabinoid effects on pertussis toxin (PTX) sensitive G proteins [44]. This was soon followed by more detailed studies showing the possible involvement of the different Gi/o subtypes. Thus, $G\alpha o$ and various $G\alpha i$ subtypes were co-immunoprecipitated with CB1 receptor from solubilized rat brain membranes [45]. In PTX-treated rat primary neurons, expression of the PTX insensitive Go, Gi2, and Gi3, but not Gi1 was able to rescue the decreased excitatory postsynaptic currents [46]. More recently, specific activation of $G\alpha i1$, $G\alpha i2$, and $G\alpha i3$ but not $G\alpha o$ protein subunits by CB1 was shown using $[^{35}S]GTP\gamma S$ scintillation proximity assay [47] in CB1-transfected HEK cells.

In the synaptic active zone ion channels predominate. GPCRs can mostly be found at the extra- or perisynaptic areas where they are strategically located to sense spillover of neurotransmitters and provide feedback. Nevertheless, GPCRs were also shown to be present in the PSD [48,49] but corresponding studies showing which particular G protein subtypes these receptors couple to in vivo are still missing. However, it is important to note that our data demonstrates no functional coupling between CB1 and any specific G protein subtype. Considering their subsynaptic localization in our experiments we can postulate that in the frontal cortex of mice perisynaptic CB1 receptors can signal through Gi1, Gi3, or Go proteins or any combination of those.

Finally, we applied a protocol described by Ostrom and Insel (2006) [13] to characterize the partition of the CB1 receptor and the Gαi/o subunits located in lipid and non-lipid rafts microdomains of cortical synaptosomal membranes. Our data show that the CB1 receptor is located both in lipid raft and non-lipid raft membrane compartments, with the possibility of coupling to different Gαi/o subunits. Unexpectedly, the immunoreactivity profile of the CB1 receptor differed using the CB1-Rb-Af380 and the CB1-Immunogenes antibodies. Thus, CB1-Rb-Af380 antibody recognized the CB1 receptor exclusively in the lipid raft, whereas the CB1-Immunogenes antibody recognized the receptor in both fractions, indicating that the CB1-Rb-Af380 antibody recognizes only a partial pool of the total plasma membrane population of CB1 receptor. These two polyclonal antibodies are designed against the same last 31 amino acids of the CB1 receptor, thus adding a further degree of complexity to the interpretation of these paradoxical results. Several phosphorylation sites exist at the C-terminal of the CB1 receptor [50,51], and phosphorylation of these residues could impact differentially the affinity of these antibodies for the epitope. Presuming that the phosphorylation status of the CB1 receptor could differ between lipid and non-lipid rafts domains could account for our data and would define these antibodies as tools for detecting different states of the total population of the CB1 receptor.

In summary, to the validity of the genetic approach used to generate the CB1-RS mouse model, our results indicated that in cortical synaptosomes, the expression levels, the subsynaptic localization, and the plasma membrane lipid rafts versus non-lipid rafts partition of the CB1 receptor and Gαi/o subunits, were indistinguishable from cortical synaptosomes of the wild-type mice. The results evidence that the rescue methodology restores the levels of the presynaptic CB1 receptor at the same endogenous plasma membrane sites. Finally, to study the Gαi/o functional coupling of the CB1 receptor located in cortical and hippocampal synaptosomal membranes, we performed [^{35}S]GTPγS binding assays. Agonist-stimulated [^{35}S]GTPγS binding showed that the wild-type and the CB1-RS mice did not differ in the efficiency of CB1 receptor coupling to Gαi/o proteins both in frontal cortical and in hippocampal synaptosomes. Thus, besides restoring the levels of the CB1 receptor at endogenous plasma membrane sites, the Gαi/o coupling was not altered by the set of genetic modifications that culminate in the rescue of the CB1 receptor. Therefore, here we provided data that corroborate previous results [5–7], supporting the wild-type phenotype of the CB1-RS mice and the suitability of the genetic approach.

In the last decade, it has been demonstrated that the functionality of the CB1 receptor depends on membrane cholesterol content and the integrity of lipid rafts [52–54]. Cholesterol negatively regulates the function of canonical signalling of the CB1 receptor through Gαi/o proteins, because cholesterol depletion procedures increase both CB1 receptor agonist high-affinity maximal binding (B_{max}) as well agonist-stimulated [^{35}S]GTPγS binding efficacy (E_{max}) [54]. Therefore, it has been proposed that lipid rafts are suitable structures for the negative regulation of the CB1 receptor function by cholesterol because in these microdomains the presence of this lipid is significantly higher than in non-raft plasma membranes. Indeed, strategies used to reduce membrane cholesterol levels, such as membrane treatment with the MβCD compound, mostly deplete cholesterol from lipid rafts, supporting this hypothesis. However, most of the information that we have about this phenomenon has been obtained in heterologous cellular models. Therefore, to assess this hypothesis, frontal cortical synaptosomes from both wild-type and CB1-RS mice were treated with 10 mM of MβCD, which induced depletion of 30% of total cholesterol from the synaptosomal plasma membrane. The increase in [^{35}S]GTPγS-specific binding to a maximal concentration of the CP 55,940 suggests that the CB1 receptor located in lipid rafts of the synaptosomal membranes is probably responsible for the functional output measured. To the best of our knowledge, this is the first time that the distribution of the presynaptically located CB1 receptor at lipid and non-lipid raft microdomains has been characterized while providing robust data on the cholesterol modulation of the cannabinoid agonist-stimulated CB1 receptor coupling to Gαi/o protein. Our experimental design does not allow us to determine the contribution of the lipid raft and non-lipid raft-located CB1 receptors to the overall response to agonists as exposing plasma membrane material to Triton X-100 (1%) abolishes coupling between GPCR and G proteins [55]. Due to such technical reasons, currently, there is little data on the functional activity of GPCR-mediated signalling in plasma membrane subdomains.

We have previously addressed the potential impact that the cellular context (glutamatergic versus GABAergic neurons) could exert in the canonical coupling of the presynaptically located CB1 receptor to Gαi/o subunits performing [^{35}S]GTPγS binding assays in hippocampal tissue homogenates of cell type-specific knockout mutants, Glu-CB1-KO and GABA-CB1-KO mice [10]. Our data showed that although the level of CB1 receptors expressed in glutamatergic neurons was significantly lower than that expressed in GABAergic neurons, it was responsible for more than 50% of the maximal responses to agonists. The results showed that in glutamatergic neurons there was a more effective CB1 receptor-dependent Gαi/o protein signalling than in GABAergic neurons [10]. However, the results could be affected by the CB1 receptor–Gαi/o coupling located in other subcellular compartments since the experiments were performed in hippocampal tissue homogenates [10]. Therefore, to study the impact that cellular context produces in the presynaptic CB1 receptor–Gαi/o protein signalling, we performed Western blots and [^{35}S]GTPγS binding assays in frontal cortical and hippocampal synaptosome-enriched fractions obtained from mice that express the CB1 receptor exclusively in dorsal telencephalic glutamatergic neurons (Glu-CB1-RS) [5] or in forebrain GABAergic neurons (GABA-CB1-RS) [6]. As expected, in frontal cortical synaptosomal membranes, the specific bands resulting from the immunodetection of the CB1 receptor in both partial rescue mice (Glu-CB1-RS and GABA-CB1-RS) represented about 45% of the corresponding total signal obtained in CB1-RS mice. In contrast, the specific CB1 receptor bands in hippocampal synaptosomes of Glu-CB1-RS and GABA-CB1-RS was about 28% and 70% of the total signal found in CB1-RS, respectively. Thus, the sum of CB1 receptor immunoreactivity in glutamatergic and GABAergic terminals of frontal cortical and hippocampal synaptosomes were found to be around 90% and 100% of the total signal of the CB1 receptors in these brain areas, respectively. Anatomical studies have shown that CB1 receptor density in GABAergic terminals is considerably higher than in glutamatergic terminals in almost all cortical areas [1,38,56,57]. However, the fact that the number of excitatory terminals (80% of pyramidal glutamatergic neurons) predominate over the inhibitory ones (20% of GABAergic interneurons) in the cerebral cortex [58] could explain the observed absence of differences in the levels of CB1 receptor expression in Western blots of frontal cortical synaptosomes derived from Glu-CB1-RS and GAB-CB1-RS mice. Agonist-stimulated [^{35}S]GTPγS binding in Glu-CB1-RS- and GABA-CB1-RS-derived frontal cortical and hippocampal synaptosomes clearly showed that the maximal response (E_{max}) to full agonists correlated with the abundance of CB1 receptors, irrespective of the terminal type (glutamatergic or GABAergic) context. In this way, in frontal cortical synaptosomes, an equal contribution of glutamatergic (Glu-CB1-RS) and GABAergic (GABA-CB1-RS) CB1 receptors to the total CB1 receptor population (CB1-RS), as defined by Western blot assays, was followed by an equal contribution to the total agonist-stimulated CB1 receptor coupling to Gαi/o proteins, as defined by [^{35}S]GTPγS binding assays. Meanwhile, in hippocampal synaptosomes where 28% of the signal of the CB1-RS was found in Glu-CB1-RS, and 70% in GABA-CB1-RS, the CB1 receptor located at GABAergic terminals was responsible for considerably more Gαi/o protein activation than the CB1 receptor located at glutamatergic terminals. A similar correlation between CB1 receptor-dependent Gαi/o protein signalling to agonists and the expression levels of the CB1 receptor was also observed when cortical and hippocampal synaptosomes were compared in each genotype. Thus, the expression of the CB1 receptor and agonist-stimulated CB1 receptor coupling to Gαi/o proteins was systematically higher in hippocampal synaptosomes than in frontal cortical synaptosomes (both in wild-type and in CB1-RS mice).

The concentration–response curves for the agonists tested (CP 55,940 and WIN 55,212-2) showed similar agonist potency (pEC_{50}) in both regions and genotypes. In addition, the similar magnitude of the negative regulation exerted by cholesterol on agonist dependent CB1 receptor coupling to $G\alpha i/o$ proteins at both types of presynaptic terminals informs us that probably there are no differences to the raft location of the CB1 receptor signalling elements related to the cellular context (glutamatergic versus GABAergic neurons). In any case, as discussed above, one of the limitations of our experimental design is that it did not allow us to determine the contribution of the lipid raft- and non-lipid raft-located CB1 receptors to the overall response to agonists. Therefore, we can only speculate about the increased agonist efficacy as exclusively linked to the activation of CB1 receptor located in lipid rafts.

In conclusion, our results demonstrated the suitability of the genetic approach and support the wild-type phenotype of the CB1-RS mice with respect to the expression level, subsynaptic distribution, raft vs non-raft compartmentalization, and $G\alpha i/o$ coupling of CB1 receptors in synaptosomes. These findings showed that the plasma membrane partitioning of the CB1 receptor and its functional coupling to $G\alpha i/o$ proteins are not biased towards the cell type of CB1 receptor rescue. In addition, we provided an updated view of the functional coupling of the CB1 receptor to $G\alpha i/o$ proteins at excitatory and inhibitory terminals, showing that the extent of the canonical $G\alpha i/o$ protein-dependent CB1 receptor signalling correlated with the abundance of CB1 receptor in the respective cell type (glutamatergic versus GABAergic neurons) both in frontal cortical and hippocampal synaptosomes. Moreover, we explored the effects of plasma membrane cholesterol abundance on CB1 receptor signalling, decreasing the membrane cholesterol level by $M\beta CD$. Pretreatment of synaptosomes from Glu-CB1-RS and GABA-CB1-RS mice with $M\beta CD$ increased the agonists efficacy to the same extent. In summary, the data infered here further substantiate the potential of our approach to unravel cell-type specific CB1 receptor signalling and highlight the utility of the CB1 receptor rescue model in studying endocannabinoid physiology on the subcellular level.

4. Materials and Methods

4.1. Animal Procedures and Brain Tissue Preparation

All experimental protocols were performed in accordance with the European Community's Council Directive of 24 November 1986 (86/609/EEC. Animals were housed in a temperature- and humidity-controlled room (22 ± 1 °C; $50 \pm 1\%$) with a 12 h light/dark cycle (lights on at 7:00 A.M.) and had access to food and water ad libitum. This study was performed on adult (16–26 weeks old) male mice from the following mouse lines: conventional CB1 receptor knockout (CB1-KO) mice and wild-type (WT) littermates, Stop-CB1 mice and their Glu-CB1 receptor rescue (Glu-CB1-RS) littermates, Stop-CB1 and their GABA-CB1 receptor rescue (GABA-CB1-RS) littermates and the CB1 receptor total rescue (CB1-RS) mice. Stop-CB1 mice and the CB1 receptor total rescue (CB1-RS) mice were produced by separate breedings as the general deleter EIIα-Cre [59] caused mosaicism in the offspring. Stop-CB1 mice were generated by heterozygous breeding of CB1stop/+ mice. CB1-RS mice, on the other hand, were obtained by homozygous breeding of mice carrying the recombined floxed Stop-CB1 allele. The reader is referred to previous studies for more detailed information on generation, breeding, and genotyping of the mice [3,5,6,60].

Mice were anesthetized with isoflurane (10 s) before decapitation. For brain extraction, the skull was cut with a scissor and the complete brain was carved out with a spatula. Then, blood clots and the meninges were removed from the sample and the piece was dissected carefully to obtain different brain regions. Initially, a sagittal incision was made in the central part of the brain to allow the separation of the cerebral hemispheres. Once those were completely separated, the hippocampus and the frontal cortex were separated from the diencephalon and basal ganglia. After removing white matter from the cortical sample as much as possible, tissue was stored at -80 °C until use.

4.2. Chemicals and Antibodies

All chemicals and reagents are described in the Supplementary file. Supplementary Table S1 contains the list of the antibodies used.

4.3. Preparation of Mouse Synaptosomal Membranes and Purification of Subsynaptic Fractions

Synaptosomal membranes from the hippocampus and frontal cortex were prepared as previously described by Dodd et al. (1981) [61] with slight modifications made by our laboratory [62]. Pooled hippocampal and cortical tissue from eight mice (about 500 mg–1 g fresh tissue weight per fractionation procedure) was thawed slowly on ice-cold 0.32 M sucrose, pH 7.4, containing 80 mM Na_2HPO_4 and 20 mM NaH_2PO_4 (sucrose phosphate buffer). The tissue was minced and homogenized in 10 volumes of sucrose/phosphate buffer, using a motor-driven Potter Teflon glass homogenizer (motor speed 800 rpm; 10 up and down strokes; mortar cooled in an ice-water mixture throughout). The homogenate was centrifuged at $1000\times g$ for 10 min. The supernatant S1 was pelleted at $15,000\times g$ for 30 min. The obtained pellet (P2 crude) was resuspended in an adequate volume of the same buffer, and a 100 μL aliquot was used for protein determination using the Bio-Rad dye reagent with bovine γ-globulin as standard. The crude membrane suspension was pelleted at $15,000\times g$ for 30 min and resuspended to obtain 16 mL of a suspension with a total protein concentration of 2.5–4 mg/mL. This P2 suspension was layered onto a centrifugation tube and 8 mL of 1.2 M sucrose phosphate buffer was added on the bottom of the tube using a Pasteur pipette, and centrifuged at $180,000\times g$ for 15 min. The material retained at the gradient interface (synaptosomes + myelin + microsomes) was carefully collected with a Pasteur pipette and diluted with ice-cold 0.32 M sucrose/phosphate buffer to a final volume of 16 mL. The diluted suspension was then layered onto 8 mL of 0.8 M sucrose buffer containing 80 mM Na_2HPO_4 and 20 mM NaH_2PO_4, and centrifuged as described above. The pellet obtained was resuspended in an adequate volume of phosphate buffer (pH 7.4) to give a synaptosome suspension with a total protein concentration of 1.5–3 mg/mL. Aliquots were then centrifuged at $40,000\times g$ for 30 min, the supernatants were aspirated and the synaptosomal pellets were frozen at −80 °C. Protein content was determined using the Bio-Rad dye reagent with bovine γ-globulin as standard.

The separation of the presynaptic active zone (PAZ), postsynaptic density (PSD) and non-synaptic fractions (extrasynaptic, EXTRA) from cortical nerve terminals was carried out as initially described by Phillips et al. (2001) [11]. Cortical synaptosomal membranes (4–5 mg total protein) were diluted in 10 mL of solubilization buffer (1% Triton X-100, 20 mM Tris, 0.1 mM $CaCl_2$, pH 6.0), and were incubated for 30 min on ice with mild agitation, and the insoluble material (synaptic junctions-PAZ+PSD) pelleted ($40,000\times g$ for 30 min at 4 °C). The supernatant (EXTRA fraction) was decanted, and the contained proteins precipitated with six volumes of acetone at −20 °C. Finally, the EXTRA fraction was recovered by centrifugation ($18,000\times g$ for 30 min at 4 °C). The synaptic junction (PAZ + PSD) pellet was resuspended in 10 mL of a solubilization buffer (1% Triton X-100, 20 mM Tris, pH 8.0). After incubation for 30 min on ice with mild agitation, the mixture was centrifuged ($18,000\times g$ for 30 min at 4 °C), and the supernatant (presynaptic fraction-PAZ) processed as described for the extrasynaptic fraction. The pellets from the supernatants and the final insoluble pellet (postsynaptic fraction-PSD) were solubilized in 5% SDS, and the total protein concentration determined by the bicinchoninic acid (BCA) protein assay following the Abcam's BCA Protein Quantification Kit procedure.

4.4. Isolation of "Lipid Rafts" from Cortical Synaptosomal Membranes

Cortical synaptosomal aliquots (6 mg total protein) were solubilized at 4 °C with 2 mL of sodium phosphate buffer containing 1% Triton X-100 by end-over-end mixing (30 min). Thereafter, the extracts are adjusted to 45% sucrose, and overlaid with 4 mL of 35% sucrose in sodium phosphate buffer, and 4 mL of 5% sucrose in sodium phosphate buffer, inside an ultracentrifugation tube. Lipid raft fractions were isolated by ultracentrifugation at $140,000\times g$, for 18 h, 4 °C. Then gradient was harvested in 12 fractions of 1 mL each plus the pellet. The analysis of lipid raft (Thy-1, and flotillin)/non-raft fractions (Na$^+$/K$^+$-ATPase) markers were carried out in different gels using the same volume per gel of samples from each of the fractions from the same separation. The Alkaline Phosphatase Assay Kit (ab83369) from Abcam was used to determine the alkaline phosphatase (ALP) activity in lipid raft and non-raft fractions derived from synaptosomes.

4.5. Immunofluorescence Assay for Frontal Cortical Synaptosomes

Immunofluorescence assays were performed as previously described with minor modifications [63]. Details of the procedure are described in the Supplementary Methods.

4.6. Treatment of Cortical Synaptosomal Fractions with Deglycosylating Enzymes

PNGase F enzymatic method (New England BioLabs) was used for removing N-linked oligosaccharides from glycoproteins. PNGase F is an amidase, which cleaves between the innermost GlcNAc and asparagine residues of high mannose, hybrid and complex oligosaccharides. Briefly, Nine parts of 2.3 µg/µL of synaptosomes re-suspended in phosphate buffer were combined with one part of 10× Glycoprotein Denaturing Buffer (5% SDS, 400 mM DTT). Glycoproteins were denatured by heating the reaction at 60 °C for 10 min. Thereafter, the denatured sample was mixed in a 1:1 ratio with 2× GlycoBuffer 2 and 2% NP-40 diluted in H$_2$O. Finally, 1 µL of PNGase F was added per each 20 µg of total protein of the denatured synaptosomal fraction, and the reaction mixture was incubated at 37 °C for 1 h. The extent of deglycosylation of the CB1 receptor was assessed by mobility shifts on SDS-PAGE gel by Western blot assays.

4.7. Treatment of Cortical Synaptosomes with Methyl-β-cyclodextrin

Methyl-β-cyclodextrin (MβCD) compound was used to directly extract cholesterol from synaptic plasma membranes. Several preliminary experiments were conducted to determine the optimal concentration of MβCD to deplete cholesterol from cortical synaptosomal membranes. Synaptosomes (1 mg protein/mL) were incubated with the indicated concentration of MβCD on 50 mM Tris-HCl buffer (pH 7.4) for 30 min at 37 °C. After treatment, the reaction was stopped by adding a large volume of cold Tris-HCl buffer without MβCD, and synaptosomes were pelleted at 15,000× *g* for 15 min at 4 °C. The obtained pellet was re-suspended again in a Tris-HCl buffer without MβCD and aliquoted in microcentrifuge tubes. Aliquots were then centrifuged at 15,000× *g* for 30 min and pellets corresponding to synaptosomes were stored at −80 °C. The Cholesterol Assay Kit (ab65359) from Abcam was used to determine total cholesterol level of synaptosomal membranes.

4.8. Western Blot Assay in Purified Fractions of Synaptosomal Membranes

Western blot experiments were performed as previously described with minor modifications [20,62]. The procedure is described in the Supplementary Methods.

4.9. Agonist Stimulated [^{35}S]GTPγS Binding Assay in Synaptosomal Membranes

The [^{35}S]GTPγS binding assays were performed following the procedure described elsewhere [64] with minor modifications. Detailed experimental protocol is presented in the Supplementary Methods.

Supplementary Materials: The following supplementary material is available online, Supplementary materials and methods [65,66]; Figure S1: Western blot of the homogenate and subcellular fractions obtained from sequential fractionation of adult mouse brain cortical homogenates; Figure S2: Double-immunofluorescence MAP2/GFAP and SNAP25/GFAP combined with the membrane marker DiIC16 in isolated cortical synaptosomes maintained in isotonic buffer and seeded on poly-L-ornitine coated coverslip; Figure S3: Migration profile of the CB1 receptor immunorreactive bands in synaptosome samples subjected to a potentially proteolytic condition and to an N-glycosidase treatment with the PNGase F enzyme; Figure S4: Subsynaptic compartmentalization of the protein selected markers in PAZ, PSD and EXTRA fractions isolated from cortical synaptosomes derived from wild-type mice; Figure S5: Subsynaptic compartmentalization of the protein selected markers in PAZ, PSD and EXTRA fractions isolated from cortical synaptosomes derived from CB1-RS mice; Figure S6: Subsynaptic compartmentalization of the CB1 receptor, the canonical transducers coupled to CB1 receptors and other proteins of the endocannabinoid system in PAZ, PSD and EXTRA fractions isolated from cortical synaptosomes derived from CB1-RS mice; Figure S7: Alkaline phosphatase enzymatic activity, total protein amount and distribution of raft and non-raft markers in lipid raft and non-lipid raft fractions isolated from frontal cortical synaptosomes derived from CB1-RS mice. Figure S8: CB1 receptor protein expression in synaptosomes obtained from frontal cortical and hippocampal tissue of wild-type, CB1-RS and Stop-CB1 mice; Figure S9: CB1 receptor coupling to Gαi/o proteins in synaptosomes obtained from frontal cortical and hippocampal tissue of wild-type and CB1-RS mice; Figure S10: CB1 receptor coupling to Gαi/o proteins in control and in 5 mM, 10 mM and 20 mM MβCD pretreated synaptosomes from frontal cortical tissue of Glu-CB1-RS and GABA-CB1-RS mice; Figure S11: Model for the phosphorylation-dependent electrophoretic mobility shift (PDEMS) phenomenon and the EMS-related motif. Adapted from Figure 4 in Lee et al. (2019) [31]; Table S1: The list of primary antibodies used; Table S2: Migration profile of the CB1 receptor immunorreactive bands in synaptosome samples obtained in the absence or the presence of protease inhibitors during the fractionation procedure; Table S3: Statistical analysis of the subsynaptic distribution of the CB1 receptor immunoreactive signals of ~50 kDa and ~35 kDa bands between CB1-RS and wild-type mice (WT); Table S4: Densitometry analysis of the specific immunorreactive signals (~50 kDa and ~35kDa bands) of the CB1 receptor (see Supplemental Figure S8), normalized to the signal of frontal cortical synaptosomes of wild-type; Table S5: Concentration–response curves for the CP 55,940-stimulated specific [^{35}S]GTPγS binding in frontal cortical and hippocampal synaptosomes derived from wild-type (WT) and CB1-RS mice; Table S6: CB1 receptor protein expression in synaptosomal membranes of frontal cortex by the slope comparison method of the lines obtained by regression analysis of the data shown in the Figure 4B–D; Table S7: Stimulation of [^{35}S]GTPγS binding by a maximal concentration (10 µM) of the cannabinoid agonist CP 55,940 in cortical synaptosomes from wild-type mice pretreated with MβCD as described in Section 4.7 from Materials and Methods; Table S8: Stimulation of [^{35}S]GTPγS binding by a maximal concentration (10 µM) of the cannabinoid agonists CP 55,940 and WIN 55,212-2 in cortical synaptosomes from Glu-CB1-RS and GABA-CB1-RS mice pretreated with MβCD (10 mM) as described in Section 4.7 from Materials and Methods. Table S9: Summary table of the CB1 receptor density and the CB1 receptor coupling to Gαi/o in cortical and hippocampal synaptosomes from wild-type (WT), CB1-RS, Glu-CB1-RS and GABA-CB1-RS mice.

Author Contributions: B.L., J.S. and K.M. designed the research. B.L., K.M., J.S. and M.A.G. were involved in funding acquisition. M.S.-E., S.B., G.G.d.C., J.S. and K.M. performed the experimental work, and acquired and analysed the data. M.S.-E., G.G.d.C., B.L., J.S. and K.M. prepared the figures and wrote the manuscript. All authors have read and agreed to the published version of the manuscript.

Funding: This research was funded by the Basque Government (IT1230-19), MINECO, Spanish Ministry of Science, Innovation and Universities (CTQ2017-85686-R).

Institutional Review Board Statement: All experimental protocols were performed in accordance with the European Community's Council Directive of 24 November 1986 (86/609/EEC) and the animal genotyping approved by the Ethical Committee on animal care and use of Rhineland-Palatinate, Germany (A19-1-002).

Informed Consent Statement: Not applicable.

Data Availability Statement: All data supporting the findings of this study are available within the article and the associated Supplementary Materials.

Acknowledgments: Miquel Saumell-Esnaola is a recipient of a PhD contract awarded by the Department of Education of the Basque Government. Joan Sallés is a member of the Societat Catalana de Biologia, a subsidiary society of the Institut d'Estudis Catalans (Barcelona, Catalonia, Spain).

Conflicts of Interest: The authors declare no conflict of interest.

Sample Availability: Cell-type-specific CB1 receptor rescue mice are available from research Group "Molecular Mechanisms of Behavior" directed by Prof. Beat Lutz (Institute of Physiological Chemistry, University Medical Centre of the Johannes Gutenberg University Mainz, Mainz, Germany).

References

1. Monory, K.; Massa, F.; Egertová, M.; Eder, M.; Blaudzun, H.; Westenbroek, R.; Kelsch, W.; Jacob, W.; Marsch, R.; Ekker, M.; et al. The Endocannabinoid System Controls Key Epileptogenic Circuits in the Hippocampus. *Neuron* **2006**, *51*, 455–466. [CrossRef] [PubMed]
2. Busquets-Garcia, A.; Bains, J.; Marsicano, G. CB1 Receptor Signaling in the Brain: Extracting Specificity from Ubiquity. *Neuropsychopharmacology* **2018**, *43*, 4–20. [CrossRef]
3. Bellocchio, L.; Lafenêtre, P.; Cannich, A.; Cota, D.; Puente, N.; Grandes, P.; Chaouloff, F.; Piazza, P.V.; Marsicano, G. Bimodal control of stimulated food intake by the endocannabinoid system. *Nat. Neurosci.* **2010**, *13*, 281–283. [CrossRef] [PubMed]
4. Rey, A.A.; Purrio, M.; Viveros, M.P.; Lutz, B. Biphasic effects of cannabinoids in anxiety responses: CB1 and GABA B receptors in the balance of gabaergic and glutamatergic neurotransmission. *Neuropsychopharmacology* **2012**, *37*, 2624–2634. [CrossRef] [PubMed]
5. Ruehle, S.; Remmers, F.; Romo-Parra, H.; Massa, F.; Wickert, M.; Wortge, S.; Haring, M.; Kaiser, N.; Marsicano, G.; Pape, H.-C.; et al. Cannabinoid CB1 Receptor in Dorsal Telencephalic Glutamatergic Neurons: Distinctive Sufficiency for Hippocampus-Dependent and Amygdala-Dependent Synaptic and Behavioral Functions. *J. Neurosci.* **2013**, *33*, 10264–10277. [CrossRef]
6. Remmers, F.; Lange, M.D.; Hamann, M.; Ruehle, S.; Pape, H.-C.; Lutz, B. Addressing sufficiency of the CB1 receptor for endocannabinoid-mediated functions through conditional genetic rescue in forebrain GABAergic neurons. *Brain Struct. Funct.* **2017**, *222*, 3431–3452. [CrossRef]
7. Gutiérrez-Rodríguez, A.; Puente, N.; Elezgarai, I.; Ruehle, S.; Lutz, B.; Reguero, L.; Gerrikagoitia, I.; Marsicano, G.; Grandes, P. Anatomical characterization of the cannabinoid CB 1 receptor in cell-type-specific mutant mouse rescue models. *J. Comp. Neurol.* **2017**, *525*, 302–318. [CrossRef]
8. De Giacomo, V.; Ruehle, S.; Lutz, B.; Häring, M.; Remmers, F. Differential glutamatergic and GABAergic contributions to the tetrad effects of Δ9-tetrahydrocannabinol revealed by cell-type-specific reconstitution of the CB1 receptor. *Neuropharmacology* **2020**, *179*, 108287. [CrossRef]
9. Monory, K.; Blaudzun, H.; Massa, F.; Kaiser, N.; Lemberger, T.; Schütz, G.; Wotjak, C.T.; Lutz, B.; Marsicano, G. Genetic Dissection of Behavioural and Autonomic Effects of Δ9-Tetrahydrocannabinol in Mice. *PLoS Biol.* **2007**, *5*, e269. [CrossRef]
10. Steindel, F.; Lerner, R.; Häring, M.; Ruehle, S.; Marsicano, G.; Lutz, B.; Monory, K. Neuron-type specific cannabinoid-mediated G protein signalling in mouse hippocampus. *J. Neurochem.* **2013**, *124*, 795–807. [CrossRef]
11. Phillips, G.R.; Huang, J.K.; Wang, Y.; Tanaka, H.; Shapiro, L.; Zhang, W.; Shan, W.S.; Arndt, K.; Frank, M.; Gordon, R.E.; et al. The presynaptic particle web: Ultrastructure, composition, dissolution, and reconstitution. *Neuron* **2001**, *32*, 63–77. [CrossRef]
12. Kofalvi, A.; Rodrigues, R.J.; Ledent, C.; Mackie, K.; Vizi, E.S.; Cunha, R.A.; Sperlágh, B. Involvement of Cannabinoid Receptors in the Regulation of Neurotransmitter Release in the Rodent Striatum: A Combined Immunochemical and Pharmacological Analysis. *J. Neurosci.* **2005**, *25*, 2874–2884. [CrossRef]
13. Ostrom, R.S.; Insel, P.A. Methods for the Study of Signaling Molecules in Membrane Lipid Rafts and Caveolae. *Methods Mol. Biol.* **2006**, *332*, 181–192. [CrossRef]
14. Yoneda, T.; Kameyama, K.; Esumi, K.; Daimyo, Y.; Watanabe, M.; Hata, Y. Developmental and Visual Input-Dependent Regulation of the CB1 Cannabinoid Receptor in the Mouse Visual Cortex. *PLoS ONE* **2013**, *8*, e53082. [CrossRef]
15. Echeazarra, L.; García del Caño, G.; Barrondo, S.; González-Burguera, I.; Saumell-Esnaola, M.; Aretxabala, X.; López de Jesús, M.; Borrega-Román, L.; Mato, S.; Ledent, C.; et al. Fit-for-purpose based testing and validation of antibodies to amino- and carboxy-terminal domains of cannabinoid receptor 1. *Histochem. Cell Biol.* **2021**, 1–24. [CrossRef]
16. Fukudome, Y.; Ohno-Shosaku, Ã.T.; Matsui, Ã.M.; Omori, Y.; Fukaya, M.; Tsubokawa, H.; Taketo, M.M.; Watanabe, M.; Manabe, T.; Kano, M. Two distinct classes of muscarinic action on hippocampal inhibitory synapses. *Eur. J. Neurosci.* **2004**, *19*, 2682–2692. [CrossRef] [PubMed]
17. Rodríguez-Cueto, C.; Hernández-Gálvez, M.; Hillard, C.J.; Maciel, P.; García-García, L.; Valdeolivas, S.; Pozo, M.A.; Ramos, J.A.; Gómez-Ruiz, M.; Fernández-Ruiz, J. Dysregulation of the endocannabinoid signaling system in the cerebellum and brainstem in a transgenic mouse model of spinocerebellar ataxia type-3. *Neuroscience* **2016**, *339*, 191–209. [CrossRef] [PubMed]
18. Peñasco, S.; Rico-Barrio, I.; Puente, N.; Fontaine, C.J.; Ramos, A.; Reguero, L.; Gerrikagoitia, I.; de Fonseca, F.R.; Suarez, J.; Barrondo, S.; et al. Intermittent ethanol exposure during adolescence impairs cannabinoid type 1 receptor-dependent long-term depression and recognition memory in adult mice. *Neuropsychopharmacology* **2020**, *45*, 309–318. [CrossRef]
19. Egaña-Huguet, J.; Bonilla-Del Río, I.; Gómez-Urquijo, S.M.; Mimenza, A.; Saumell-Esnaola, M.; Borrega-Roman, L.; García del Caño, G.; Sallés, J.; Puente, N.; Gerrikagoitia, I.; et al. The Absence of the Transient Receptor Potential Vanilloid 1 Directly Impacts on the Expression and Localization of the Endocannabinoid System in the Mouse Hippocampus. *Front. Neuroanat.* **2021**, *15*, 1–17. [CrossRef]
20. De Jesús, M.L.; Sallés, J.; Meana, J.J.; Callado, L.F. Characterization of CB1 cannabinoid receptor immunoreactivity in postmortem human brain homogenates. *Neuroscience* **2006**, *140*, 635–643. [CrossRef] [PubMed]
21. Song, C.; Howlett, A.C. Rat brain cannabinoid receptors are N-linked glycosylated proteins. *Life Sci.* **1995**, *56*, 1983–1989. [CrossRef]
22. Esteban, P.F.; Garcia-Ovejero, D.; Paniagua-Torija, B.; Moreno-Luna, R.; Arredondo, L.F.; Zimmer, A.; Arevalo-Martin, A.; Molina-Holgado, E. Revisiting CB1 cannabinoid receptor detection and the exploration of its interacting partners. *J. Neurosci. Methods* **2020**, *337*, 108680. [CrossRef]

23. Nordström, R.; Andersson, H. Amino-Terminal Processing of the Human Cannabinoid Receptor 1. *J. Recept. Signal Transduct.* **2006**, *26*, 259–267. [CrossRef]

24. Ruehle, S.; Wager-Miller, J.; Straiker, A.; Farnsworth, J.; Murphy, M.N.; Loch, S.; Monory, K.; Mackie, K.; Lutz, B. Discovery and characterization of two novel CB1 receptor splice variants with modified N-termini in mouse. *J. Neurochem.* **2017**, *142*, 521–533. [CrossRef]

25. Egertová, M.; Elphick, M.R. Localisation of cannabinoid receptors in the rat brain using antibodies to the intracellular C-terminal tail of CB1. *J. Comp. Neurol.* **2000**, *422*, 159–171. [CrossRef]

26. Wager-Miller, J.; Westenbroek, R.; Mackie, K. Dimerization of G protein-coupled receptors: CB1 cannabinoid receptors as an example. *Chem. Phys. Lipids* **2002**, *121*, 83–89. [CrossRef]

27. Mukhopadhyay, S.; Howlett, A.C. CB 1 receptor-G protein association. *Eur. J. Biochem.* **2001**, *268*, 499–505. [CrossRef]

28. Diniz, C.R.A.F.; Biojone, C.; Joca, S.R.L.; Rantamäki, T.; Castrén, E.; Guimarães, F.S.; Casarotto, P.C. Dual mechanism of TRKB activation by anandamide through CB1 and TRPV1 receptors. *PeerJ* **2019**, *7*, e6493. [CrossRef]

29. Plummer, T.H.; Tarentino, A.L. Purification of the oligosaccharide-cleaving enzymes of *Flavobacterium meningosepticum*. *Glycobiology* **1991**, *1*, 257–263. [CrossRef] [PubMed]

30. Tarentino, A.L.; Plummer, T.H. [4] Enzymatic deglycosylation of asparagine-linked glycans: Purification, properties, and specificity of oligosaccharide-cleaving enzymes from Flavobacterium meningosepticum. *Methods Enzymol.* **1994**, *230*, 44–57. [PubMed]

31. Lee, C.R.; Park, Y.H.; Min, H.; Kim, Y.R.; Seok, Y.J. Determination of protein phosphorylation by polyacrylamide gel electrophoresis. *J. Microbiol.* **2019**, *57*, 93–100. [CrossRef] [PubMed]

32. Nyíri, G.; Cserép, C.; Szabadits, E.; Mackie, K.; Freund, T.F. CB1 cannabinoid receptors are enriched in the perisynaptic annulus and on preterminal segments of hippocampal GABAergic axons. *Neuroscience* **2005**, *136*, 811–822. [CrossRef]

33. Niehaus, J.L.; Liu, Y.; Wallis, K.T.; Egertová, M.; Bhartur, S.G.; Mukhopadhyay, S.; Shi, S.; He, H.; Selley, D.E.; Howlett, A.C.; et al. CB 1 Cannabinoid Receptor Activity Is Modulated by the Cannabinoid Receptor Interacting Protein CRIP 1a. *Mol. Pharmacol.* **2007**, *72*, 1557–1566. [CrossRef]

34. Guggenhuber, S.; Alpar, A.; Chen, R.; Schmitz, N.; Wickert, M.; Mattheus, T.; Harasta, A.E.; Purrio, M.; Kaiser, N.; Elphick, M.R.; et al. Cannabinoid receptor-interacting protein Crip1a modulates CB1 receptor signaling in mouse hippocampus. *Brain Struct. Funct.* **2016**, *221*, 2061–2074. [CrossRef]

35. Rodríguez, J.J.; Mackie, K.; Pickel, V.M. Ultrastructural Localization of the CB1 Cannabinoid Receptor in μ-Opioid Receptor Patches of the Rat Caudate Putamen Nucleus. *J. Neurosci.* **2001**, *21*, 823–833. [CrossRef] [PubMed]

36. Katona, I.; Sperlágh, B.; Maglóczky, Z.; Sántha, E.; Köfalvi, A.; Czirják, S.; Mackie, K.; Vizi, E.; Freund, T. GABAergic interneurons are the targets of cannabinoid actions in the human hippocampus. *Neuroscience* **2000**, *100*, 797–804. [CrossRef]

37. Katona, I.; Sperlágh, B.; Sík, A.; Käfalvi, A.; Vizi, E.S.; Mackie, K.; Freund, T.F. Presynaptically Located CB1 Cannabinoid Receptors Regulate GABA Release from Axon Terminals of Specific Hippocampal Interneurons. *J. Neurosci.* **1999**, *19*, 4544–4558. [CrossRef]

38. Katona, I.; Urban, G.M.; Wallace, M.; Ledent, C.; Jung, K.-M.; Piomelli, D.; Mackie, K.; Freund, T.F. Molecular Composition of the Endocannabinoid System at Glutamatergic Synapses. *J. Neurosci.* **2006**, *26*, 5628–5637. [CrossRef] [PubMed]

39. Yoshida, T.; Fukaya, M.; Uchigashima, M.; Miura, E.; Kamiya, H.; Kano, M.; Watanabe, M. Localization of diacylglycerol lipase-α around postsynaptic spine suggests close proximity between production site of an endocannabinoid, 2-arachidonoyl-glycerol, and presynaptic cannabinoid CB1 receptor. *J. Neurosci.* **2006**, *26*, 4740–4751. [CrossRef]

40. Fukaya, M.; Uchigashima, M.; Nomura, S.; Hasegawa, Y.; Kikuchi, H.; Watanabe, M. Predominant expression of phospholipase Cβ1 in telencephalic principal neurons and cerebellar interneurons, and its close association with related signaling molecules in somatodendritic neuronal elements. *Eur. J. Neurosci.* **2008**, *28*, 1744–1759. [CrossRef]

41. Montaña, M.; García del Caño, G.; López de Jesús, M.; González-Burguera, I.; Echeazarra, L.; Barrondo, S.; Sallés, J. Cellular neurochemical characterization and subcellular localization of phospholipase C β1 in rat brain. *Neuroscience* **2012**, *222*, 239–268. [CrossRef]

42. García del Caño, G.; Aretxabala, X.; González-Burguera, I.; Montaña, M.; López de Jesús, M.; Barrondo, S.; Barrio, R.J.; Sampedro, C.; Goicolea, M.A.; Sallés, J. Nuclear diacylglycerol lipase-α in rat brain cortical neurons: Evidence of 2-arachidonoylglycerol production in concert with phospholipase C-β activity. *J. Neurochem.* **2015**, *132*, 489–503. [CrossRef] [PubMed]

43. Jung, K.-M.; Astarita, G.; Zhu, C.; Wallace, M.; Mackie, K.; Piomelli, D. A Key Role for Diacylglycerol Lipase-α in Metabotropic Glutamate Receptor-Dependent Endocannabinoid Mobilization. *Mol. Pharmacol.* **2007**, *72*, 612–621. [CrossRef]

44. Howlett, A.C.; Qualy, J.M.; Khachatrian, L.L. Involvement of G(i) in the inhibition of adenylate cyclase by cannabimimetic drugs. *Mol. Pharmacol.* **1986**, *29*, 307–313.

45. Mukhopadhyay, S.; McIntosh, H.H.; Houston, D.B.; Howlett, A.C. The CB1 cannabinoid receptor juxtamembrane C-terminal peptide confers activation to specific G proteins in brain. *Mol. Pharmacol.* **2000**, *57*, 162–170.

46. Straiker, A.J.; Borden, C.R.; Sullivan, J.M. G-Protein α Subunit Isoforms Couple Differentially to Receptors that Mediate Presynaptic Inhibition at Rat Hippocampal Synapses. *J. Neurosci.* **2002**, *22*, 2460–2468. [CrossRef]

47. Costas-Insua, C.; Moreno, E.; Maroto, I.B.; Ruiz-Calvo, A.; Bajo-Grañeras, R.; Martín-Gutiérrez, D.; Diez-Alarcia, R.; Vilaró, M.T.; Cortés, R.; García-Font, N.; et al. Identification of BiP as a CB 1 Receptor-Interacting Protein That Fine-Tunes Cannabinoid Signaling in the Mouse Brain. *J. Neurosci.* **2021**, *41*, 7924–7941. [CrossRef] [PubMed]

48. Techlovská, Š.; Chambers, J.N.; Dvořáková, M.; Petralia, R.S.; Wang, Y.-X.; Hájková, A.; Nová, A.; Franková, D.; Prezeau, L.; Blahos, J. Metabotropic glutamate receptor 1 splice variants mGluR1a and mGluR1b combine in mGluR1a/b dimers in vivo. *Neuropharmacology* **2014**, *86*, 329–336. [CrossRef]

49. Akama, K.T.; Thompson, L.I.; Milner, T.A.; McEwen, B.S. Post-synaptic Density-95 (PSD-95) Binding Capacity of G-protein-coupled Receptor 30 (GPR30), an Estrogen Receptor That Can Be Identified in Hippocampal Dendritic Spines. *J. Biol. Chem.* **2013**, *288*, 6438–6450. [CrossRef] [PubMed]

50. Daigle, T.L.; Kwok, M.L.; Mackie, K. Regulation of CB 1 cannabinoid receptor internalization by a promiscuous phosphorylation-dependent mechanism. *J. Neurochem.* **2008**, *106*, 70–82. [CrossRef] [PubMed]

51. Straiker, A.; Wager-Miller, J.; Mackie, K. The CB1 cannabinoid receptor C-terminus regulates receptor desensitization in autaptic hippocampal neurones. *Br. J. Pharmacol.* **2012**, *165*, 2652–2659. [CrossRef] [PubMed]

52. Bari, M.; Battista, N.; Fezza, F.; Finazzi-Agrò, A.; Maccarrone, M. Lipid Rafts Control Signaling of Type-1 Cannabinoid Receptors in Neuronal Cells. *J. Biol. Chem.* **2005**, *280*, 12212–12220. [CrossRef] [PubMed]

53. Sarnataro, D.; Grimaldi, C.; Pisanti, S.; Gazzerro, P.; Laezza, C.; Zurzolo, C.; Bifulco, M. Plasma membrane and lysosomal localization of CB1 cannabinoid receptor are dependent on lipid rafts and regulated by anandamide in human breast cancer cells. *FEBS Lett.* **2005**, *579*, 6343–6349. [CrossRef]

54. Oddi, S.; Dainese, E.; Fezza, F.; Lanuti, M.; Barcaroli, D.; De Laurenzi, V.; Centonze, D.; Maccarrone, M. Functional characterization of putative cholesterol binding sequence (CRAC) in human type-1 cannabinoid receptor. *J. Neurochem.* **2011**, *116*, 858–865. [CrossRef] [PubMed]

55. Sýkora, J.; Bouřová, L.; Hof, M.; Svoboda, P. The effect of detergents on trimeric G-protein activity in isolated plasma membranes from rat brain cortex: Correlation with studies of DPH and Laurdan fluorescence. *Biochim. Biophys. Acta-Biomembr.* **2009**, *1788*, 324–332. [CrossRef]

56. Marsicano, G.; Lutz, B. Expression of the cannabinoid receptor CB1 in distinct neuronal subpopulations in the adult mouse forebrain. *Eur. J. Neurosci.* **1999**, *11*, 4213–4225. [CrossRef]

57. Kawamura, Y. The CB1 Cannabinoid Receptor Is the Major Cannabinoid Receptor at Excitatory Presynaptic Sites in the Hippocampus and Cerebellum. *J. Neurosci.* **2006**, *26*, 2991–3001. [CrossRef]

58. Kawaguchi, Y.; Kubota, Y. GABAergic cell subtypes and their synaptic connections in rat frontal cortex. *Cereb. Cortex* **1997**, *7*, 476–486. [CrossRef]

59. Lakso, M.; Pichel, J.G.; Gorman, J.R.; Sauer, B.; Okamoto, Y.; Lee, E.; Alt, F.W.; Westphal, H. Efficient in vivo manipulation of mouse genomic sequences at the zygote stage. *Proc. Natl. Acad. Sci. USA* **1996**, *93*, 5860–5865. [CrossRef]

60. Marsicano, G.; Wotjak, C.T.; Azad, S.C.; Bisogno, T.; Rammes, G.; Cascio, M.G.; Hermann, H.; Tang, J.; Hofmann, C.; Zieglgänsberger, W.; et al. The endogenous cannabinoid system controls extinction of aversive memories. *Nature* **2002**, *418*, 530–534. [CrossRef]

61. Dodd, P.R.; Hardy, J.A.; Oakley, A.E.; Edwardson, J.A.; Perry, E.K.; Delaunoy, J.-P. A rapid method for preparing synaptosomes: Comparison, with alternative procedures. *Brain Res.* **1981**, *226*, 107–118. [CrossRef]

62. Garro, M.A.; López de Jesús, M.; Ruíz de Azúa, I.; Callado, L.F.; Javier Meana, J.; Sallés, J. Regulation of phospholipase Cβ activity by muscarinic acetylcholine and 5-HT2 receptors in crude and synaptosomal membranes from human cerebral cortex. *Neuropharmacology* **2001**, *40*, 686–695. [CrossRef]

63. Stigliani, S.; Zappettini, S.; Raiteri, L.; Passalacqua, M.; Melloni, E.; Venturi, C.; Tacchetti, C.; Diaspro, A.; Usai, C.; Bonanno, G. Glia re-sealed particles freshly prepared from adult rat brain are competent for exocytotic release of glutamate. *J. Neurochem.* **2006**, *96*, 656–668. [CrossRef] [PubMed]

64. Barrondo, S.; Sallés, J. Allosteric modulation of 5-HT1A receptors by zinc: Binding studies. *Neuropharmacology* **2009**, *56*, 455–462. [CrossRef] [PubMed]

65. Fleming, W.W.; Westfall, D.P.; De la Lande, I.S.; Jellett, L.B. Log-normal distribution of equiefective doses of norepinephrine and acetylcholine in several tissues. *J. Pharmacol. Exp. Ther.* **1972**, *181*, 339–345.

66. Christopoulos, A. Assessing the distribution of parameters in models of ligand–receptor interaction: To log or not to log. *Trends Pharmacol. Sci.* **1998**, *19*, 351–357. [CrossRef]

Review

CB1 Cannabinoid Receptor Signaling and Biased Signaling

Luciana M. Leo and Mary E. Abood *

Center for Substance Abuse Research, Lewis Katz School of Medicine, Temple University, Philadelphia, PA 19140, USA; luciana.leo@temple.edu
* Correspondence: mabood@temple.edu

Abstract: The CB1 cannabinoid receptor is a G-protein coupled receptor highly expressed throughout the central nervous system that is a promising target for the treatment of various disorders, including anxiety, pain, and neurodegeneration. Despite the wide therapeutic potential of CB1, the development of drug candidates is hindered by adverse effects, rapid tolerance development, and abuse potential. Ligands that produce biased signaling—the preferential activation of a signaling transducer in detriment of another—have been proposed as a strategy to dissociate therapeutic and adverse effects for a variety of G-protein coupled receptors. However, biased signaling at the CB1 receptor is poorly understood due to a lack of strongly biased agonists. Here, we review studies that have investigated the biased signaling profile of classical cannabinoid agonists and allosteric ligands, searching for a potential therapeutic advantage of CB1 biased signaling in different pathological states. Agonist and antagonist bound structures of CB1 and proposed mechanisms of action of biased allosteric modulators are used to discuss a putative molecular mechanism for CB1 receptor activation and biased signaling. Current studies suggest that allosteric binding sites on CB1 can be explored to yield biased ligands that favor or hinder conformational changes important for biased signaling.

Keywords: cannabinoid; CB1; biased signaling; functional selectivity; G-protein; β-arrestin

Citation: Leo, L.M.; Abood, M.E. CB1 Cannabinoid Receptor Signaling and Biased Signaling. *Molecules* **2021**, *26*, 5413. https://doi.org/10.3390/molecules26175413

Academic Editor: Mauro Maccarrone

Received: 29 July 2021
Accepted: 3 September 2021
Published: 6 September 2021

Publisher's Note: MDPI stays neutral with regard to jurisdictional claims in published maps and institutional affiliations.

1. Introduction

The cannabinoid receptor type 1 (CB1) is a class A G-protein coupled receptor (GPCR) that was first discovered as the main target for Δ^9-tetrahydrocannabinol (THC), the psychoactive compound in *Cannabis*. CB1 was first described in rat [1,2] and later cloned from a human brain cDNA library [3]. At the protein level, rat and human CB1 share 97% sequence identity, with only two amino acid substitutions within the transmembrane domains. Of these, one is found on the extracellular (EC) end of transmembrane helix 2 (TMH2)—position 2.62 in Ballesteros–Weinstein nomenclature [4], Ile175 in human and Val176 in rat—and one on the EC end of TMH3—position 3.22, Arg186 in human and Pro187 in rat. Interestingly, the C-terminal tail of CB1 forms an extra α-helix between residues Ala440 and Met461 (amino acid numbers for human CB1), termed Helix 9 (Hx9), that associates with the plasma membrane.

In addition to THC, other exogenous ligands for CB1 have been described. Notably, THC analogs and other synthetic cannabinoids are widely used as CB1 agonists, such as HU-210, CP55940, and WIN55212 [5–7]. Endogenous ligands for CB1 are derived from arachidonic acid, which is metabolized by diacylglycerol lipase into 2-arachidonoylacylglycerol (2-AG) and by N-acyl-phosphatidylethanolamine-hydrolyzing phospholipase D into anandamide (AEA) [8–10]. AEA and 2-AG are primarily degraded by fatty acid amide hydrolase and monoacylglycerol lipase, respectively [11]. However, 2-AG and AEA also bind to other targets, such as the CB2 receptor and transient receptor potential vanilloid 1 (TRPV1) [11]. These endogenous ligands, their receptors, and their synthesis and degradation enzymes form the endocannabinoid system [11].

Here, we will review the role of CB1 in physiological and pathological conditions and explore its various signaling mechanisms. CB1 has been investigated as a source of biased signaling, a process by which a given ligand can preferentially elicit signaling via one signal transducer to the detriment of another [12]. However, the physiological role of CB1-biased signaling is poorly understood. Therefore, studies that suggest a therapeutic advantage for CB1 biased ligands will be discussed. Finally, considering the solved molecular structures of agonist bound CB1, along with the proposed mechanisms of action of certain biased allosteric modulators, we will analyze the potential molecular mechanism of CB1 biased signaling.

2. The CB1 Receptor

2.1. Therapeutic Potential

Dysregulation of the endocannabinoid system in physiological aging and in brain pathologies along with the prevalence of CB1 in a variety of CNS circuits make it an attractive target for the treatment of multiple neurological conditions. In fact, *Cannabis* and cannabinoid formulations are already approved for certain medicinal uses in several countries and in most US states. Dronabinol and nabilone are synthetic THC analogs approved by the US Food and Drug Administration as antiemetics and orexigenics for patients undergoing chemotherapy and patients with acquired immunodeficiency syndrome. Nabiximols (Sativex®) are *Cannabis* extracts containing THC and cannabidiol at a near 1:1 ratio approved in the United Kingdom, Spain, Brazil, Colombia, Chile, Australia, among several countries, for mitigation of symptoms, including spasticity, of treatment-resistant multiple sclerosis. In the United States, Sativex® is currently under investigation in a phase 3 clinical trial for the treatment of neuropathic pain (NCT00711880) with promising preliminary results [13]. There are also several currently active clinical trials investigating the efficacy of medicinal *Cannabis* use in the treatment of acute and chronic pain. The analgesic properties of cannabinoids are well known, and enhancing CB1 activity has been proposed as a treatment for various forms of pain [14] due to its ability to suppress nociception at dorsal root ganglia [15,16], spinal cord [17–19], and the descending pain modulatory system, such as in the periaqueductal gray (PAG) [20–22].

Although cannabinoid use is generally associated with cognitive impairment [23], a recent study showed that, while in young mice a chronic low dose THC treatment acts through CB1 to impair memory, it has the opposite effect in aged mice [24]. This result, along with findings of reduced CB1 expression and function in aged mice [25] and of early onset cognitive dysfunction in mice with CB1 deletion [26], suggests that CB1 agonists may have a beneficial effect in the treatment of age-related cognitive impairment.

CB1 agonists have also been shown to reduce anxiety-like behavior [27,28] and depressive-like behavior [29,30] in preclinical models, showing promise for the treatment of generalized anxiety and major depression disorders. The anxiolytic effect of cannabinoids, along with their negative modulation of hypothalamus–pituitary–adrenal axis activity mediated stress responses and facilitation of extinction learning in fear memory, led cannabinoid agonists to be investigated in the treatment of posttraumatic stress disorder (PTSD). In this context, positive results have been reported from CB1 and CB2 agonists in preclinical models [31], and a current phase 2 clinical study is underway to investigate the effect of *Cannabis* on symptoms of PTSD in war veterans (NCT02759185). Further, the anticonvulsant action of cannabinoids in preclinical models makes CB1 a possible target for the treatment of epilepsy [32,33].

Neuroprotection has been suggested as a function of the endocannabinoid system, and findings that CB1 agonists protect against cerebral ischemia and that CB1 deletion enhances the severity of ischemia–reperfusion injury in mice [34–36] suggest that it could also be targeted for the treatment of stroke. Finally, there is evidence that CB1 activity is beneficial for the treatment of Huntington's disease (HD), a genetic neurodegenerative disorder marked by expression of mutant Huntingtin (mHTT) protein with polyglutamine repeats, which forms aggregates that lead to striatal neurodegeneration and progressive motor dysfunction [37]. Loss of CB1 receptors in basal nuclei was reported in HD mouse models [38,39] and in the brains of HD patients [40]. These findings suggest that CB1 function is impaired in HD, and therefore, restoring CB1 signaling could have a beneficial effect in the treatment of HD. Indeed, Chairlone et al. found that deletion of CB1 receptors from glutamatergic corticostriatal neurons exacerbates striatal neuron cell death and motor dysfunction in a mouse model of HD [41]. Therefore, CB1 agonists may mitigate HD progression and motor symptoms.

2.2. CB1 Physiology

CB1 is the main endocannabinoid system GPCR in the nervous system and is one of the most highly expressed GPCRs in the central nervous system (CNS). Neurons are the primary source of CB1 expression in the CNS, where a high density of CB1 is found in axons, especially at presynaptic terminals [42]. In presynaptic terminals, endocannabinoids act as retrograde neuromodulators, that is, synaptic transmission triggers endocannabinoid synthesis at the postsynaptic terminal, which activate presynaptic CB1 receptors that, in turn, inhibit neurotransmitter release [43]. Since CB1 is found in both GABAergic and glutamatergic synapses, endocannabinoids induce short-term synaptic plasticity via depolarization-induced suppression of inhibition (DSI in GABAergic terminals) or depolarization-induced suppression of excitation (DSE in glutamatergic terminals) [44]. However, CB1 does not act only in presynaptic terminals but also regulates somatodendritic excitability, such as in low-threshold spiking cortical interneurons, where 2-AG promotes slow self-inhibition [45]. A putative role for CB1 in neuronal mitochondria has been proposed, where it could contribute to suppression of neurotransmitter release by negatively regulating mitochondrial respiration and adenosine triphosphate (ATP) generation [46]. To a lower extent, CB1 is expressed in astrocytes, where it regulates gliotransmitter release, glucose metabolism, and the release of inflammatory mediators [47–50]. CB1 is not found at the protein level in resting microglia but has been detected in activated microglial cells in primary cultures from mollusk, mouse, and rat but not human tissue [51]. Additionally, CB1 is found in neurons of the dorsal root ganglia (DRG), in peripheral nerve terminals, and in neurons of the enteric nervous system [52]. At low levels, CB1 is also expressed in some peripheral tissues, such as adipose tissue, testis, prostate, adrenal glands, thymus, bone marrow, and heart [42].

2.3. Toxicity and Adverse Effects

Cannabinoids are generally well tolerated; however, acute and chronic toxicity is known to occur after consumption of *Cannabis* or, more frequently, synthetic cannabinoids. In the CNS, cannabinoids can induce cognitive and psychomotor impairment. In more severe cases, and especially with synthetic cannabinoids, agitation and acute psychosis may occur [53]. Regulation of neurotransmitter release by CB1 receptors is likely responsible for these effects. Overactivation of peripheral CB1 can also contribute to the development and progression of cardiovascular and metabolic diseases. Notably, the endocannabinoid system can affect cardiovascular function in a complex manner. CB1 activation reduces cardiac contractility likely via sympathetic inhibition and reduced Na^+ and Ca^{2+} influx in myocytes [54]. Further, CB1 activation causes hypotension in healthy individuals, but CB1 antagonism reduced blood pressure in obese and diabetic patients with hypertension [54]. Similarly, CB1 activation may exacerbate myocardial injury in the context of cardiac pathology [55]. CB1 can also contribute to diet-induced obesity. In addition to regulating feeding behavior in the CNS [56], peripheral CB1 enhances lipogenesis [57–59], inhibits lipolysis [59–61], and promotes leptin resistance [62–64]. Peripheral CB1 also promotes the development of nonalcoholic fatty liver disease [65–67], pancreatic β-cell death, and peripheral insulin resistance [68–70]. Therefore, therapies aiming at CB1 agonism may not be suitable for patients that already suffer from cardiovascular and metabolic diseases. In these contexts, CB1 antagonism and inverse agonism could be viable therapeutic strategies. However, the CNS effects of CB1 blockade have proven to be hazardous, as exemplified by the Rimonabant clinical trials, a CB1 inverse agonist that promoted depression and suicide [71,72]. Peripherally restricted CB1 antagonists and inverse agonists are currently being pursued to avoid CNS effects [59,60,63,67,70].

3. CB1 Mechanism of Activation

X-ray crystal structures of inactive CB1 bound to an antagonist/inverse agonist [73,74], canonical active CB1 bound to a potent agonist [75], and cryoelectron microscopy (cryo-EM) structures of CB1 in complex with heterotrimeric G_i protein [76,77] have been determined. In the inactive structure studies, the ionic lock between Arg214$^{3.50}$ and Asp338$^{6.30}$ (Ballesteros–Weinstein nomenclature [4] in superscript) was present in CB1, and the antagonist/inverse agonist compounds were deduced to enter the binding pocket via a gap between TMH1 and TMH7 [73]. In the active state structure study, novel washout resistant agonists were generated to enable crystallography. These were of similar potency and efficacy to CP55940 in a cAMP inhibition assay [75], but non-$G\alpha_{i/o}$ signaling, receptor internalization, or β-arrestin recruitment were not evaluated. The ligand-binding pocket was formed mainly by hydrophobic interactions with residues on extracellular loop 2 (ECL2), TMH3, TMH5, TMH6, and TMH7 [74,75], apart from a hydrogen bond formed between the phenolic hydroxyl of the agonist AM11542 and Ser383$^{7.39}$. Importantly, a previous study showed that mutating Ser383$^{7.39}$ to Ala resulted in severely reduced binding of several CB1 ligands [78], further supporting the role of Ser383$^{7.39}$ in ligand-binding interactions. Comparing the structures of antagonist-bound and agonist-bound CB1 revealed important features that likely participate in the molecular mechanism of receptor activation. The most noticeable conformational change in the transmembrane helices is the outward movement of the intracellular (IC) domain of TMH6. In the CWXP motif, a "twin toggle switch" mechanism is formed between Trp356$^{6.48}$ and Phe200$^{3.36}$. In the inactive state, the side chains of Phe200$^{3.36}$ and Trp356$^{6.48}$ point away and toward the ligand-binding pocket, respectively, forming an aromatic stacking interaction that maintains the inactive state. Upon agonist binding, the rotation of TMH3 causes the Phe200$^{3.36}$ side chain to flip, facing the binding pocket and disrupting the interaction with Trp356$^{6.48}$. Now released, Trp356$^{6.48}$ rotates inward, which results in the relaxation of the kink at Pro358$^{6.50}$, causing TMH6 to straighten, moving its IC end away from the receptor core [75,76]. This "twin toggle switch" mechanism was previously demonstrated using mutagenesis and molecular dynamics (MD) simulations [79] and was confirmed by the crystal structure. Conformational changes important to receptor activation also occur in the DRY motif, where Arg214$^{3.50}$ adopts an extended conformation, leading to disruption of the hydrogen bonding network with Asp213$^{3.49}$ and Asp338$^{6.30}$ (ionic lock). With the ionic lock broken, TMH6 moves outward, exposing sites for interaction with the G-protein. In the NPXXY motif, TMH7 unwinds around Tyr397$^{7.53}$. Further interactions formed by amino acids in this motif are not shown. Although the crystal structure elucidated many important features of agonist binding and molecular mechanisms of activation, there is an issue with the receptor used in the study. Four amino acid mutations (T210A, E273K, T283V, R340E) were introduced to improve expression and thermostability, thus allowing crystallography to be performed. This could have an impact on the overall structure of the activated receptor. In fact, the T210A mutation reduced cAMP inhibition in response to three different agonists, and Hua et al. state that the modified receptor construct cannot induce signaling [75,80].

Nonetheless, a cryo-EM structure of human CB1 bound to the highly potent agonist MDMB-FUBINACA and in complex with G_i showed that the two structures highly match, with a broad overlap in the ligand-binding site and "twin toggle switch" mechanism [76]. Differences were found in a more extended outward movement of the IC end of TMH6 and rotation of Arg214$^{3.50}$ toward the α5-helix of the $G\alpha_i$ protein. Further, they found a weaker interaction between the intracellular loop 2 (ICL2) of CB1 and the Ras domain of the $G\alpha_i$, which could explain the G-protein coupling promiscuity observed with CB1 [76]. These studies provided much valuable information on the mechanism behind G-protein signaling at CB1. However, in the crystal structures, the agonist's ability to promote β-arrestin recruitment is unknown. Further, the receptor was truncated at the C-terminus for crystallization, which precludes β-arrestin binding. In the cryo-EM structure, the receptor is stabilized by forming a complex with G_i, so this conformation is most likely to resemble the one responsible for G-protein-biased signaling. Therefore, the active CB1 structures available to date do not provide clues to a molecular mechanism behind β-arrestin-biased signaling.

Arrestins bind to GPCRs in two locations, the phosphorylated C-terminus and the cytoplasmic end of the activated GPCR transmembrane core [81]. One of these β-arrestin-GPCR core interactions occurs with Arg214[3.50] in the highly conserved DRY motif [82]. A mutational study focused on the role of the DRY motif in CB1 G-protein signaling and β-arrestin recruitment [83]. They found that mutating both Arg214[3.50] and Tyr215[3.51] to Ala (DAA) yielded a CB1 receptor with a G-protein-biased signaling profile. Although G-protein signaling was partially reduced, β-arrestin recruitment was eliminated. In contrast, mutating Asp213[3.49] and Arg214[3.50] to Ala (AAY) yielded a CB1 receptor with a β-arrestin-biased signaling profile. While G-protein signaling was reduced, β-arrestin recruitment was enhanced. Both mutated receptors also have increased constitutive activity. These mutations impacted both G-protein and β-arrestin signaling, consistent with the roles of the DRY motif in the interaction between the α5-helix of the G-protein [76] and the finger loop of β-arrestin [82]. However, it is possible that increases in β-arrestin recruitment are due to impaired G-protein coupling and reduced competition for GPCR binding. Nonetheless, the intramolecular interactions that promote β-arrestin-biased signaling at the CB1 receptor remain elusive.

Biophysical studies of other GPCRs show that while G-protein-biased ligands induce movement of TMH6, β-arrestin-biased ligands favor movement of TMH7 [84–86]. Unfortunately, these studies cannot identify precise structural modifications or intramolecular interactions. Crystal structures of the 5-HT2B receptor bound to the β-arrestin-biased ligand ergotamine [87] and visual arrestin bound Rhodopsin [88] showed a reduced movement of TMH6, compared to the canonical G-protein active state and structural modifications on TMH7 and Hx8. Although the structure of β-arrestin bound CB1 has not been solved yet, the studies from other class A GPCRs indicate that a similar molecular mechanism involving TMH7/Hx8 regulates β-arrestin-biased signaling in CB1.

4. CB1 Signaling

4.1. G-Proteins

Canonical GPCR signaling depends on the coupling to heterotrimeric G-proteins, composed of α, β, and γ subunits. These are classified according to the type of Gα subunit, which will activate or inhibit specific second messengers, leading to different downstream signaling events. $G\alpha_s$ proteins stimulate the activity of adenylyl cyclase (AC), enhancing cAMP levels, while $G\alpha_{i/o}$ proteins inhibit AC, suppressing cAMP production (Figure 1). This second messenger binds to and activates protein kinase A (PKA), the exchange protein directly activated by cAMP (Epac) and cyclic nucleotide-gated ion channels, stimulating intracellular signaling cascades that regulate a variety of essential cellular functions, such as metabolism, gene expression, cell growth and differentiation, apoptosis and neurotransmission [89]. On the other hand, $G\alpha_{q/11}$ proteins stimulate the activity of phospholipase C (PLC) β, an enzyme that catalyzes the hydrolysis of phosphatidylinositol 4,5-biphosphate (PIP_2), releasing diacylglycerol (DAG) and inositol 1,4,5-triphosphate (IP_3). The latter binds to IP_3 receptors in the endoplasmic reticulum (ER), releasing Ca^{2+} from the ER to the cytoplasm. The increase in cytosolic calcium levels leads to activation of various signaling cascades, including activation of protein kinase C (PKC) [90]. $G\alpha_{12/13}$ proteins recruit RhoGEFs to the membrane, leading to the activation of RhoA, which, in turn, activates Rho-associated protein kinase (ROCK). ROCK catalyzes the phosphorylation of focal adhesion kinase (FAK), stimulating the formation of actin stress fibers. ROCK also inhibits myosin light chain phosphatase, promoting cell contractility, and activates serum response factors [91]. These signaling pathways are also known to modulate the activation of extracellular signal-regulated kinase homologs 1 and 2 (ERK1/2).

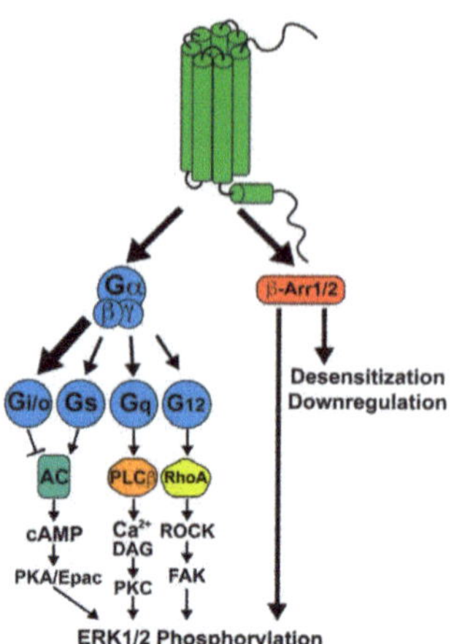

Figure 1. Signaling activated by the CB1 receptor. The CB1 receptor primarily couples to $G_{i/o}$ proteins (large arrow), but also to G_s, $G_{q/11}$, and $G_{12/13}$ (smaller arrows) to a lower extent. CB1 also recruits both β-arrestin1 and β-arrestin2. These proteins mediate receptor desensitization, endocytosis, and pERK1/2 signaling. The latter can also be induced by downstream signaling events stemming from the G-protein pathways. AC: adenylyl cyclase; β-Arr1/2: β-arrestin1/2; cAMP: cyclic adenosine monophosphate; DAG: diacylglycerol; Epac: exchange protein directly activated by cAMP; ERK1/2: extracellular-signal-regulated kinase $\frac{1}{2}$; FAK: focal adhesion kinase; PKA: protein kinase A; PKC: protein kinase C; ROCK: Rho-associated protein kinase.

Presynaptic membrane CB1 receptors induce DSI and DSE through suppression of neurotransmitter release via G-protein activity. CB1 mainly couples to $G_{i/o}$ proteins (Figure 1) [5], leading to reduced cAMP levels via $G\alpha_{i/o}$ and inhibition of voltage-gated Ca^{2+} channels via $G\beta\gamma$, both of which suppress neurotransmitter release [44]. Although $G_{i/o}$ proteins account for most of CB1-stimulated G-protein activity, low efficacy coupling to $G\alpha_s$ (Figure 1), about 10% of total $G\alpha_{i/o}$ coupling, has been described in N18TG2 neuroblastoma cells in response to CP55,940 [92]. As a result, in conditions where $G\alpha_{i/o}$ proteins are suppressed, such as under Pertussis toxin (PTx) treatment, CB1 agonists stimulate cAMP accumulation [93–95]. Under physiological conditions, however, since $G\alpha_{i/o}$ coupling is much more significant compared to $G\alpha_s$, the net effect of CB1 agonists is to suppress AC activity and cAMP production. Coupling to $G\alpha_{q/11}$ has been reported in human embryonic kidney (HEK293) cells transfected with CB1 receptor (Figure 1), but WIN55212-2 was the only agonist capable of eliciting $G\alpha_{q/11}$ mediated Ca^{2+} signaling, suggesting ligand specificity for this response [96]. CB1 $G\alpha_{12/13}$ coupling has been suggested (Figure 1) due to AEA-induced B103 neuroblastoma cell rounding, which was found to be dependent on ROCK and independent of $G\alpha_{i/o}$ [97]. An evaluation of $[^{35}S]GTP\gamma S$ binding in N18TG2 cells demonstrated that $G\alpha_{12/13}$ activity accounts for about 7 to 10% of G-protein activity in unstimulated and CP55940 stimulated cells, respectively [98]. Further, WIN55212-2 was found to induce growth cone retraction in primary hippocampal neurons, and this effect was disrupted by suppression of $G\alpha_{12}$ and $G\alpha_{13}$ expression using small interfering ribonucleic acid (siRNA), which suggests that cannabinoid receptors induce $G\alpha_{12/13}$ to regulate neurite growth [99]. Studies supporting $G\alpha_{12/13}$ signaling by CB1 remain limited, and this pathway, therefore, still requires further characterization.

4.2. β-Arrestins

Not unlike other class A GPCRs, CB1 is capable of recruiting β-arrestins. Ligand-induced interaction with both β-arrestin1 and β-arrestin2 has been demonstrated [100,101]. These are known to induce receptor desensitization and internalization. Therefore, chronic exposure to cannabinoids leads to tolerance and downregulation of CB1 receptor activity in the brain [102–105], which could underlie *Cannabis* dependence. In addition, β-arrestin recruitment can promote ERK1/2 phosphorylation (pERK1/2) via the scaffolding of mitogen-activated protein kinases [106]. Although pERK1/2 is induced by either heterotrimeric G-proteins or β-arrestins (Figure 1), these responses differ in magnitude, kinetics, and likely physiological function. The G-protein-mediated pERK response was found to be strong, fast, and transient, while the β-arrestin-mediated pERK response is of lower magnitude, slow, and longer lasting [107]. Further, the subcellular location of pERK differs depending on the originating signal. G-protein mediated pERK1/2 is largely translocated to the nucleus, where it promotes gene transcription and cell proliferation. Conversely, β-arrestin-induced pERK1/2 concentrates on endosomes, inhibiting gene transcription and phosphorylating cytoplasmic substrates that regulate protein translation, cytoskeleton dynamics, apoptosis, cell migration, and cross talk with other signaling cascades [107–110].

Interestingly, β-arrestin deletion studies have shown that β-arrestin recruitment and signaling can have different effects on cannabinoid-induced behaviors. When administered systemically, CB1 receptor agonists produce four typical behaviors that are used in a battery of tests for preclinical models to assess cannabinoid response, known as the cannabinoid tetrad—analgesia, hypothermia, catalepsy, and hypolocomotion [111]. Cannabinoid tetrad tests were used to investigate cannabinoid responsiveness in mice lacking either β-arrestin1 or β-arrestin2. Mice with deletion of β-arrestin1 showed reduced analgesia and hypothermia in response to CP55940 under acute treatment but not in response to THC. This occurred despite the fact that β-arrestin1 knockout (KO) enhanced [^{35}S]GTPγS binding induced by CP55,940 in cortex membranes, indicating a loss of G-protein desensitization [105]. This finding suggests that receptor desensitization, pERK1/2 signaling, or both β-arrestin1 functions together contribute to antinociception in mice. In contrast, antinociception or hypothermia induced by acute CP55940 treatment was not influenced by deletion of β-arrestin2, while THC-mediated antinociception and hypothermia were increased in β-arrestin2 KO mice [112]. Interestingly, a follow-up study found that despite increasing cannabinoid radioligand binding and availability in whole-brain P2 subcellular fraction—crude synaptosomes—β-arrestin2 KO actually decreased basal and agonist-stimulated [^{35}S]GTPγS binding in hippocampus and cortex, while [^{35}S]GTPγS binding in the cerebellum was unchanged [104]. This finding suggests that the increased antinociception and hypothermic effects of THC in β-arrestin2 KO mice are not due to increased G-protein signaling but may reflect a role of β-arrestin1 in mediating these cannabinoid-induced behaviors and a negative regulatory role for β-arrestin2 on the effects of β-arrestin1 signaling by cannabinoid receptors. Unfortunately, the authors did not evaluate G-protein signaling in the hypothalamus, midbrain, and spinal cord, where CNS regions involved in these responses are found. Taken together, these studies show that under acute treatment, β-arrestin1, and β-arrestin2 can have diverging effects on cannabinoid-induced antinociception and hypothermia. Catalepsy and hypolocomotion were not investigated in these studies; therefore, the role of β-arrestin1 and β-arrestin2 in these behaviors remains unknown. In addition, cannabinoid ligands, under certain conditions, may preferentially recruit β-arrestin1 or β-arrestin2, with CP55940 favoring β-arrestin1 and THC favoring β-arrestin2 [105,112].

Another study investigated the role of β-arrestin2 deletion on chronic cannabinoid exposure and tolerance development and found that β-arrestin2 downregulates CB1 receptor activity in a brain region-specific manner [103]. In accordance with Breivogel et al. [112], this study found that β-arrestin2 KO increased antinociception and hypothermia in response to acute THC treatment; however, no difference was found in cannabinoid-induced G-protein activity in CNS regions associated with antinociception, i.e., PAG and spinal cord, or hypothermia—the preoptic area of the hypothalamus. In contrast, the catalepsy response to acute THC was not affected by the β-arrestin2 deletion. After repeated THC administration, on the other hand, wild-type (WT) and β-arrestin2 KO mice developed different degrees of tolerance to THC antinociception, hypothermia, and catalepsy [103]. Although both genotypes develop a similar level of tolerance to hypothermia, tolerance to antinociception was attenuated in β-arrestin2 KO mice. Correspondingly, agonist-stimulated [^{35}S]GTPγS binding in the PAG and spinal cord was reduced by chronic THC treatment in WT but not in β-arrestin2 KO mice, while no changes were found in the preoptic area of the hypothalamus for either genotype. These findings indicate that β-arrestin2 regulates desensitization of CB1 induced G-protein activity in PAG, spinal cord, and preoptic area of the hypothalamus, and that desensitization by β-arrestin2 is the underlying mechanism behind the development of tolerance to cannabinoid antinociception and hypothermia. Interestingly, the development of tolerance to THC catalepsy was enhanced in β-arrestin2 KO mice, and agonist-stimulated [^{35}S]GTPγS binding was reduced in basal nuclei—globus pallidus and substantia nigra—after chronic THC treatment in β-arrestin2 KO but not in WT mice [103]. Since basal nuclei have been implicated in cannabinoid-induced catalepsy [113,114], these findings indicate that G-protein desensitization in CB1 receptors located in the basal nuclei confers tolerance to catalepsy, but the mechanism for tolerance development, in this case, is not due to CB1 interaction with β-arrestin2 but may instead be due to β-arrestin1.

It has been suggested that, for the CB1 receptor, β-arrestin1 and β-arrestin2 have different roles in signaling and endocytosis, with β-arrestin1 responsible for pERK1/2 signaling and β-arrestin2 responsible for receptor internalization [115]. Since β-arrestin recruitment is preceded by G-Protein-coupled receptor kinase (GRK)-mediated phosphorylation of Ser/Thr residues on the C-terminus, studies investigated the impact of mutations on the C-terminal putative GRK3 phosphorylation sites Ser426 and Ser430 on CB1 receptor desensitization, internalization, and β-arrestin-mediated signaling. The S426A/S430A CB1 receptor shows attenuated desensitization and receptor internalization [116,117]. Further, when compared to WT, S426A/S430A elicits a more prolonged pERK1/2 response, which is independent of receptor internalization but also insensitive to inhibition of Gα$_{i/o}$ and Gα$_s$ with PTx and cholera toxin, respectively [117]. Delgado-Peraza et al. [118] showed that suppressing β-arrestin1 translation eliminated 2-AG and WIN55212-2 induced pERK1/2 signaling by S426A/S430A CB1, while suppressing β-arrestin2 translation had no effect on early pERK1/2 and only partially reduced sustained pERK1/2. They also showed that at 20 min after treatment with WIN55212-2, S426A/S430A highly colocalizes with β-arrestin1, while WT CB1 does not. Further, by performing coimmunoprecipitation, they found that S426A/S430A CB1 shows greatly enhanced association with β-arrestin1 after 5 min WIN55212-2 treatment. In contrast, association with β-arrestin2 was present in WT CB1 but greatly reduced in S426A/S430A. This finding indicates that GRK3 phosphorylation at Ser426 and Ser430 (Ser425 and Ser429 in human CB1) switches the receptor's preference from recruitment of β-arrestin1 to the recruitment of β-arrestin2. Indeed, suppressing GRK3 translation, which likely inhibits CB1 internalization, promoted sustained pERK1/2 signaling at the WT CB1 receptor [118]. The observation that S426A/S430A highly recruits β-arrestin1 instead of β-arrestin2 and shows enhanced pERK1/2 suggests that β-arrestin1 mostly mediates arrestin-dependent pERK1/2 signaling. In contrast, the finding that S426A/S430A is resistant to internalization and shows reduced β-arrestin2 recruitment indicates that β-arrestin2 mostly mediates receptor internalization. Nevertheless, the fact that CB1 downregulation, as measured by radioligand binding, still occurs in the brains of β-arrestin2 KO mice, albeit in a brain-region-specific manner [103], shows that β-arrestin1 is capable of internalization under certain conditions. Further, the fact that β-arrestin2 siRNA knockdown partially reduced pERK1/2 from S426A/S430A CB1, although only at later time points [118], shows that β-arrestin2 is capable of inducing pERK1/2 signaling to a lower extent. In conclusion, although CB1 recruits both β-arrestin1 and β-arrestin2, there are brain region- and ligand-specific differences in the roles of each of these proteins regarding CB1 internalization and signaling that may translate to different roles on cannabinoid-induced effects.

5. CB1-Biased Signaling

5.1. Orthosteric Ligands

As mentioned above, CB1 ligands show potential therapeutic effects in numerous neurological disorders. However, the development of CB1 targeted pharmacotherapeutics remains hindered by concerns about adverse effects, rapid tolerance, and abuse potential. Since CB1 can activate both heterotrimeric G-proteins and β-arrestins, novel drug discovery efforts have focused on exploring biased signaling to mitigate some of these issues while maintaining therapeutic effects, as has been reported for several other GPCR systems [119].

CB1 ligands are capable of functional selectivity, but clear biased-signaling profiles have been challenging to characterize reliably across different studies. Laprairie et al. [101] compared β-arrestin1 signaling from CB1 orthosteric agonists in a mouse striatal derived cell line (STHdh). In this study, the rank order of potency for β-arrestin1 recruitment was as follows: THC > CP55940 > WIN55212-2 » 2-AG » AEA. Efficacy for β-arrestin1 recruitment was similar among THC, CP55940, and 2-AG but lower with AEA. WIN55212-2 efficacy was lower than that of THC, CP55940, and 2-AG but did not reach statistical significance. Furthermore, pERK1/2 signaling was sensitive to PTx treatment in an early time point for AEA, 2-AG, CP55940, and WIN55212-2 but not for THC. These data indicate that THC has a more β-arrestin1 biased signaling profile than the other ligands tested, while AEA shows more sensitivity to G-protein inhibition. Since the pERK1/2 response is used to estimate G-protein signaling, affirmations on G-protein-biased signaling should be confirmed by analysis of G-protein activation, with cAMP inhibition, or [^{35}S]GTPγS binding, for instance. In another study, Laprairie et al. [120] investigated CB1-biased signaling in STHdh cells expressing WT or mHTT, and calculated bias factors using the operational model [121] with WIN55212-2 as the reference ligand. When comparing the $G\alpha_{i/o}$-dependent pERK1/2 response and β-arrestin1 recruitment, they found that THC and CP55940 show β-arrestin1 biased signaling, while the endocannabinoids 2-AG and AEA show $G\alpha_{i/o}$ biased signaling. Since pERK was used to assess $G\alpha_{i/o}$ signaling, it is necessary to exercise caution when analyzing these data, as pERK1/2 is a response that can be elicited by other G-proteins as well as by β-arrestins, which may be a confounding factor. In addition, β-arrestin2 recruitment was not investigated since STHdh cells do not express β-arrestin2. Nonetheless, CP55940 and THC were detrimental to cell viability, while 2-AG, AEA, and WIN55212-2 improved viability in cells expressing mHTT. This finding suggests that CB1 G-protein signaling is neuroprotective in HD.

In a different study, Khajehali et al. [122] investigated cAMP inhibition and pERK1/2 in Chinese hamster ovary (CHO) cells stably expressing human CB1 receptor and calculated the bias factor using the operational model with 2-AG as the reference ligand. In this case, WIN55212-2 showed a similar signaling profile to 2-AG, while CP55940, THC, and AEA showed a tendency toward cAMP inhibition bias, although that difference was not statistically significant. HU-210 and methanandamide, on the other hand, showed a significant bias toward cAMP inhibition. Since this study did not assess PTx sensitivity or β-arrestin recruitment, it is difficult to ascertain the origin of the pERK1/2 response and whether it could be used to estimate relative levels of β-arrestin bias.

More recently, Zhu et al. [123] evaluated cAMP inhibition, pERK1/2 response and receptor internalization in HEK293 cells stably expressing human CB1 receptor and calculated ligand bias factors using a kinetic model with 2-AG as the reference ligand. In this study, WIN55212-2 also showed a similar signaling profile to 2-AG. On the other hand, THC showed a strong bias toward pERK1/2 and receptor internalization over cAMP inhibition, while CP55940 and AEA showed bias toward receptor internalization but only moderate bias toward pERK1/2. These findings suggest that CP55940, AEA, and THC show a β-arrestin-biased signaling profile; however, this should be confirmed by β-arrestin recruitment assays. As previously mentioned, pERK1/2 can be stimulated by multiple transducers, and although receptor internalization is generally a good proxy for β-arrestin recruitment, it is possible that β-arrestin1 and β-arrestin2 exert different functions, in which case, ligand-specific preference for either β-arrestin could be a confounding factor.

In contrast, Ibsen et al. [100] investigated CB1 mediated β-arrestin1 and β-arrestin2 translocation to the plasma membrane in HEK293 cells with different results. In this study, only 2-AG and WIN produced an amount of β-arrestin1 translocation that was different from control. In β-arrestin2 translocation; however, the rank order of potency was CP55940 > WIN55212-2 > AEA > 2-AG, while the rank order of efficacy was 2-AG > WIN55212-2 > CP55940 > AEA. In this case, THC did not significantly stimulate β-arrestin2 translocation. All in all, the studies that have sought to compare ligand bias among orthosteric CB1 agonists have failed to reliably identify biased ligands, with conflicting results under different experimental conditions, even in the same cellular background. This could indicate that all of these ligands are relatively balanced when it comes to shifting the conformational dynamics to a state that favors G-protein coupling or a state that favors β-arrestin recruitment, and that strongly biased CB1 orthosteric ligands have not yet been described. Understanding the molecular mechanism behind biased signaling will be of paramount importance for the design of novel CB1 ligands with a better biased-signaling profile.

5.2. Allosteric Ligands

Orthosteric agonists, antagonists, and inverse agonists bind to the primary, orthosteric, binding pocket and compete for binding with endogenous ligands. On the other hand, allosteric ligands bind to an allosteric site, which is topologically distinct from the orthosteric binding pocket and do not compete for binding with orthosteric/endogenous ligands [124]. A ligand can be a negative allosteric modulator (NAM), inhibiting signaling from an orthosteric agonist, or a positive allosteric modulator (PAM), enhancing signaling from an orthosteric agonist. Neither NAMs nor PAMs produce signaling in the absence of an orthosteric agonist. However, some ligands are allosteric agonists, promoting signaling in the absence of an orthosteric ligand, and some compounds can have both PAM and allosteric agonist effects (ago-PAM), inducing signaling when administered alone, as well as potentiating signaling from an orthosteric agonist. Pharmacologists are increasingly seeking allosteric ligands as a strategy to develop improved small molecule therapeutics to target GPCRs. These drugs may produce fewer side effects, given that NAMs and PAMs can alter signaling from endogenous agonists in a time-specific and site-specific manner. Another advantage is that amino acid residues in allosteric sites are less conserved across different GPCRs, which would contribute to target specificity. Further, some allosteric modulators can confer biased signaling properties to otherwise balanced agonists [124]. Endogenous and exogenous allosteric modulators have been described for the CB1 receptor (Figure 2). Some of these allosteric ligands display a biased signaling profile. Since orthosteric agonists promote balanced levels of G-protein signaling and β-arrestin recruitment (Figure 3A), the mechanism of action of biased allosteric ligands may shed light on the conformational changes that are required for CB1 mediated β-arrestin-biased signaling and make allosteric binding pockets better candidates for the development of novel CB1 biased ligands.

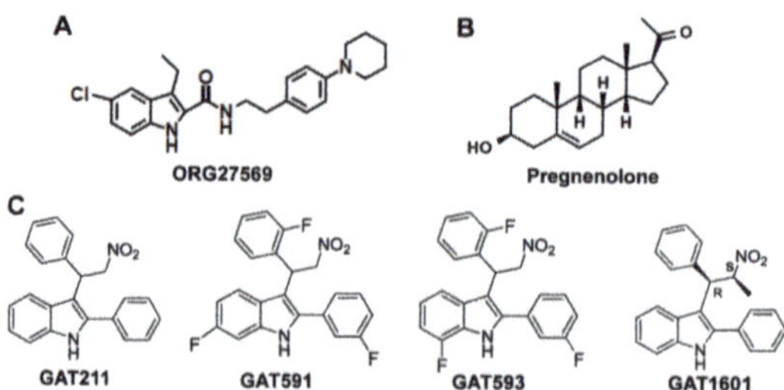

Figure 2. Biased allosteric ligands of CB1: (**A**) molecular structure of ORG27569; (**B**) molecular structure of pregnenolone; (**C**) molecular structures of GAT211 and its analogs, two fluorinated analogs—GAT591 and GAT593—and one methylated analog—GAT1601.

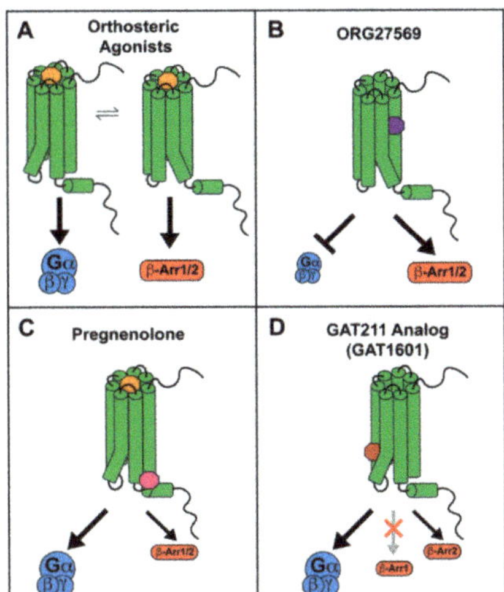

Figure 3. Biased signaling by CB1 allosteric ligands: (**A**) orthosteric ligands, represented in yellow, promote G-protein signaling and β-arrestin recruitment in a balanced manner, stimulating conformational changes important for both; (**B**) ORG27569, in purple, inhibits G-protein signaling and stimulates β-arrestin signaling, whether alone or in the presence of an orthosteric ligand, generating β-arrestin-biased signaling. This occurs due to inhibition of TMH6 movement and stimulation of TMH7/Hx8 movement, respectively; (**C**) pregnenolone, in magenta, inhibits β-arrestin signaling in the presence of an orthosteric ligand, generating G-protein-biased signaling. This occurs due to inhibition of TMH7/Hx8 movement while allowing TMH6 movement; (**D**) GAT1601, in dark red, stimulates G-protein signaling and β-arrestin2 recruitment to a smaller extent but not β-arrestin1 recruitment, generating G-protein-biased signaling. This occurs due to the facilitation of TMH6 movement.

5.2.1. ORG27569 as a Biased Allosteric Modulator of CB1

The first allosteric modulator described for the CB1 receptor was ORG27569 (ORG), a 1*H*-indole-2-carboxamide analog (Figure 2A) that was first described as a NAM for CB1. ORG was found to enhance binding and slow the dissociation rate of CP55940 but inhibit G-protein activation [125–127]. ORG also antagonized inhibition of cAMP by CP55940, WIN55212-2 and AEA [126,128]. In the absence of an orthosteric agonist, ORG inhibited CB1 constitutive activity [126,127]. These findings indicate that ORG promotes desensitization of G-protein signaling at the CB1 receptor. In accordance, ORG inhibited DSE in primary hippocampal neurons [129], indicating that it can also negatively regulate 2-AG mediated G-protein signaling.

Remarkably, ORG enhanced CP55940-mediated pERK1/2 signaling, and also stimulated this signaling pathway when administered alone [115,126,127]. This effect was abolished by suppressing β-arrestin1 translation but not by suppression of β-arrestin2 [115]. Further, β-arrestin1 colocalized with CB1 receptor under fluorescence microscopy after treatment with ORG [115]. Agonist-induced receptor internalization and β-arrestin2 recruitment, however, were inhibited by ORG [126,128,130]. In contrast, Ahn et al. [115,127] reported increased receptor internalization via a β-arrestin2-dependent mechanism when ORG was administered alone. This disparity could be attributed to the fact that Ahn et al. used a mutant CB1 receptor (T210A) for their internalization assays to enhance the presence of CB1 on the plasma membrane at baseline. The enhanced presence of CB1 on the membrane or differences in the receptor structure caused by the mutation could impact the way ORG influences CB1 signaling, switching from β-arrestin1 to β-arrestin2 recruitment. Likewise, differences in GRK expression levels among different in vitro experimental systems could affect ORG-induced β-arrestin recruitment, where preference for β-arrestin1 results in ORG-stimulated pERK1/2 signaling and reduced internalization, but no discrimination between β-arrestin isoforms leads to stimulation of both pERK1/2 and internalization. A similar mechanism could also explain divergent results from different labs, where ORG enhanced [115,126,127] or inhibited [122,130] agonist-stimulated pERK1/2. If ORG preferentially recruits β-arrestin1, differences in GRK isoform expression or β-arrestin1 expression levels may occlude β-arrestin1 pERK1/2 signaling and produce an antagonistic effect on G-protein mediated pERK1/2 signaling. All in all, the body of evidence suggests that ORG is a β-arrestin biased ligand at the CB1 receptor, functioning as an allosteric inverse agonist and NAM for G-protein signaling and an allosteric agonist and PAM for β-arrestin1-mediated pERK1/2 signaling (Figure 3B).

Studies on the impact of ORG on receptor conformation have begun to shed light on the molecular mechanism for β-arrestin recruitment and signaling at the CB1 receptor. Using site-directed fluorescence labeling, Fay and Farrens [86] showed that ORG enhances conformational changes at TMH7/Hx8, in the absence of TMH6 movement. Although this is an important finding, biophysical methods such as these are unable to define conformational changes in such a way as to determine a molecular mechanism. Recently, a crystal structure of CB1 bound to ORG and CP55940 was reported [131]. In this structure, the outward movement of TMH6 that would normally be induced by an orthosteric agonist such as CP55,940, was greatly inhibited [131], which explains the NAM effect of ORG on CP55,940 mediated G-protein signaling [125]. Further, a slight outward movement of the intracellular domain of TMH7 is seen in ORG- and CP55940-bound CB1, when compared to MDMB-FUBINACA-bound CB1 [76,131]. However, the presence of five thermostabilizing mutations (T210A, E273K, T283V, R340E, and S203K) within the transmembrane helices and a truncated C-terminus that were applied to aid in crystallization [131] may affect overall receptor structure and preclude observation of further conformational changes and intramolecular interactions that may be induced by ORG on the TMH7/Hx8 elbow. Further, ORG was shown to also induce β-arrestin signaling in the absence of CP55940 [115,127]. Since the CB1 crystal structure was obtained with both ORG and CP55940 [131], how ORG may affect CB1 conformational dynamics in the absence of an orthosteric agonist is still poorly understood. Using MD simulations, Lynch et al. [132] indicated that after ORG binds to CB1, it promotes an outward movement of the IC domain of TMH7. This does not open the site for interaction with the G-protein at the TMH3/5/6 region but opens a site for interaction with β-arrestin at the TMH7/1/2 region. However, the findings have yet to be confirmed experimentally. These studies attribute to the TMH7/Hx8 region the role of promoting the alternative active state of CB1 that promotes β-arrestin-biased signaling.

When administered in vivo, ORG had no effect on CP55940-mediated antinociception or catalepsy [133,134]. Interestingly, ORG had no effect on CP55940-mediated hypothermia in C57BL/6J mice but attenuated this response in Sprague Dawley rats [133,134]. Further, ORG reduced AEA-induced hypothermia [134], an effect that is not blocked by the selective CB1 inverse agonist SR141716A [135,136], indicating that ORG antagonizes AEA hypothermia via a non-CB1 mechanism. Interestingly, ORG administered alone decreases body weight and food intake in mice [133] and rats [134]. However, this effect was also observed in mice with genetic deletion of CB1, indicating that the effect is not mediated by CB1. These findings suggest that ORG has at least one non-CB1 target in vivo, and this could be an additional possible explanation for diverging results with ORG in different studies.

5.2.2. Pregnenolone as a Biased Allosteric Modulator of CB1

An endogenous NAM for CB1 has also been described. Pregnenolone, 3α-hydroxy-5β-pregnan-20-one (Figure 2B), is a steroid hormone that was found to be a signaling specific NAM for CB1 [137]. Exposure to THC upregulates pregnenolone synthesis via a pERK1/2-induced increase in the levels of cytochrome P450scc [137]. Pregnenolone then antagonizes the effects of THC on synaptic transmission and on the cannabinoid tetrad, forming a negative feedback loop. The CB1 signaling profile of THC in the presence of pregnenolone was evaluated in vitro, showing that pregnenolone effectively antagonizes THC-mediated pERK1/2 signaling and suppression of cellular and mitochondrial respiration, without influencing cAMP inhibition. Using the Force-Biased Metropolis Monte Carlo (MMC) simulated annealing program, Vallée et al. [137] showed that the pregnenolone binding site on CB1 lies on the cytoplasmic end, where pregnenolone forms hydrogen bonds with Glu133$^{1.49}$ and Arg409$^{7.65}$. This was confirmed by mutational analysis, as in an E133G CB1 mutant, pregnenolone has no effect on THC-mediated suppression of cellular respiration [137]. Binding at this site would tether TMH7 near TMH1, restricting conformational changes on TMH7 that are believed to be important for the β-arrestin-biased signaling state [86,132]. These findings are consistent with a role for pregnenolone as a biased NAM for CB1 β-arrestin signaling (Figure 3C). However, changes to β-arrestin recruitment in the presence of pregnenolone should be directly measured to confirm this effect.

In rodents, pregnenolone prevented THC-induced increases in food intake and memory impairment. Further, neuronal firing in the ventral tegmental area and dopamine release in the nucleus accumbens induced by THC were reduced by pregnenolone. Accordingly, pregnenolone also reduced WIN55212-2 self-administration [137]. These findings indicate that CB1-biased signaling mediated by pregnenolone may be useful for the treatment of *Cannabis* intoxication and to reduce *Cannabis* abuse potential. Importantly, pregnenolone was also shown to block a cannabinoid-induced psychotic-like state in mice [138]. As medicinal and recreational *Cannabis* become more popular, pregnenolone or potential novel analogs could become an important tool in the clinic. However, its efficacy for inhibiting signaling from endogenous cannabinoids has not yet been investigated and could be a source of adverse effects.

5.2.3. GAT211 as a Positive Allosteric Modulator of CB1

GAT211 is a compound derived from 2-phenylindole (Figure 2C) that has been described as an allosteric ligand for the CB1 receptor [139]. GAT211 increased binding and slowed the dissociation rate of CP55940 from CB1 and reduced the binding of SR141716A. In functional assays, GAT211 enhanced the effect of CP55940, 2-AG, and AEA on both G-protein signaling and β-arrestin1 recruitment to similar degrees. When compared to β-arrestin2 recruitment, on the other hand, GAT211 significantly favored cAMP inhibition in CHO cells [140], suggesting a G-protein-biased signaling profile when using CP55940 as a reference ligand. In the absence of an orthosteric agonist, GAT211 is also capable of eliciting G-protein signaling and β-arrestin1 recruitment, demonstrating an ago-PAM effect at CB1. GAT211 is a racemic mixture of GAT228 (R-(+) enantiomer) and GAT229 (S-(-) enantiomer). Interestingly, GAT229 is responsible for the PAM effect on agonist-mediated signaling and shows no effect when administered alone, while GAT228 stimulated signal transduction on its own, showing an allosteric agonist profile [139]. In hippocampal neurons in vitro, GAT228 inhibited excitatory postsynaptic currents (EPSCs) on its own, further demonstrating its allosteric agonist effect [141]. While GAT229 had no effect on EPSCs alone, it enhanced DSE, supporting its role as a PAM for endocannabinoid signaling [141]. This enantiospecific effect is possible because GAT228 and GAT229 likely bind to two different allosteric sites on the CB1 receptor. Using Force-Biased MMC Simulated Annealing, Hurst et al. [142] found that GAT228 binds at an IC exosite, forming interactions with residues on TMH1, TMH2, TMH4, and ICL1, while GAT229 binds at an EC site, forming interactions with residues on TMH2, TMH3, and ECL1. These findings support the existence of separate allosteric agonist and PAM binding sites for GAT211 enantiomers.

The therapeutic potential of GAT211 and its enantiomers has been shown in several preclinical models. GAT211 and enantiomers enhanced cell viability in a striatal cell line expressing mHTT, and improved motor coordination and prevented motor impairment in the R6/2 mouse model of HD [143]. In a preclinical model of glaucoma, GAT229 reduced intraocular pressure [144]. GAT211 and enantiomers also reduced seizures in a preclinical model of childhood epilepsy [145]. Further, GAT211 induced antinociception in preclinical models of inflammatory and neuropathic pain without affecting motor coordination or body temperature, and without inducing tolerance, conditioned place preference, or antagonist precipitated withdrawal symptoms [146]. Additionally, in a preclinical model of neuropathic pain, GAT211 enhanced morphine analgesia and prevented opioid tolerance development [147]. These studies suggest that a CB1 ago-PAM, such as GAT211, could have therapeutic effects in the contexts of HD, epilepsy, and pathological pain. Importantly, GAT211 could be useful as an opioid-sparing treatment, which is especially relevant in the face of the current opioid epidemic. GAT211, as a CB1 ago-PAM, may also have therapeutic potential in disorders associated with an impaired endocannabinoid system, such as during aging and neurodegeneration, as it would be able to counteract reduced CB1 expression and boost endogenous cannabinoid signaling [24–26,148].

However, GAT211 is a probe compound not intended to be developed for the clinic, due to its low affinity for CB1 and rapid metabolic clearance. To address these issues, fluorinated analogs of GAT211 were developed—GAT591 and GAT593 (Figure 2C). These showed significantly enhanced potency and greater metabolic stability as measured by a microsomal stability assay [140]. The analogs did not improve upon the moderate G-protein-biased signaling profile of GAT211 but maintained a similar slight preference for cAMP inhibition over β-arrestin2 recruitment, compared to CP55940 [140]. When administered in vivo, the fluorinated analogs suppressed mechanical allodynia in a preclinical model of inflammatory pain with a much longer duration of action than previously reported for GAT211 [140], likely due to enhanced metabolic stability. Remarkably, the fluorinated analogs also produced antinociception in naïve mice, without affecting catalepsy or hypothermia [140], which are frequently observed with orthosteric agonists. This may be attributed to the preference for G-protein signaling over β-arrestin2 recruitment since β-arrestin2 KO studies suggest that cannabinoid antinociception is hindered, while catalepsy is enhanced by the presence of β-arrestin2 [103], as previously discussed. Interestingly, methylated GAT211 analogs modified biased signaling in a diastereomer-specific manner. One diastereomer, GAT1601 (Figure 2C), was an effective ago-PAM for cAMP inhibition and β-arrestin2 recruitment, but it did not enhance β-arrestin1 recruitment [149]. In a preclinical model of glaucoma, this compound was more effective and had a longer-lasting effect than the more balanced diastereomer or the parent compound GAT211 [149]. These results suggest that this "anti-β-arrestin1" signaling bias (Figure 3D) may also present a therapeutic advantage in some pathological conditions.

6. Conclusions

CB1 is a GPCR that signals primarily via $G_{i/o}$ proteins. However, signaling promiscuity is reported throughout the literature. In addition to $G_{i/o}$, CB1 has been shown to couple to G_s [92], $G_{q/11}$ [96], and $G_{12/13}$ [98]. Although the fraction of non-$G_{i/o}$ protein activation is reportedly small, it is possible that activation of different G-protein subtypes is associated with different CB1 functions. For instance, while $G_{i/o}$ is responsible for the suppression of synaptic neurotransmitter release [44], $G_{12/13}$ activation may be responsible for CB1 mediated regulation of neurite growth [99]. Consequently, this poorly studied aspect of CB1 signaling could have a fundamental impact on the role of CB1 during brain development. How non-$G_{i/o}$ signaling affects cannabinoid-induced physiological effects must be studied further to ascertain whether shifting G-protein subtype preference could pose a therapeutic advantage when targeting CB1. Regardless of which G-protein is coupled by CB1, the mechanism of activation culminates in an outward movement of the intracellular domain of TMH6, while the G-protein subtype flexibility is likely due to weak interactions of the receptor ICL2 with the G_α on the intracellular surface [76]. Mutations on the ICL2, therefore, may increase or decrease CB1 mediated signaling via non-$G_{i/o}$ proteins, which could elucidate the role of these signaling pathways on cannabinoid function. This activation mechanism is now well understood for CB1 orthosteric agonists and it shares similarities with other class A GPCRs [150–152].

On the other hand, the mechanism and function of β-arrestin recruitment by CB1 is less well understood. At the functional level, β-arrestins regulate CB1 desensitization and downregulation in a brain-region-specific manner, which can result in differential tolerance development to cannabinoid effects [103]. An important distinction seems to exist between the functions of β-arrestin1 and β-arrestin2, where cannabinoid antinociception may be enhanced by β-arrestin1 [105] but hindered by β-arrestin2 [103,112]. Further, switching between β-arrestin1 and β-arrestin2 via the absence or presence of phosphorylation by GRK3 affects early and sustained pERK1/2 responses [118], which could have different roles on the downstream effects of CB1 activity. However, little is known about the consequences of favoring CB1 recruitment of β-arrestin1 or β-arrestin2. Most known agonists stimulate both β-arrestins and specificity may stem from a cell-specific context. One recently discovered exception is GAT1601, which enhances β-arrestin2 recruitment but not β-arrestin1 [149]. Interestingly, this compound showed stronger therapeutic potential than more balanced compounds in a preclinical model of glaucoma, suggesting that dissociating β-arrestin1 from β-arrestin2 recruitment could be beneficial when targeting CB1 in this context. More studies are required to better understand the differences between the functions of both β-arrestins downstream of CB1 activation under different cellular backgrounds so this potential can be exploited in CB1 targeted therapeutics.

The biased-signaling properties of CB1 orthosteric agonists have been investigated in different in vitro systems. However, common CB1 ligands show little preference for either G-protein or β-arrestin signaling, with conflicting results across different studies [100, 101,120,122,123]. This suggests that classical cannabinoids are fairly unbiased and that differential signaling depends largely on the cellular background. Allosteric CB1 ligands, on the other hand, have been successfully used to selectively trigger or inhibit specific signal transducers [115,126,127,137,140,149]. This suggests that allosteric-binding sites hold better promise for the development of strongly biased CB1 ligands. In fact, their proposed mechanism of action is to stimulate or inhibit conformational changes on the TMH7/Hx8 elbow of the receptor to stimulate or inhibit β-arrestin signaling [86,132,137], which is consistent with the putative molecular mechanism of β-arrestin-biased signaling [84,86–88,153]. As an exception, GAT211 and analogs do not affect the conformation of TMH7 but instead facilitate the movement of TMH6 [142], leading to G-protein-biased signaling. In fact, when administered alone, in agonist mode, GAT211 analogs showed little to no stimulation of β-arrestin recruitment [149]. As allosteric agonists, these compounds bind to a TMH1-2-4 exosite, stimulating the movement of TMH3 toward TMH4 and stretching the ionic lock until it is broken, which facilitates the outward movement of TMH6 [142]. Therefore, exploration of this allosteric agonist site can potentially produce novel strongly G-protein-biased CB1 agonists.

In conclusion, CB1 is highly expressed throughout the CNS in excitatory and inhibitory neurons as well as astrocytes, giving it the potential to impact a myriad of CNS physiological functions and disease states. However, this broad expression also limits its utility due to adverse effects and abuse potential. Biased signaling has been suggested as a strategy to dissociate therapeutic effects from the undesired effects of CB1 activity. However, the functional consequences of CB1-biased signaling are still poorly understood due to the lack of signaling specificity of know orthosteric agonists. The development of biased allosteric ligands may be a viable strategy to dissociate the activation of G-proteins, β-arrestin1, or β-arrestin2 and refine CB1 targeted therapeutics.

Author Contributions: Literature review, L.M.L.; writing—original draft preparation, L.M.L.; writing—review and editing, M.E.A.; funding acquisition, M.E.A. Both authors have read and agreed to the published version of the manuscript.

Funding: Funded by National Institutes of Health/National Institute on Drug Abuse (NIH/NIDA) grants: R01 DA045698, T32 DA007237, and P30 DA013429.

Institutional Review Board Statement: Not applicable.

Informed Consent Statement: Not applicable.

Data Availability Statement: Not applicable.

Conflicts of Interest: The authors declare no conflict of interest. The funders had no role in the design of the study; in the collection, analyses, or interpretation of data; in the writing of the manuscript, or in the decision to publish the results.

Sample Availability: Not applicable.

References

1. Devane, W.A.; Dysarz, F.A.; Johnson, M.R.; Melvin, L.S.; Howlett, A.C. Determination and Characterization of a Canna-Binoid Receptor in Rat Brain. *Mol. Pharmacol.* **1988**, *34*, 605–613.
2. Matsuda, L.A.; Lolait, S.J.; Brownstein, M.J.; Young, A.C.; Bonner, T.I. Structure of a cannabinoid receptor and functional expression of the cloned cDNA. *Nat. Cell Biol.* **1990**, *346*, 561–564. [CrossRef]
3. Gérard, C.M.; Mollereau, C.; Vassart, G.; Parmentier, M. Molecular cloning of a human cannabinoid receptor which is also expressed in testis. *Biochem. J.* **1991**, *279*, 129–134. [CrossRef]
4. Ballesteros, J.A.; Weinstein, H. Integrated methods for the construction of three-dimensional models and computational probing of structure-function relations in G protein-coupled receptors. In *Methods in Neurosciences*; Sealfon, S.C., Ed.; Academic Press: Cambridge, MA, USA, 1995; Volume 25, pp. 366–428.
5. Howlett, A.C. Pharmacology of Cannabinoid Receptors. *Annu. Rev. Pharmacol. Toxicol.* **1995**, *35*, 607–634. [CrossRef] [PubMed]
6. Xie, X.-Q.; Melvin, L.S.; Makriyannis, A. The Conformational Properties of the Highly Selective Cannabinoid Receptor Ligand CP-55,940. *J. Biol. Chem.* **1996**, *271*, 10640–10647. [CrossRef] [PubMed]
7. Eissenstat, M.A.; Bell, M.R.; D'Ambra, T.E.; Alexander, E.J.; Daum, S.J.; Ackerman, J.H.; Gruett, M.D.; Kumar, V.; Estep, K.G. Aminoalkylindoles: Structure-Activity Relationships of Novel Cannabinoid Mimetics. *J. Med. Chem.* **1995**, *38*, 3094–3105. [CrossRef] [PubMed]
8. Devane, W.A.; Hanus, L.; Breuer, A.; Pertwee, R.G.; Stevenson, L.A.; Griffin, G.; Gibson, D.; Mandelbaum, A.; Etinger, A.; Mechoulam, R. Isolation and structure of a brain constituent that binds to the cannabinoid receptor. *Science* **1992**, *258*, 1946–1949. [CrossRef] [PubMed]
9. Sugiura, T.; Kondo, S.; Sukagawa, A.; Nakane, S.; Shinoda, A.; Itoh, K.; Yamashita, A.; Waku, K. 2-Arachidonoylgylcerol: A Possible Endogenous Cannabinoid Receptor Ligand in Brain. *Biochem. Biophys. Res. Commun.* **1995**, *215*, 89–97. [CrossRef]
10. Mechoulam, R.; Ben-Shabat, S.; Hanus, L.; Ligumsky, M.; Kaminski, N.E.; Schatz, A.R.; Gopher, A.; Almog, S.; Martin, B.R.; Compton, D.R.; et al. Identification of an endogenous 2-monoglyceride, present in canine gut, that binds to cannabinoid receptors. *Biochem. Pharmacol.* **1995**, *50*, 83–90. [CrossRef]
11. Di Marzo, V. New approaches and challenges to targeting the endocannabinoid system. *Nat. Rev. Drug Discov.* **2018**, *17*, 623–639. [CrossRef]
12. Smith, J.; Lefkowitz, R.J.; Rajagopal, S. Biased signalling: From simple switches to allosteric microprocessors. *Nat. Rev. Drug Discov.* **2018**, *17*, 243–260. [CrossRef] [PubMed]
13. Nurmikko, T.J.; Serpell, M.G.; Hoggart, B.; Toomey, P.J.; Morlion, B.J.; Haines, D. Sativex successfully treats neuropathic pain characterised by allodynia: A randomised, double-blind, placebo-controlled clinical trial. *Pain* **2007**, *133*, 210–220. [CrossRef] [PubMed]
14. Woodhams, S.G.; Chapman, V.; Finn, D.P.; Hohmann, A.G.; Neugebauer, V. The cannabinoid system and pain. *Neuropharmacology* **2017**, *124*, 105–120. [CrossRef]
15. Hohmann, A.; Herkenham, M. Localization of central cannabinoid CB1 receptor messenger RNA in neuronal subpopulations of rat dorsal root ganglia: A double-label in situ hybridization study. *Neuroscience* **1999**, *90*, 923–931. [CrossRef]
16. Ahluwalia, J.; Urban, L.; Capogna, M.; Bevan, S.J.; Nagy, I. Cannabinoid 1 receptors are expressed in nociceptive primary sensory neurons. *Neuroscience* **2000**, *100*, 685–688. [CrossRef]
17. Rahn, E.J.; Makriyannis, A.; Hohmann, A.G. Activation of cannabinoid CB1 and CB2 receptors suppresses neuropathic nociception evoked by the chemotherapeutic agent vincristine in rats. *Br. J. Pharmacol.* **2007**, *152*, 765–777. [CrossRef] [PubMed]
18. Pernía-Andrade, A.J.; Kato, A.; Witschi, R.; Nyilas, R.; Katona, I.; Freund, T.F.; Watanabe, M.; Filitz, J.; Koppert, W.; Schüttler, J.; et al. Spinal Endocannabinoids and CB1 Receptors Mediate C-Fiber-Induced Heterosynaptic Pain Sensitization. *Science* **2009**, *325*, 760–764. [CrossRef]
19. Yang, F.; Xu, Q.; Shu, B.; Tiwari, V.; He, S.-Q.; Vera-Portocarrero, L.P.; Dong, X.; Linderoth, B.; Raja, S.N.; Wang, Y.; et al. Activation of cannabinoid CB1 receptor contributes to suppression of spinal nociceptive transmission and inhibition of mechanical hypersensitivity by Aβ-fiber stimulation. *Pain* **2016**, *157*, 2582–2593. [CrossRef]
20. Lichtman, A.H.; Cook, S.A.; Martin, B.R. Investigation of brain sites mediating cannabinoid-induced antinociception in rats: Evidence supporting periaqueductal gray involvement. *J. Pharmacol. Exp. Ther.* **1996**, *276*, 585–593.
21. Finn, D.; Jhaveri, M.; Beckett, S.; Roe, C.; Kendall, D.; Marsden, C.; Chapman, V. Effects of direct periaqueductal grey administration of a cannabinoid receptor agonist on nociceptive and aversive responses in rats. *Neuropharmacology* **2003**, *45*, 594–604. [CrossRef]
22. Hohmann, A.G.; Suplita, R.L.; Bolton, N.M.; Neely, M.H.; Fegley, D.; Mangieri, R.; Krey, J.F.; Walker, J.M.; Holmes, P.V.; Crystal, J.D.; et al. An endocannabinoid mechanism for stress-induced analgesia. *Nat. Cell Biol.* **2005**, *435*, 1108–1112. [CrossRef] [PubMed]

23. Broyd, S.J.; Van Hell, H.H.; Beale, C.; Yücel, M.; Solowij, N. Acute and Chronic Effects of Cannabinoids on Human Cognition—A Systematic Review. *Biol. Psychiatry* **2016**, *79*, 557–567. [CrossRef]
24. Bilkei-Gorzo, A.; Albayram, O.; Draffehn, A.; Michel, K.; Piyanova, A.; Oppenheimer, H.; Dvir-Ginzberg, M.; Rácz, I.; Ulas, T.; Imbeault, S.; et al. A chronic low dose of Δ9-tetrahydrocannabinol (THC) restores cognitive function in old mice. *Nat. Med.* **2017**, *23*, 782–787. [CrossRef]
25. Bilkei-Gorzo, A. The endocannabinoid system in normal and pathological brain ageing. *Philos. Trans. R. Soc. B Biol. Sci.* **2012**, *367*, 3326–3341. [CrossRef]
26. Bilkei-Gorzo, A.; Racz, I.; Valverde, O.; Otto, M.; Michel, K.; Sarstre, M.; Zimmer, A. Early age-related cognitive impairment in mice lacking cannabinoid CB1 receptors. *Proc. Natl. Acad. Sci. USA* **2005**, *102*, 15670–15675. [CrossRef]
27. Haller, J.; Varga, B.; Ledent, C.; Freund, T.F. CB1 cannabinoid receptors mediate anxiolytic effects: Convergent genetic and pharmacological evidence with CB1-specific agents. *Behav. Pharmacol.* **2004**, *15*, 299–304. [CrossRef]
28. Naderi, N.; Haghparast, A.; Saber-Tehrani, A.; Rezaii, N.; Alizadeh, A.-M.; Khani, A.; Motamedi, F. Interaction between cannabinoid compounds and diazepam on anxiety-like behaviour of mice. *Pharmacol. Biochem. Behav.* **2008**, *89*, 64–75. [CrossRef] [PubMed]
29. Hill, M.N.; Carrier, E.J.; McLaughlin, R.; Morrish, A.C.; Meier, S.E.; Hillard, C.J.; Gorzalka, B.B. Regional alterations in the endocannabinoid system in an animal model of depression: Effects of concurrent antidepressant treatment. *J. Neurochem.* **2008**, *106*, 2322–2336. [CrossRef] [PubMed]
30. Shen, C.-J.; Zheng, D.; Li, K.-X.; Yang, J.-M.; Pan, H.-Q.; Yu, X.-D.; Fu, J.-Y.; Zhu, Y.; Sun, Q.-X.; Tang, M.-Y.; et al. Cannabinoid CB1 receptors in the amygdalar cholecystokinin glutamatergic afferents to nucleus accumbens modulate depressive-like behavior. *Nat. Med.* **2019**, *25*, 337–349. [CrossRef]
31. Sbarski, B.; Akirav, I. Cannabinoids as therapeutics for PTSD. *Pharmacol. Ther.* **2020**, *211*, 107551. [CrossRef]
32. Wallace, M.; Wiley, J.; Martin, B.R.; DeLorenzo, R.J. Assessment of the role of CB1 receptors in cannabinoid anticonvulsant effects. *Eur. J. Pharmacol.* **2001**, *428*, 51–57. [CrossRef]
33. Bahremand, A.; Nasrabady, S.E.; Shafaroodi, H.; Ghasemi, M.; Dehpour, A.R. Involvement of nitrergic system in the anticonvulsant effect of the cannabinoid CB1 agonist ACEA in the pentylenetetrazole-induced seizure in mice. *Epilepsy Res.* **2009**, *84*, 110–119. [CrossRef]
34. Parmentier-Batteur, S.; Jin, K.; Mao, X.O.; Xie, L.; Greenberg, D.A. Increased Severity of Stroke in CB1 Cannabinoid Receptor Knock-Out Mice. *J. Neurosci.* **2002**, *22*, 9771–9775. [CrossRef]
35. Hayakawa, K.; Mishima, K.; Nozako, M.; Hazekawa, M.; Ogata, A.; Fujioka, M.; Harada, K.; Mishima, S.; Orito, K.; Egashira, N.; et al. Δ9-tetrahydrocannabinol (Δ9-THC) prevents cerebral infarction via hypothalamic-independent hypothermia. *Life Sci.* **2007**, *80*, 1466–1471. [CrossRef] [PubMed]
36. Ma, L.; Jia, J.; Niu, W.; Jiang, T.; Zhai, Q.; Yang, L.; Bai, F.; Wang, Q.; Xiong, L. Mitochondrial CB1 receptor is involved in ACEA-induced protective effects on neurons and mitochondrial functions. *Sci. Rep.* **2015**, *5*, 12440. [CrossRef] [PubMed]
37. Ross, C.A.; Tabrizi, S. Huntington's disease: From molecular pathogenesis to clinical treatment. *Lancet Neurol.* **2011**, *10*, 83–98. [CrossRef]
38. Denovan-Wright, E.M.; Robertson, H.A. Cannabinoid receptor messenger RNA levels decrease in a subset of neurons of the lateral striatum, cortex and hippocampus of transgenic Huntington's disease mice. *Neuroscience* **2000**, *98*, 705–713. [CrossRef]
39. Lastres-Becker, I.; Berrendero, F.; Lucas, J.J.; Martín-Aparicio, E.; Yamamoto, A.; Ramos, J.; Fernández-Ruiz, J.J. Loss of mRNA levels, binding and activation of GTP-binding proteins for cannabinoid CB1 receptors in the basal ganglia of a transgenic model of Huntington's disease. *Brain Res.* **2002**, *929*, 236–242. [CrossRef]
40. Glass, M.; Faull, R.; Dragunow, M. Loss of cannabinoid receptors in the substantia nigra in huntington's disease. *Neuroscience* **1993**, *56*, 523–527. [CrossRef]
41. Chiarlone, A.; Bellocchio, L.; Blázquez, C.; Resel, E.; Soria-Gomez, E.; Cannich, A.; Ferrero, J.J.; Sagredo, O.; Benito, C.; Romero, J.; et al. A restricted population of CB1 cannabinoid receptors with neuroprotective activity. *Proc. Natl. Acad. Sci. USA* **2014**, *111*, 8257–8262. [CrossRef]
42. Howlett, A. International Union of Pharmacology. XXVII. Classification of Cannabinoid Receptors. *Pharmacol. Rev.* **2002**, *54*, 161–202. [CrossRef]
43. Wilson, R.I.; Nicoll, R.A. Endogenous cannabinoids mediate retrograde signalling at hippocampal synapses. *Nat. Cell Biol.* **2001**, *410*, 588–592. [CrossRef] [PubMed]
44. Araque, A.; Castillo, P.E.; Manzoni, O.J.; Tonini, R. Synaptic functions of endocannabinoid signaling in health and disease. *Neuropharmacology* **2017**, *124*, 13–24. [CrossRef] [PubMed]
45. Marinelli, S.; Pacioni, S.; Bisogno, T.; Di Marzo, V.; Prince, D.A.; Huguenard, J.R.; Bacci, A. The Endocannabinoid 2-Arachidonoylglycerol Is Responsible for the Slow Self-Inhibition in Neocortical Interneurons. *J. Neurosci.* **2008**, *28*, 13532–13541. [CrossRef]
46. Bénard, G.; Massa, F.; Puente, N.; Lourenço, J.; Bellocchio, L.; Soria-Gomez, E.; Matias, I.; Delamarre, A.; Metna-Laurent, M.; Cannich, A.; et al. Mitochondrial CB1 receptors regulate neuronal energy metabolism. *Nat. Neurosci.* **2012**, *15*, 558–564. [CrossRef]
47. Han, J.; Kesner, P.; Metna-Laurent, M.; Duan, T.; Xu, L.; Georges, F.; Koehl, M.; Abrous, N.; Mendizabal-Zubiaga, J.; Grandes, P.; et al. Acute Cannabinoids Impair Working Memory through Astroglial CB1 Receptor Modulation of Hippocampal LTD. *Cell* **2012**, *148*, 1039–1050. [CrossRef]

48. Metna-Laurent, M.; Marsicano, G. Rising stars: Modulation of brain functions by astroglial type-1 cannabinoid receptors. *Glia* **2014**, *63*, 353–364. [CrossRef]

49. Robin, L.M.; da Cruz, J.F.O.; Langlais, V.C.; Martin-Fernandez, M.; Metna-Laurent, M.; Busquets-Garcia, A.; Bellocchio, L.; Soria-Gomez, E.; Papouin, T.; Varilh, M.; et al. Astroglial CB1 Receptors Determine Synaptic D-Serine Availability to Enable Recognition Memory. *Neuron* **2018**, *98*, 935–944.e5. [CrossRef]

50. Jimenez-Blasco, D.; Busquets-Garcia, A.; Hebert-Chatelain, E.; Serrat, R.; Vicente-Gutierrez, C.; Ioannidou, C.; Gómez-Sotres, P.; Lopez-Fabuel, I.; Resch-Beusher, M.; Resel, E.; et al. Glucose metabolism links astroglial mitochondria to cannabinoid effects. *Nat. Cell Biol.* **2020**, *583*, 603–608. [CrossRef]

51. Stella, N. Cannabinoid and cannabinoid-like receptors in microglia, astrocytes, and astrocytomas. *Glia* **2010**, *58*, 1017–1030. [CrossRef]

52. Mackie, K. Distribution of Cannabinoid Receptors in the Central and Peripheral Nervous System. In *Cannabinoids. Handbook of Experimental Pharmacology*; Pertwee, R.G., Ed.; Springer: Berlin/Heidelberg, Germany, 2005; Volume 168, pp. 299–325.

53. Kelly, B.F.; Nappe, T.M. Cannabinoid Toxicity. In *StatPearls*; StatPearls Publishing: Treasure Island, FL, USA, 2021.

54. O'Sullivan, S.E. Endocannabinoids and the Cardiovascular System in Health and Disease. In *Handbook of Experimental Pharmacology*; Springer: Berlin/Heidelberg, Germany, 2015; Volume 231, pp. 393–422.

55. Tang, X.; Liu, Z.; Li, X.; Wang, J.; Li, L. Cannabinoid Receptors in Myocardial Injury: A Brother Born to Rival. *Int. J. Mol. Sci.* **2021**, *22*, 6886. [CrossRef]

56. Koch, M.; Varela, L.; Kim, J.G.; Kim, J.D.; Hernández-Nuño, F.; Simonds, S.; Castorena, C.M.; Vianna, C.R.; Elmquist, J.K.; Morozov, Y.; et al. Hypothalamic POMC neurons promote cannabinoid-induced feeding. *Nat. Cell Biol.* **2015**, *519*, 45–50. [CrossRef]

57. Cota, D.; Marsicano, G.; Tschöp, M.; Grübler, Y.; Flachskamm, C.; Schubert, M.; Auer, D.; Yassouridis, A.; Thöne-Reineke, C.; Ortmann, S.; et al. The endogenous cannabinoid system affects energy balance via central orexigenic drive and peripheral lipogenesis. *J. Clin. Investig.* **2003**, *112*, 423–431. [CrossRef]

58. Osei-Hyiaman, D.; DePetrillo, M.; Pacher, P.; Liu, J.; Radaeva, S.; Bátkai, S.; Harvey-White, J.; Mackie, K.; Offertáler, L.; Wang, L.; et al. Endocannabinoid activation at hepatic CB1 receptors stimulates fatty acid synthesis and contributes to diet-induced obesity. *J. Clin. Investig.* **2005**, *115*, 1298–1305. [CrossRef] [PubMed]

59. Ma, H.; Zhang, G.; Mou, C.; Fu, X.; Chen, Y. Peripheral CB1 Receptor Neutral Antagonist, AM6545, Ameliorates Hypometabolic Obesity and Improves Adipokine Secretion in Monosodium Glutamate Induced Obese Mice. *Front. Pharmacol.* **2018**, *9*, 156. [CrossRef]

60. Paszkiewicz, R.L.; Bergman, R.N.; Santos, R.S.; Frank, A.P.; Woolcott, O.O.; Iyer, M.S.; Stefanovski, D.; Clegg, D.J.; Kabir, M. A Peripheral CB1R Antagonist Increases Lipolysis, Oxygen Consumption Rate, and Markers of Beiging in 3T3-L1 Adipocytes Similar to RIM, Suggesting that Central Effects Can Be Avoided. *Int. J. Mol. Sci.* **2020**, *21*, 6639. [CrossRef]

61. Müller, G.A.; Herling, A.W.; Wied, S.; Müller, T.D. CB1 Receptor-Dependent and Independent Induction of Lipolysis in Primary Rat Adipocytes by the Inverse Agonist Rimonabant (SR141716A). *Molecules* **2020**, *25*, 896. [CrossRef]

62. Tam, J.; Cinar, R.; Liu, J.; Godlewski, G.; Wesley, D.; Jourdan, T.; Szanda, G.; Mukhopadhyay, B.; Chedester, L.; Liow, J.-S.; et al. Peripheral Cannabinoid-1 Receptor Inverse Agonism Reduces Obesity by Reversing Leptin Resistance. *Cell Metab.* **2012**, *16*, 167–179. [CrossRef]

63. Tam, J.; Szanda, G.; Drori, A.; Liu, Z.; Cinar, R.; Kashiwaya, Y.; Reitman, M.L.; Kunos, G. Peripheral cannabinoid-1 receptor blockade restores hypothalamic leptin signaling. *Mol. Metab.* **2017**, *6*, 1113–1125. [CrossRef] [PubMed]

64. Drori, A.; Gammal, A.; Azar, S.; Hinden, L.; Hadar, R.; Wesley, D.; Nemirovski, A.; Szanda, G.; Salton, M.; Tirosh, B.; et al. CB1R regulates soluble leptin receptor levels via CHOP, contributing to hepatic leptin resistance. *eLife* **2020**, *9*, 60771. [CrossRef] [PubMed]

65. Osei-Hyiaman, U.; Liu, J.; Zhou, L.; Godlewski, G.; Harvey-White, J.; Jeong, W.-I.; Bátkai, S.; Marsicano, G.; Lutz, B.; Buettner, C.; et al. Hepatic CB1 receptor is required for development of diet-induced steatosis, dyslipidemia, and insulin and leptin resistance in mice. *J. Clin. Investig.* **2008**, *118*, 3160–3169. [CrossRef] [PubMed]

66. Azar, S.; Udi, S.; Drori, A.; Hadar, R.; Nemirovski, A.; Vemuri, K.V.; Miller, M.; Sherill-Rofe, D.; Arad, Y.; Gur-Wahnon, D.; et al. Reversal of diet-induced hepatic steatosis by peripheral CB1 receptor blockade in mice is p53/miRNA-22/SIRT1/PPARα dependent. *Mol. Metab.* **2020**, *42*, 101087. [CrossRef] [PubMed]

67. Khan, N.; Laudermilk, L.; Ware, J.; Rosa, T.; Mathews, K.; Gay, E.; Amato, G.; Maitra, R. Peripherally Selective CB1 Receptor Antagonist Improves Symptoms of Metabolic Syndrome in Mice. *ACS Pharmacol. Transl. Sci.* **2021**, *4*, 757–764. [CrossRef] [PubMed]

68. Kim, W.; Lao, Q.; Shin, Y.-K.; Carlson, O.D.; Lee, E.K.; Gorospe, M.; Kulkarni, R.N.; Egan, J.M. Cannabinoids Induce Pancreatic -Cell Death by Directly Inhibiting Insulin Receptor Activation. *Sci. Signal.* **2012**, *5*, ra23. [CrossRef]

69. Jourdan, T.; Nicoloro, S.M.; Zhou, Z.; Shen, Y.; Liu, J.; Coffey, N.J.; Cinar, R.; Godlewski, G.; Gao, B.; Aouadi, M.; et al. Decreasing CB 1 receptor signaling in Kupffer cells improves insulin sensitivity in obese mice. *Mol. Metab.* **2017**, *6*, 1517–1528. [CrossRef]

70. Eid, B.; Neamatallah, T.; Hanafy, A.; El-Bassossy, H.; Aldawsari, H.; Vemuri, K.; Makriyannis, A. Effects of the CB1 Receptor Antagonists AM6545 and AM4113 on Insulin Resistance in a High-Fructose High-Salt Rat Model of Metabolic Syndrome. *Medicina* **2020**, *56*, 573. [CrossRef]

71. Christensen, R.; Kristensen, P.K.; Bartels, E.M.; Bliddal, H.; Astrup, A. Efficacy and safety of the weight-loss drug rimonabant: A meta-analysis of randomised trials. *Lancet* **2007**, *370*, 1706–1713. [CrossRef]

72. Topol, E.J.; Bousser, M.-G.; Fox, K.; Creager, M.A.; Despres, J.-P.; Easton, J.D.; Hamm, C.W.; Montalescot, G.; Steg, P.G.; Pearson, T.A.; et al. Rimonabant for prevention of cardiovascular events (CRESCENDO): A randomised, multicentre, placebo-controlled trial. *Lancet* **2010**, *376*, 517–523. [CrossRef]

73. Shao, Z.; Yin, J.; Chapman, K.; Grzemska, M.; Clark, L.; Wang, J.; Rosenbaum, D.M. High-resolution crystal structure of the human CB1 cannabinoid receptor. *Nat. Cell Biol.* **2016**, *540*, 602–606. [CrossRef]

74. Hua, T.; Vemuri, K.; Pu, M.; Qu, L.; Han, G.W.; Wu, Y.; Zhao, S.; Shui, W.; Li, S.; Korde, A.; et al. Crystal Structure of the Human Cannabinoid Receptor CB1. *Cell* **2016**, *167*, 750–762. [CrossRef]

75. Hua, T.; Vemuri, K.; Nikas, S.P.; LaPrairie, R.B.; Wu, Y.; Qu, L.; Pu, M.; Korde, A.; Jiang, S.; Ho, J.-H.; et al. Crystal structures of agonist-bound human cannabinoid receptor CB1. *Nat. Cell Biol.* **2017**, *547*, 468–471. [CrossRef]

76. Kumar, K.K.; Shalev-Benami, M.; Robertson, M.J.; Hu, H.; Banister, S.; Hollingsworth, S.A.; Latorraca, N.R.; Kato, H.; Hilger, D.; Maeda, S.; et al. Structure of a Signaling Cannabinoid Receptor 1-G Protein Complex. *Cell* **2019**, *176*, 448–458. [CrossRef]

77. Hua, T.; Li, X.; Wu, L.; Iliopoulos-Tsoutsouvas, C.; Wang, Y.; Wu, M.; Shen, L.; Brust, C.A.; Nikas, S.P.; Song, F.; et al. Activation and Signaling Mechanism Revealed by Cannabinoid Receptor-Gi Complex Structures. *Cell* **2020**, *180*, 655–665. [CrossRef] [PubMed]

78. Kapur, A.; Hurst, D.P.; Fleischer, D.; Whitnell, R.; Thakur, G.A.; Makriyannis, A.; Reggio, P.H.; Abood, M.E. Mutation Studies of Ser7.39 and Ser2.60 in the Human CB1 Cannabinoid Receptor: Evidence for a Serine-Induced Bend in CB1 Transmem-brane Helix 7. *Mol. Pharm.* **2007**, *71*, 1512–1524. [CrossRef] [PubMed]

79. McAllister, S.D.; Hurst, D.P.; Barnett-Norris, J.; Lynch, D.; Reggio, P.H.; Abood, M. Structural Mimicry in Class A G Protein-coupled Receptor Rotamer Toggle Switches. *J. Biol. Chem.* **2004**, *279*, 48024–48037. [CrossRef] [PubMed]

80. D'Antona, A.M.; Ahn, K.H.; Kendall, D.A. Mutations of CB1 T210 Produce Active and Inactive Receptor Forms: Correlations with Ligand Affinity, Receptor Stability, and Cellular Localization. *Biochemistry* **2006**, *45*, 5606–5617. [CrossRef]

81. Cahill, T.J.; Thomsen, A.; Tarrasch, J.T.; Plouffe, B.; Nguyen, A.; Yang, F.; Huang, L.-Y.; Kahsai, A.W.; Bassoni, D.L.; Gavino, B.J.; et al. Distinct conformations of GPCR–β-arrestin complexes mediate desensitization, signaling, and endocytosis. *Proc. Natl. Acad. Sci. USA* **2017**, *114*, 2562–2567. [CrossRef]

82. Staus, D.P.; Hu, H.; Robertson, M.J.; Kleinhenz, A.L.W.; Wingler, L.M.; Capel, W.D.; Latorraca, N.R.; Lefkowitz, R.J.; Skiniotis, G. Structure of the M2 muscarinic receptor–β-arrestin complex in a lipid nanodisc. *Nat. Cell Biol.* **2020**, *579*, 297–302. [CrossRef]

83. Gyombolai, P.; Tóth, A.D.; Tímár, D.; Turu, G.; Hunyady, L. Mutations in the 'DRY' motif of the CB1 cannabinoid receptor result in biased receptor variants. *J. Mol. Endocrinol.* **2014**, *54*, 75–89. [CrossRef]

84. Liu, J.J.; Horst, R.; Katritch, V.; Stevens, R.C.; Wüthrich, K. Biased Signaling Pathways in 2-Adrenergic Receptor Characterized by 19F-NMR. *Science* **2012**, *335*, 1106–1110. [CrossRef]

85. Rahmeh, R.; Damian, M.; Cottet, M.; Orcel, H.; Mendre, C.; Durroux, T.; Sharma, K.S.; Durand, G.; Pucci, B.; Trinquet, E.; et al. Structural insights into biased G protein-coupled receptor signaling revealed by fluorescence spectroscopy. *Proc. Natl. Acad. Sci. USA* **2012**, *109*, 6733–6738. [CrossRef]

86. Fay, J.F.; Farrens, D.L. Structural dynamics and energetics underlying allosteric inactivation of the cannabinoid receptor CB1. *Proc. Natl. Acad. Sci. USA* **2015**, *112*, 8469–8474. [CrossRef]

87. Wacker, D.; Wang, C.; Katritch, V.; Han, G.W.; Huang, X.-P.; Vardy, E.; McCorvy, J.D.; Jiang, Y.; Chu, M.; Siu, F.Y.; et al. Structural Features for Functional Selectivity at Serotonin Receptors. *Science* **2013**, *340*, 615–619. [CrossRef] [PubMed]

88. Kang, Y.; Zhou, X.E.; Gao, X.; He, Y.; Liu, W.; Ishchenko, A.; Barty, A.; White, T.A.; Yefanov, O.; Han, G.W.; et al. Crystal structure of rhodopsin bound to arrestin by femtosecond X-ray laser. *Nat. Cell Biol.* **2015**, *523*, 561–567. [CrossRef] [PubMed]

89. Cheng, X.; Ji, Z.; Tsalkova, T.; Mei, F. Epac and PKA: A tale of two intracellular cAMP receptors. *Acta Biochim. Biophys. Sin.* **2008**, *40*, 651–662. [CrossRef] [PubMed]

90. Mizuno, N.; Itoh, H. Functions and Regulatory Mechanisms of Gq-Signaling Pathways. *Neurosignals* **2009**, *17*, 42–54. [CrossRef] [PubMed]

91. Siehler, S. Regulation of RhoGEF proteins by G12/13-coupled receptors. *Br. J. Pharmacol.* **2009**, *158*, 41–49. [CrossRef] [PubMed]

92. Eldeeb, K.; Leone-Kabler, S.; Howlett, A.C. CB1 cannabinoid receptor-mediated increases in cyclic AMP accumulation are correlated with reduced $G_{i/o}$ function. *J. Basic Clin. Physiol. Pharmacol.* **2016**, *27*, 311–322. [CrossRef]

93. Bonhaus, D.W.; Chang, L.K.; Kwan, J.; Martin, G.R. Dual activation and inhibition of adenylyl cyclase by cannabinoid receptor agonists: Evidence for agonist-specific trafficking of intracellular responses. *J. Pharmacol. Exp. Ther.* **1998**, *287*, 884–888.

94. Calandra, B.; Portier, M.; Kernéis, A.; Delpech, M.; Carillon, C.; Le Fur, G.; Ferrara, P.; Shire, D. Dual intracellular signaling pathways mediated by the human cannabinoid CB1 receptor. *Eur. J. Pharmacol.* **1999**, *374*, 445–455. [CrossRef]

95. Glass, M.; Felder, C.C. Concurrent Stimulation of Cannabinoid CB1 and Dopamine D2 Receptors Augments cAMP Accumulation in Striatal Neurons: Evidence for a GsLinkage to the CB1 Receptor. *J. Neurosci.* **1997**, *17*, 5327–5333. [CrossRef]

96. Lauckner, J.E.; Hille, B.; Mackie, K. The cannabinoid agonist WIN55,212-2 increases intracellular calcium via CB1 receptor coupling to Gq/11 G proteins. *Proc. Natl. Acad. Sci. USA* **2005**, *102*, 19144–19149. [CrossRef]

97. Ishii, I.; Chun, J. Anandamide-induced neuroblastoma cell rounding via the CB1 cannabinoid receptors. *NeuroReport* **2002**, *13*, 593–596. [CrossRef]

98. Eldeeb, K.; Leone-Kabler, S.; Howlett, A.C. Mouse Neuroblastoma CB1 Cannabinoid Receptor-Stimulated [^{35}S]GTPGS Binding: Total and Antibody-Targeted Gα Protein-Specific Scintillation Proximity Assays. In *Methods in Enzymology*; Reggio, P.H., Ed.; Academic Press: Cambridge, MA, USA, 2017; Volume 593, pp. 1–21.

99. Roland, A.; Ricobaraza, A.; Carrel, D.; Jordan, B.M.; Rico, F.; Simon, A.; Humbert-Claude, M.; Ferrier, J.; McFadden, M.H.; Scheuring, S.; et al. Cannabinoid-induced actomyosin contractility shapes neuronal morphology and growth. *eLife* **2014**, *3*, e03159. [CrossRef]

100. Ibsen, M.S.; Finlay, D.; Patel, M.; Javitch, J.A.; Glass, M.; Grimsey, N.L. Cannabinoid CB1 and CB2 Receptor-Mediated Arrestin Translocation: Species, Subtype, and Agonist-Dependence. *Front. Pharmacol.* **2019**, *10*, 350. [CrossRef] [PubMed]

101. Laprairie, R.B.; Bagher, A.M.; Kelly, M.E.; Dupre, D.J.; Denovan-Wright, E.M. Type 1 Cannabinoid Receptor Ligands Display Functional Selectivity in a Cell Culture Model of Striatal Medium Spiny Projection Neurons. *J. Biol. Chem.* **2014**, *289*, 24845–24862. [CrossRef]

102. Breivogel, C.; Childers, S.R.; Deadwyler, S.A.; Hampson, R.E.; Vogt, L.J.; Sim-Selley, L.J. Chronic delta9-Tetrahydrocannabinol Treatment Produces a Time-Dependent Loss of Cannabinoid Receptors and Cannabinoid Receptor-Activated G Proteins in Rat Brain. *J. Neurochem.* **2002**, *73*, 2447–2459. [CrossRef]

103. Nguyen, P.T.; Schmid, C.L.; Raehal, K.M.; Selley, D.E.; Bohn, L.M.; Sim-Selley, L.J. β-Arrestin2 Regulates Cannabinoid CB1 Receptor Signaling and Adaptation in a Central Nervous System Region–Dependent Manner. *Biol. Psychiatry* **2012**, *71*, 714–724. [CrossRef]

104. Breivogel, C.S.; Puri, V.; Lambert, J.M.; Hill, D.K.; Huffman, J.W.; Razdan, R.K. The influence of beta-arrestin2 on cannabinoid CB1receptor coupling to G-proteins and subcellular localization and relative levels of beta-arrestin1 and 2 in mouse brain. *J. Recept. Signal Transduct.* **2013**, *33*, 367–379. [CrossRef]

105. Breivogel, C.S.; Vaghela, M.S. The effects of beta-arrestin1 deletion on acute cannabinoid activity, brain cannabinoid receptors and tolerance to cannabinoids in mice. *J. Recept. Signal Transduct.* **2014**, *35*, 98–106. [CrossRef]

106. Lefkowitz, R.J. Transduction of Receptor Signals by -Arrestins. *Science* **2005**, *308*, 512–517. [CrossRef] [PubMed]

107. Ahn, S.; Shenoy, S.K.; Wei, H.; Lefkowitz, R.J. Differential Kinetic and Spatial Patterns of β-Arrestin and G Protein-mediated ERK Activation by the Angiotensin II Receptor. *J. Biol. Chem.* **2004**, *279*, 35518–35525. [CrossRef] [PubMed]

108. Luttrell, L.M.; Roudabush, F.L.; Choy, E.W.; Miller, W.; Field, M.E.; Pierce, K.L.; Lefkowitz, R.J. Activation and targeting of extracellular signal-regulated kinases by -arrestin scaffolds. *Proc. Natl. Acad. Sci. USA* **2001**, *98*, 2449–2454. [CrossRef]

109. Tohgo, A.; Pierce, K.L.; Choy, E.W.; Lefkowitz, R.J.; Luttrell, L. β-Arrestin Scaffolding of the ERK Cascade Enhances Cytosolic ERK Activity but Inhibits ERK-mediated Transcription following Angiotensin AT1a Receptor Stimulation. *J. Biol. Chem.* **2002**, *277*, 9429–9436. [CrossRef]

110. Caunt, C.J.; Finch, A.R.; Sedgley, K.R.; McArdle, C.A. Seven-transmembrane receptor signalling and ERK compartmentalization. *Trends Endocrinol. Metab.* **2006**, *17*, 276–283. [CrossRef]

111. Metna-Laurent, M.; Mondésir, M.; Grel, A.; Vallée, M.; Piazza, P. Cannabinoid-Induced Tetrad in Mice. *Curr. Protoc. Neurosci.* **2017**, *80*, 9–59. [CrossRef]

112. Breivogel, C.S.; Lambert, J.M.; Gerfin, S.; Huffman, J.W.; Razdan, R.K. Sensitivity to Δ9-tetrahydrocannabinol is selectively enhanced in beta-arrestin2−/− mice. *Behav. Pharmacol.* **2008**, *19*, 298–307. [CrossRef]

113. Pertwee, R.; Wickens, A. Enhancement by chlordiazepoxide of catalepsy induced in rats by intravenous or intrapallidal injections of enantiomeric cannabinoids. *Neuropharmacology* **1991**, *30*, 237–244. [CrossRef]

114. Wallmichrath, I.; Szabo, B. Cannabinoids inhibit striatonigral GABAergic neurotransmission in the mouse. *Neuroscience* **2002**, *113*, 671–682. [CrossRef]

115. Ahn, K.H.; Mahmoud, M.; Shim, J.-Y.; Kendall, D.A. Distinct Roles of β-Arrestin 1 and β-Arrestin 2 in ORG27569-induced Biased Signaling and Internalization of the Cannabinoid Receptor 1 (CB1). *J. Biol. Chem.* **2013**, *288*, 9790–9800. [CrossRef]

116. Jin, W.; Brown, S.; Roche, J.P.; Hsieh, C.; Celver, J.P.; Kovoor, A.; Chavkin, C.; Mackie, K. Distinct Domains of the CB1 Cannabinoid Receptor Mediate Desensitization and Internalization. *J. Neurosci.* **1999**, *19*, 3773–3780. [CrossRef]

117. Daigle, T.L.; Kearn, C.S.; Mackie, K. Rapid CB1 cannabinoid receptor desensitization defines the time course of ERK1/2 MAP kinase signaling. *Neuropharmacology* **2008**, *54*, 36–44. [CrossRef]

118. Delgado-Peraza, F.; Ahn, K.H.; Nogueras-Ortiz, C.; Mungrue, I.; Mackie, K.; Kendall, D.A.; Yudowski, G.A. Mechanisms of Biased β-Arrestin-Mediated Signaling Downstream from the Cannabinoid 1 Receptor. *Mol. Pharmacol.* **2016**, *89*, 618–629. [CrossRef] [PubMed]

119. Whalen, E.J.; Rajagopal, S.; Lefkowitz, R.J. Therapeutic potential of β-arrestin- and G protein-biased agonists. *Trends Mol. Med.* **2011**, *17*, 126–139. [CrossRef] [PubMed]

120. LaPrairie, R.B.; Bagher, A.M.; Kelly, M.E.M.; Denovan-Wright, E.M. Biased Type 1 Cannabinoid Receptor Signaling Influences Neuronal Viability in a Cell Culture Model of Huntington Disease. *Mol. Pharmacol.* **2015**, *89*, 364–375. [CrossRef] [PubMed]

121. Kenakin, T.; Watson, C.; Muniz-Medina, V.; Christopoulos, A.; Novick, S. A Simple Method for Quantifying Functional Selectivity and Agonist Bias. *ACS Chem. Neurosci.* **2012**, *3*, 193–203. [CrossRef]

122. Khajehali, E.; Malone, D.T.; Glass, M.; Sexton, P.; Christopoulos, A.; Leach, K. Biased Agonism and Biased Allosteric Modulation at the CB1 Cannabinoid Receptor. *Mol. Pharmacol.* **2015**, *88*, 368–379. [CrossRef]

123. Zhu, X.; Finlay, D.B.; Glass, M.; Duffull, S.B. Evaluation of the profiles of CB 1 cannabinoid receptor signalling bias using joint kinetic modelling. *Br. J. Pharmacol.* **2020**, *177*, 3449–3463. [CrossRef]

124. Wold, E.A.; Chen, J.; Cunningham, K.A.; Zhou, J. Allosteric Modulation of Class A GPCRs: Targets, Agents, and Emerging Concepts. *J. Med. Chem.* **2019**, *62*, 88–127. [CrossRef]

125. Price, M.R.; Baillie, G.L.; Thomas, A.; Stevenson, L.A.; Easson, M.; Goodwin, R.; McLean, A.; McIntosh, L.; Goodwin, G.; Walker, G.; et al. Allosteric Modulation of the Cannabinoid CB1 Receptor. *Mol. Pharmacol.* **2005**, *68*, 1484–1495. [CrossRef]

126. Baillie, G.L.; Horswill, J.G.; Anavi-Goffer, S.; Reggio, P.H.; Bolognini, D.; Abood, M.; McAllister, S.D.; Strange, P.G.; Stephens, G.J.; Pertwee, R.; et al. CB1 Receptor Allosteric Modulators Display Both Agonist and Signaling Pathway Specificity. *Mol. Pharmacol.* **2013**, *83*, 322–338. [CrossRef]

127. Ahn, K.H.; Mahmoud, M.M.; Kendall, D.A. Allosteric Modulator ORG27569 Induces CB1 Cannabinoid Receptor High Affinity Agonist Binding State, Receptor Internalization, and Gi Protein-Independent ERK1/2 Kinase Activation. *J. Bioloical Chem.* **2012**, *287*, 12070–12082. [CrossRef]

128. Cawston, E.E.; Redmond, W.J.; Breen, C.M.; Grimsey, N.; Connor, M.; Glass, M. Real-time characterization of cannabinoid receptor 1 (CB1) allosteric modulators reveals novel mechanism of action. *Br. J. Pharmacol.* **2013**, *170*, 893–907. [CrossRef] [PubMed]

129. Straiker, A.; Mitjavila, J.; Yin, D.; Gibson, A.; Mackie, K. Aiming for allosterism: Evaluation of allosteric modulators of CB1 in a neuronal model. *Pharmacol. Res.* **2015**, *99*, 370–376. [CrossRef] [PubMed]

130. Gamage, T.F.; Anderson, J.C.; Abood, M.E. CB1 Allosteric Modulator Org27569 Is an Antagonist/Inverse Agonist of ERK1/2 Signaling. *Cannabis Cannabinoid Res.* **2016**, *1*, 272–280. [CrossRef] [PubMed]

131. Shao, Z.; Yan, W.; Chapman, K.; Ramesh, K.; Ferrell, A.J.; Yin, J.; Wang, X.; Xu, Q.; Rosenbaum, D.M. Structure of an allosteric modulator bound to the CB1 cannabinoid receptor. *Nat. Chem. Biol.* **2019**, *15*, 1199–1205. [CrossRef]

132. Lynch, D.L.; Hurst, W.P.; Shore, D.M.; Pitman, M.C.; Reggio, P.H. Molecular Dynamics Methodologies for Probing Cannabinoid Ligand/Receptor Interaction. *Methods Enzym.* **2017**, *593*, 449–490. [CrossRef]

133. Gamage, T.F.; Ignatowska-Jankowska, B.; Wiley, J.; Abdelrahman, M.; Trembleau, L.; Greig, I.; Thakur, G.A.; Tichkule, R.; Poklis, J.; Ross, R.A.; et al. In-vivo pharmacological evaluation of the CB1-receptor allosteric modulator Org-27569. *Behav. Pharmacol.* **2014**, *25*, 182–185. [CrossRef]

134. Ding, Y.; Qiu, Y.; Jing, L.; Thorn, D.A.; Zhang, Y.; Li, J.-X. Behavioral effects of the cannabinoid CB1receptor allosteric modulator ORG27569 in rats. *Pharmacol. Res. Perspect.* **2014**, *2*, e00069. [CrossRef]

135. Giuffrida, A.; McMahon, L.R. In vivo pharmacology of endocannabinoids and their metabolic inhibitors: Therapeutic implications in Parkinson's disease and abuse liability. *Prostaglandins Other Lipid Mediat.* **2010**, *91*, 90–103. [CrossRef]

136. Singh, H.; Schulze, D.R.; McMahon, L.R. Tolerance and cross-tolerance to cannabinoids in mice: Schedule-controlled responding and hypothermia. *Psychopharmacology* **2011**, *215*, 665–675. [CrossRef]

137. Vallee, M.; Vitiello, S.; Bellocchio, L.; Hébert-Chatelain, E.; Monlezun, S.; Martín-García, E.; Kasanetz, F.; Baillie, G.L.; Panin, F.; Cathala, A.; et al. Pregnenolone Can Protect the Brain from Cannabis Intoxication. *Science* **2014**, *343*, 94–98. [CrossRef]

138. Busquets-Garcia, A.; Soria-Gomez, E.; Redon, B.; Mackenbach, Y.; Vallee, M.; Chaouloff, F.; Varilh, M.; Ferreira, G.; Piazza, P.-V.; Marsicano, G. Pregnenolone blocks cannabinoid-induced acute psychotic-like states in mice. *Mol. Psychiatry* **2017**, *22*, 1594–1603. [CrossRef]

139. Laprairie, R.B.; Kulkarni, P.M.; Deschamps, J.R.; Kelly, M.E.M.; Janero, D.R.; Cascio, M.G.; Stevenson, L.A.; Pertwee, R.G.; Kenakin, T.P.; Denovan-Wright, E.M.; et al. Enantiospecific Allosteric Modulation of Cannabinoid 1 Receptor. *ACS Chem. Neurosci.* **2017**, *8*, 1188–1203. [CrossRef] [PubMed]

140. Garai, S.; Kulkarni, P.M.; Schaffer, P.C.; Leo, L.M.; Brandt, A.L.; Zagzoog, A.; Black, T.; Lin, X.; Hurst, D.P.; Janero, D.R.; et al. Application of Fluorine- and Nitrogen-Walk Approaches: Defining the Structural and Functional Diversity of 2-Phenylindole Class of Cannabinoid 1 Receptor Positive Allosteric Modulators. *J. Med. Chem.* **2020**, *63*, 542–568. [CrossRef] [PubMed]

141. Mitjavila, J.; Yin, D.; Kulkarni, P.M.; Zanato, C.; Thakur, G.A.; Ross, R.; Greig, I.; Mackie, K.; Straiker, A. Enantiomer-specific positive allosteric modulation of CB1 signaling in autaptic hippocampal neurons. *Pharmacol. Res.* **2018**, *129*, 475–481. [CrossRef] [PubMed]

142. Hurst, D.P.; Garai, S.; Kulkarni, P.M.; Schaffer, P.C.; Reggio, P.H.; Thakur, G.A. Identification of CB1 Receptor Allosteric Sites Using Force-Biased MMC Simulated Annealing and Validation by Structure–Activity Relationship Studies. *ACS Med. Chem. Lett.* **2019**, *10*, 1216–1221. [CrossRef]

143. Laprairie, R.B.; Bagher, A.M.; Rourke, J.L.; Zrein, A.; Cairns, E.A.; Kelly, M.E.; Sinal, C.J.; Kulkarni, P.M.; Thakur, G.A.; Denovan-Wright, E.M. Positive allosteric modulation of the type 1 cannabinoid receptor reduces the signs and symptoms of Huntington's disease in the R6/2 mouse model. *Neuropharmacology* **2019**, *151*, 1–12. [CrossRef]

144. Cairns, E.A.; Szczesniak, A.-M.; Straiker, A.J.; Kulkarni, P.M.; Pertwee, R.G.; Thakur, G.A.; Baldridge, W.H.; Kelly, M.E. The In Vivo Effects of the CB1-Positive Allosteric Modulator GAT229 on Intraocular Pressure in Ocular Normotensive and Hypertensive Mice. *J. Ocul. Pharmacol. Ther.* **2017**, *33*, 582–590. [CrossRef]

145. Roebuck, A.J.; Greba, Q.; Smolyakova, A.-M.; Alaverdashvili, M.; Marks, W.N.; Garai, S.; Baglot, S.L.; Petrie, G.; Cain, S.M.; Snutch, T.P.; et al. Positive allosteric modulation of type 1 cannabinoid receptors reduces spike-and-wave discharges in Genetic Absence Epilepsy Rats from Strasbourg. *Neuropharmacology* **2021**, *190*, 108553. [CrossRef]

146. Slivicki, R.A.; Xu, Z.; Kulkarni, P.M.; Pertwee, R.G.; Mackie, K.; Thakur, G.A.; Hohmann, A.G. Positive Allosteric Modulation of Cannabinoid Receptor Type 1 Suppresses Pathological Pain Without Producing Tolerance or Dependence. *Biol. Psychiatry* **2018**, *84*, 722–733. [CrossRef]

147. Slivicki, R.A.; Iyer, V.; Mali, S.S.; Garai, S.; Thakur, G.A.; Crystal, J.D.; Hohmann, A.G. Positive Allosteric Modulation of CB1 Cannabinoid Receptor Signaling Enhances Morphine Antinociception and Attenuates Morphine Tolerance Without Enhancing Morphine- Induced Dependence or Reward. *Front. Mol. Neurosci.* **2020**, *13*, 54. [CrossRef] [PubMed]

148. Cristino, L.; Bisogno, T.; Di Marzo, V. Cannabinoids and the expanded endocannabinoid system in neurological disorders. *Nat. Rev. Neurol.* **2020**, *16*, 9–29. [CrossRef]
149. Garai, S.; Leo, L.M.; Szczesniak, A.-M.; Hurst, D.P.; Schaffer, P.C.; Zagzoog, A.; Black, T.; Deschamps, J.R.; Miess, E.; Schulz, S.; et al. Discovery of a Biased Allosteric Modulator for Cannabinoid 1 Receptor: Preclinical Anti-Glaucoma Efficacy. *J. Med. Chem.* **2021**, *64*, 8104–8126. [CrossRef] [PubMed]
150. Ballesteros, J.A.; Jensen, A.D.; Liapakis, G.; Rasmussen, S.; Shi, L.; Gether, U.; Javitch, J. Activation of the β2-Adrenergic Receptor Involves Disruption of an Ionic Lock between the Cytoplasmic Ends of Transmembrane Segments 3 and 6. *J. Biol. Chem.* **2001**, *276*, 29171–29177. [CrossRef] [PubMed]
151. Rosenbaum, D.M.; Zhang, C.; Lyons, J.; Holl, R.; Aragao, D.; Arlow, D.H.; Rasmussen, S.; Choi, H.-J.; DeVree, B.; Sunahara, R.K.; et al. Structure and function of an irreversible agonist-β2 adrenoceptor complex. *Nat. Cell Biol.* **2011**, *469*, 236–240. [CrossRef]
152. Carpenter, B.; Nehmé, R.; Warne, T.; Leslie, A.G.W.; Tate, C.G. Structure of the adenosine A2A receptor bound to an engineered G protein. *Nat. Cell Biol.* **2016**, *536*, 104–107. [CrossRef]
153. Suomivuori, C.-M.; Latorraca, N.R.; Wingler, L.M.; Eismann, S.; King, M.C.; Kleinhenz, A.L.W.; Skiba, M.A.; Staus, D.P.; Kruse, A.C.; Lefkowitz, R.J.; et al. Molecular mechanism of biased signaling in a prototypical G protein–coupled receptor. *Science* **2020**, *367*, 881–887. [CrossRef]

Article

Cannabinoid Receptor 2 Alters Social Memory and Microglial Activity in an Age-Dependent Manner

Joanna Agnieszka Komorowska-Müller [1,2], Tanushka Rana [1], Bolanle Fatimat Olabiyi [1], Andreas Zimmer [1,*] and Anne-Caroline Schmöle [1]

[1] Institute for Molecular Psychiatry, Medical Faculty, University of Bonn, 53127 Bonn, Germany; jkimp@uni-bonn.de (J.A.K.-M.); tanushka.rana@mdc-berlin.de (T.R.); olabiyi@uni-bonn.de (B.F.O.); anne.schmoele@uni-bonn.de (A.-C.S.)

[2] International Max Planck Research School for Brain and Behavior, Ludwig-Erhard-Allee 2, 53175 Bonn, Germany

[*] Correspondence: a.zimmer@uni-bonn.de; Tel.: +49-228-6885300

Abstract: Physiological brain aging is characterized by gradual, substantial changes in cognitive ability, accompanied by chronic activation of the neural immune system. This form of inflammation, termed inflammaging, in the central nervous system is primarily enacted through microglia, the resident immune cells. The endocannabinoid system, and particularly the cannabinoid receptor 2 (CB_2R), is a major regulator of the activity of microglia and is upregulated under inflammatory conditions. Here, we elucidated the role of the CB_2R in physiological brain aging. We used $CB_2R^{-/-}$ mice of progressive ages in a behavioral test battery to assess social and spatial learning and memory. This was followed by detailed immunohistochemical analysis of microglial activity and morphology, and of the expression of pro-inflammatory cytokines in the hippocampus. CB_2R deletion decreased social memory in young mice, but did not affect spatial memory. In fact, old $CB_2R^{-/-}$ mice had a slightly improved social memory, whereas in WT mice we detected an age-related cognitive decline. On a cellular level, CB_2R deletion increased lipofuscin accumulation in microglia, but not in neurons. $CB_2R^{-/-}$ microglia showed an increase of activity markers Iba1 and CD68, and minor upregulation in *tnfa* and *il6* expression and downregulation of *ccl2* with age. This was accompanied by a change in morphology as $CB_2R^{-/-}$ microglia had smaller somas and lower polarity, with increased branching, cell volume, and tree length. We present that CB_2Rs are involved in cognition and age-induced microglial activity, but may also be important for microglial activation itself.

Keywords: cannabinoid receptor 2 (CB_2R); microglia; inflammaging; memory; lipofuscin

Citation: Komorowska-Müller, J.A.; Rana, T.; Olabiyi, B.F.; Zimmer, A.; Schmöle, A.-C. Cannabinoid Receptor 2 Alters Social Memory and Microglial Activity in an Age-Dependent Manner. *Molecules* **2021**, *26*, 5984. https://doi.org/10.3390/molecules26195984

Academic Editor: Mauro Maccarone

Received: 2 August 2021
Accepted: 28 September 2021
Published: 2 October 2021

1. Introduction

Inflammaging, low-grade age-dependent inflammation, has been named one of the seven pillars of aging [1–3] and is one of the main causes of altered intracellular communication. In this type of inflammation, accumulating molecular signals produced throughout life act as the primary stimuli that activate macrophages and microglia [2,3]. These molecular signals can include a dysfunctional immune system that fails to efficiently clean pathogens, enhanced pro-inflammatory tissue damage, cellular senescence, enhanced NF-kB activation, or a defective autophagy response [4].

In the brain, inflammaging affects the activity of the resident innate immune cells—microglia. In young mice, microglia scan their surroundings to react to changes in the environment. Upon detection of neuronal damage or assault, they travel to the site of injury to phagocytose debris and to potentially induce a neuroinflammatory signaling cascade. However, this process is disturbed with aging. Aged microglia are less motile and have deficits in their phagocytic capacity, but show increased secretion of pro-inflammatory cytokines [5]. This age-induced priming of microglia is thought to influence their responses to infections or even stress [6,7]. Aged microglia also frequently become senescent, which further hinders their protective functions [7].

Many studies indicated that the endocannabinoid system (ECS) is an important regulator of microglial activity [8–10] and age-related cellular and molecular changes. ECS consists of two main receptors, the endocannabinoid receptors 1 and 2 (CB_1R and CB_2R); their ligands the endocannabinoids (ECs) 2-arachidonoylglycerol and anandamide, as well as EC-synthesizing and -degrading enzymes. Presynaptic CB_1Rs are an integral part of a synaptic feedback mechanism [11], whereas CB_2Rs modulate immune cell functions and microglia activity. Under basal conditions, CB_2R expression in the brain is low and not readily detectable with most conventional methods [12–16]. However, it is upregulated under inflammation [17]. Moreover, recent findings also support the presence of functional CB_2Rs on neurons [13,18–22].

Mice lacking CB_1R exhibit accelerated age-related cognitive decline, gliosis, and increased expression of inflammatory cytokines in the brain [23–26]. At the same time, overall endocannabinoid tone decreases with age, as 2-AG level and DAGLα expression declined in 12-month old versus 2-month old mice alongside with CB_1R binding to G-protein [27–29]. A chronic low-dose treatment of 18-month old mice with Δ^9-THC, a CB_1/CB_2 agonist, resulted in recovery of their cognitive impairment to the levels of 2-month old mice [30]. The change in cognition was accompanied by an increase in synaptic proteins and in dendritic spine density in the hippocampus [30]. While these data suggest that the ECS is an important player in brain aging, its precise function remains unclear. In particular, it remains unknown if and how CB_2Rs contribute to brain aging.

In this study, we characterized the role of CB_2R in physiological brain aging, focusing on cognition and inflammaging. We investigated cognitive performance of young, adult, and old $CB_2R^{-/-}$ mice, and subsequently analyzed age-induced changes in microglial morphology and activity.

2. Results

2.1. CB_2R Deletion Has a Moderate Age-Dependent Effect on Cognition

To investigate the age-related cognitive performance in $CB_2R^{-/-}$ mice, we used the partner recognition (PR) and Morris water maze (MWM). Anxiety-related behaviors were analyzed in the o-maze test (Figure 1A). All experiments were performed with young (3-months), adult (12-months), and old (18-months) male mice.

In the PR test, all groups showed intact sociability (Figure 1B), as evidenced by a significantly increased preference for the caged mouse of the metal can (one sample-test **** $p < 0.0001$ for each group). We detected no significant effects of genotype or age with regard to sociability. WT mice recognized their previous partner after 30 min separation and showed a preference for the novel partner in the 3-months and 12-months, but not in the 18-months group (one sample *t*-test against a hypothetical mean (50%): 3-months WT mice $p = 0.0468$; 12-months WT mice $p = 0.0035$; 18-months old $CB_2R^{-/-}$ mice $p = 0.0472$) (Figure 1C). We also detected a significant decrease in preference between 3-months and 18-months indicating an age-related cognitive decline (two-way ANOVA age x genotype effect: $F_{2,71} = 17$; $p = 0.0336$). In contrast, the preference in $CB_2R^{-/-}$ mice was higher than the chance level exclusively in the 18-month group. Consistently, after 1 h separation, we determined a preference for the novel partner in 3-month, but not in 12-months or 18-months old WT mice (one sample *t*-test against a hypothetical mean (50%): 3-months WT mice $p = 0.0021$) (Supplementary Figure S1A). We also revealed a significant age-related decrease in preference between 3-months and 12-months and 3-months and 18-months WT mice. In comparison, preference of the $CB_2R^{-/-}$ did not differ from the chance level in any of the investigated age groups, but it was increased in the 18-months old group in comparison to WT mice. Thus, CB_2R deletion caused a moderate age-dependent change in social memory.

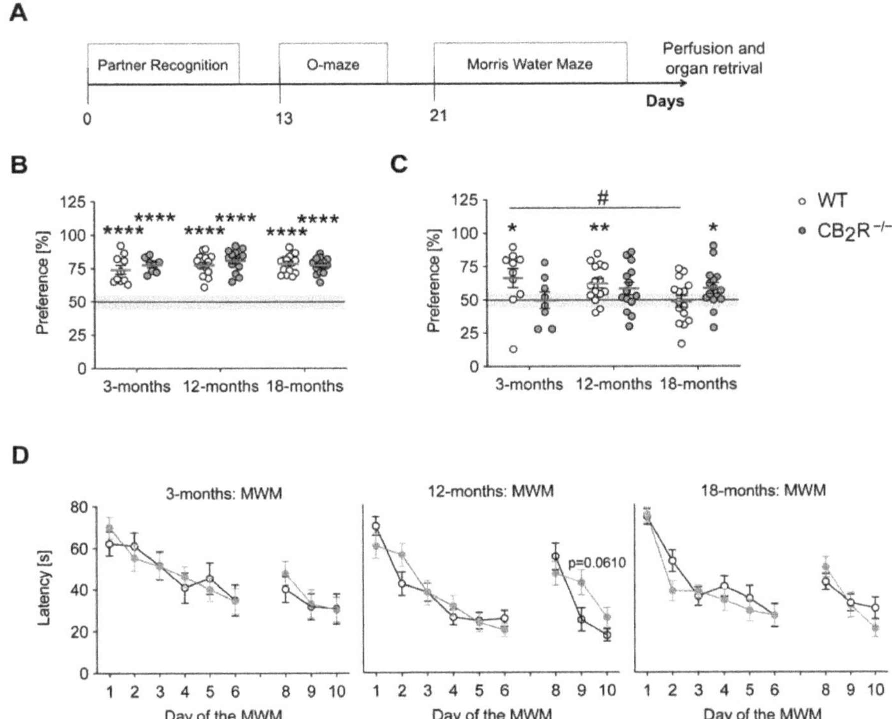

Figure 1. CB$_2$R deletion has a moderate age-dependent effect on cognition. (**A**) Experimental timeline: Partner Recognition (PR), O-maze, and Morris Water Maze (MWM). (**B**) Sociability in the PR test was calculated as interaction time with a partner mouse over total interaction time. All groups were social as indicated by mean sociability >50%. (**C**) Preference for the novel partner (after 30 min separation) was calculated as time with the novel partner mouse over total interaction time. Each group was analyzed individually by one-sample t-test (hypothetical mean = 50). Significant difference from the 50% chance level indicated learning (one sample t-test against a hypothetical mean (50%): 3-months WT mice $p = 0.0021$). * $p < 0.05$; ** $p < 0.01$; **** $p < 0.0001$. Each point represents a single mouse. Red line indicates the mean value ± SEM. Line indicates a 50% chance level. Grey box indicates 5% variance around the chance level. Two-way ANOVA followed by Sidak's multiple comparison test with # $p < 0.05$ significance between age groups within the same genotype. (**D**) Acquisition and reversal phase of the MWM. Panels from left to right: 3-, 12- and 18-months old mice. Decrease in the average latency was detected in all groups. RM ANOVA followed by Sidak's multiple comparison test with exact p-value reported between genotypes within the same age group. WT mice—white circle; CB$_2$R$^{-/-}$ mice—grey circles. N = 14–15 mice/genotype/age group.

In the MWM test, all age groups of CB$_2$R$^{-/-}$ mice and WT controls showed a similar improvement during the acquisition phase and a similar performance during the reversal phase of the test (RM ANOVA: 3-months acquisition time: $F_{5,135} = 12.29$; $p < 0.0001$, reversal time: $F_{2,54} = 8.366$; $p = 0.0007$; 12-months acquisition time: $F_{5,135} = 33.06$; $p < 0.0001$, reversal interaction: $F_{2,54} = 3.92$; $p = 0.0257$, time: $F_{2,54} = 20.04$; $p < 0.0001$, 18-months acquisition time: $F_{5,135} = 31.62$; $p < 0.0001$, reversal time: $F_{2,54} = 12.82$; $p < 0.0001$) (Figure 1D). Additionally, all groups showed preference for the target quadrant during the probe trial (one sample t-test against a hypothetical mean (22.5 s): WT mice: 3-months $p = 0.0088$; 12-months $p = 0.0002$; 18-months $p = 0.0421$; CB$_2$R$^{-/-}$ mice: 3-months $p = 0.0219$; 12-months $p = 0.0008$; 18-months $p = 0.0018$) (Supplementary Figure S1B). Moreover, changes in memory performance cannot be explained by changes in motility, as we did not detect any significant genotype effect in distance travelled or velocity in any of the tests (Supplementary Figure S2B). Taken together, these provide no evidence for age-related, CB$_2$R-mediated effects on cognitive performance.

2.2. CB$_2$R Deletion Decreases Anxiety in an Age-Independent Manner

In the o-maze test, we determined a significant age and genotype effect for the time spent in the open compartment (two-way ANOVA age effect: $F_{2,82} = 10.26$; $p = 0.0001$; genotype effect: $F_{1,82} = 4.427$; $p = 0.0384$) (Figure 2A). Post hoc testing showed that the time spent in the open compartments decreased significantly in adult and old mice. We did not detect any significant genotype differences within the same age-group (Supplementary Table S1). Furthermore, we measured a significant age and genotype effect for the distance travelled in the open compartment. We revealed a significant decrease between 3- and 18-month old WT mice and 3- and 12-months old as well as 3- and 18-months old CB$_2$R$^{-/-}$ mice (two-way ANOVA age effect: $F_{2,82} = 17$; $p < 0.0001$; genotype effect: $F_{1,82} = 7.009$; $p = 0.0097$). Post hoc analysis did not reveal any significant differences between genotypes within the same age group, but we noted a trend for increased distance travelled in the open compartment in CB$_2$R$^{-/-}$ mice.

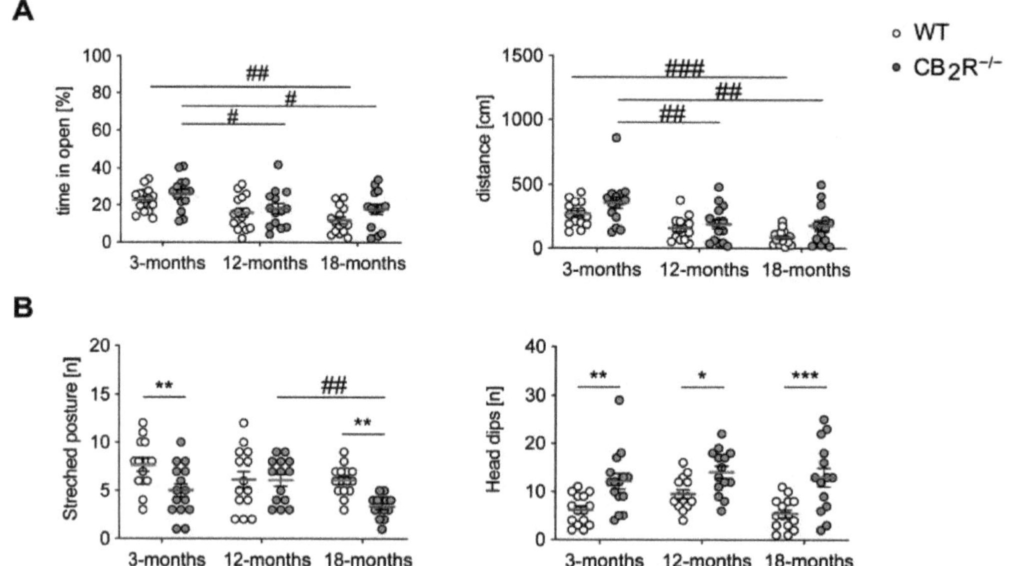

Figure 2. CB$_2$R deletion results in a decreased anxiety phenotype in O-maze. (**A**) Left panel: % of time spent in the open compartment depended on the age and genotype of the mice. Decreased % of time indicates higher anxiety. Right panel: distance travelled in the open compartment depended on the age and genotype of the mice. Decreased distance indicates higher anxiety. (**B**) Left panel: number of stretched posture behaviors was dependent on genotype and age. Increased number of stretched postures indicates higher anxiety. Right panel: number of head dipping behaviors in the open compartment was increased in CB$_2$R$^{-/-}$ mice independent of age. Decreased number of head dips indicates higher anxiety. WT mice—white circle; CB$_2$R$^{-/-}$ mice—grey circles. N = 14–15 mice/genotype/age group. Each point represents a single mouse. Red line indicates the mean value $\pm$ SEM. Two-way ANOVA followed by Sidak's multiple comparison test with * $p < 0.05$, ** $p < 0.01$, *** $p < 0.001$ significance between genotypes within the same age group; # $p < 0.05$, ## $p < 0.01$, ### $p < 0.001$ significance between age groups within the same genotype.

Then, we assessed behaviors associated with anxiety and risk assessment (Figure 2B). An increased number of stretched postures and a decreased amount of head-dipping is interpreted as increased anxiety-like behavior. In contrast, we detected a genotype and age effects for the number of stretched postures (two-way ANOVA age effect: $F_{2,81} = 4.152$; $p = 0.0192$; genotype effect: $F_{1,81} = 13.24$; $p = 0.0005$). Additionally, we measured a genotype effect and a significant increase of head-dipping behavior in CB$_2$R$^{-/-}$ mice in all age groups (two-way ANOVA genotype effect: $F_{1,81} = 33.24$; $p < 0.0001$).

2.3. Age-Dependent Increase of Lipofuscin Affected by CB$_2$R Deletion in Microglia, but Not in Neurons in the Hippocampus

We next measured the accumulation of lipofuscin in hippocampal pyramidal neurons as age-related lipofuscin accumulation is associated with neuronal loss [31].

The age-related accumulation of lipofuscin, as measured by the area covered and particle density (Figure 3A,B), was similar in WT and CB$_2$R mice (two-way ANOVA area covered, age effect: $F_{2,66} = 81.41$, $p < 0.0001$, particle density age effect: $F_{2,65} = 82.98$, $p < 0.0001$) (Supplementary Table S1).

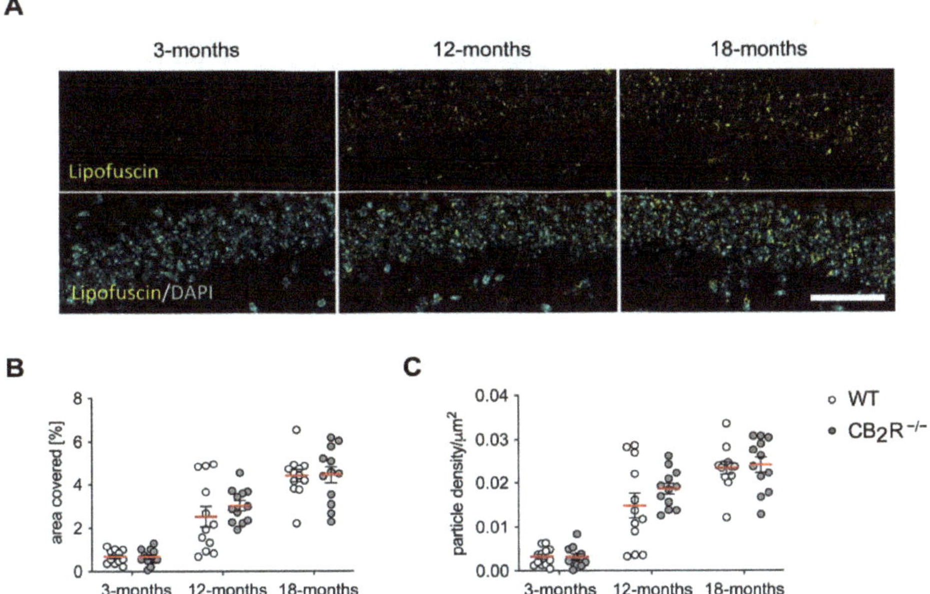

Figure 3. Accumulation of Lipofuscin in hippocampal pyramidal neurons during aging is not altered by CB2 deletion. Representative microscopy images of Lipofuscin accumulation in hippocampal pyramidal neurons with a scale bar of 50 μm (**A**). The area covered with lipofuscin (**B**) and the particle density (**C**) show a significant increase but is not different between WT and CB$_2$R$^{-/-}$ mice. WT mice—white circle; CB$_2$R$^{-/-}$ mice—grey circles. N = 6 mice/genotype/age group, two substacks per animal. Each point represents a single substack. Red line indicates the mean value ± SEM. Two-way ANOVA.

We next analyzed the accumulation of lipofuscin in microglia from hippocampal stratum radiatum as age-related lipofuscin accumulation is also associated with microglial functional decline [32,33].

Lipofuscin accumulation in hippocampal radial microglia as measured by the area covered increased significantly with age in both WT and CB$_2$R$^{-/-}$ (two-way ANOVA area covered, age effect: $F_{2,419} = 47.06$, $p < 0.0001$) (Figure 4A,B). In WT mice lipofuscin increased from on average 0.68% (3-months) to 1.06% (12-months) and reached 3.72% (18-months), while in microglia from CB$_2$R$^{-/-}$ mice increased from 0.21% (3-months) to 3.35% (12-months) and reached 4.4% (18-months). The age-related increase of lipofuscin-accumulation was higher in CB$_2$R$^{-/-}$ mice (two-way ANOVA area covered, genotype effect: $F_{1,419} = 7.857$, $p = 0.0053$) which resulted in an interaction effect (two-way ANOVA area covered $F_{2,419} = 7.514$; $p = 0.0006$) (Figure 4B). Aligned with the enhanced covered area, the particle density significantly increased with age in both genotypes (two-way ANOVA particle density, age effect: $F_{2,429} = 48.34$, $p < 0.0001$) (Figure 4C) but was not significantly different between WT and CB$_2$R$^{-/-}$ mice (Supplementary Table S1).

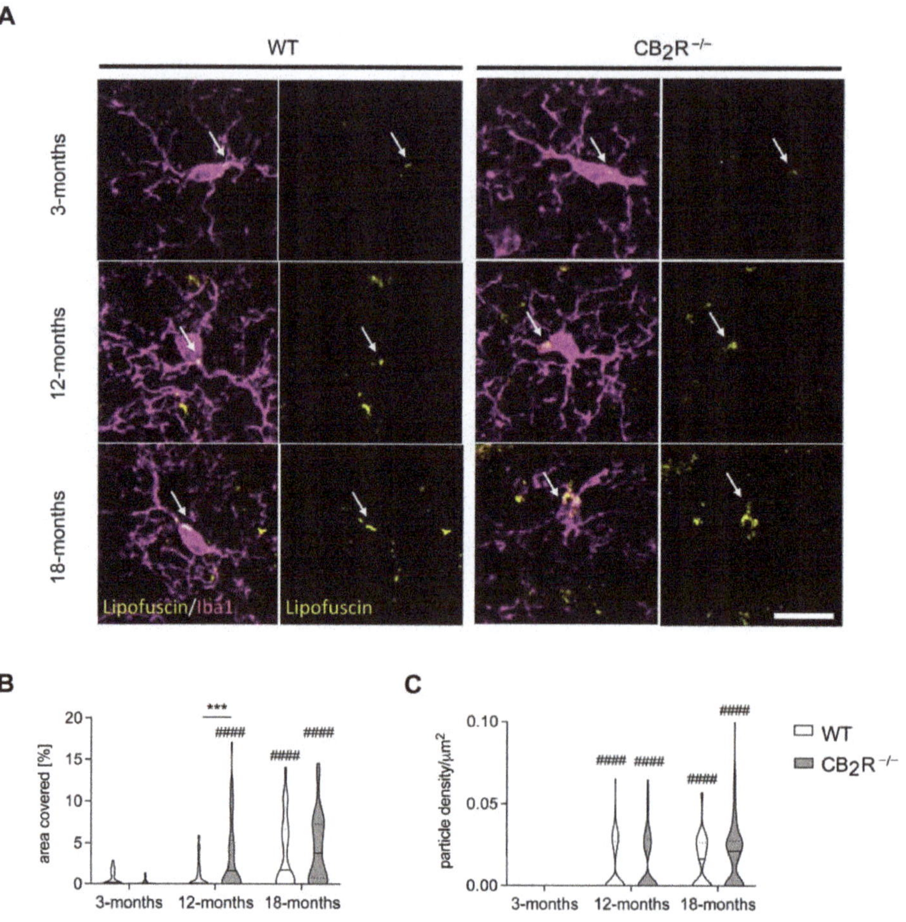

Figure 4. Accumulation of Lipofuscin in microglia is increased after CB2 deletion. Representative microscopy images of Lipofuscin accumulation in microglia in the stratum radiatum of the hippocampus. Scale bar represents 50 μm (**A**). The microglial somatic area covered with lipofuscin (**B**) and the particle density (**C**) show a significant increase with age in both genotypes. Microglia from CB$_2$R$^{-/-}$ mice show enhanced lipofuscin accumulation in comparison to WT mice, which resulted in an interaction effect. WT mice—white; CB$_2$R$^{-/-}$ mice—grey. N = 6 mice/genotype/age group. Data displayed as median (full line) with 25 and 75 percentiles (dotted lines). Two-way ANOVA followed by Sidak's multiple comparisons with *** $p < 0.001$ significance between genotypes within the same age group; #### $p < 0.0001$ significance in relation to 3-months old group within the same genotype.

2.4. Age-Induced Microglial Activity Is Altered in CB$_2$R$^{-/-}$ Microglia

Next, we analyzed Iba1 intensity and CD68 area fraction in the somas of hippocampal radial microglia from WT and CB$_2$R$^{-/-}$ mice to characterize microglial activity.

Iba1 intensity increased in both WT and CB$_2$R$^{-/-}$ microglia with age (two-way ANOVA MGV, age effect: $F_{2,345} = 11.43$, $p < 0.0001$) (Figure 5A,B). This was accompanied by an age-induced increase in CD68 expression in both WT and CB$_2$R$^{-/-}$ microglia (two-way ANOVA area covered, age effect: $F_{2,334} = 18.11$, $p < 0.0001$) (Figure 5C). Interestingly, CB$_2$R$^{-/-}$ microglia showed significantly enhanced Iba1 intensity when compared with WT microglia (two-way ANOVA MGV, genotype effect: $F_{1,345} = 36.48$, $p < 0.0001$, interaction effect: $F_{2,345} = 3.525$, $p = 0.0305$) (Figure 5B). This was further supported by enhanced CD68 content in CB$_2$R$^{-/-}$ microglia from 18-months old mice (Figure 5C).

We also analyzed the expression of the inflammatory mediators: *tnfa, il6, ccl2, arg1* and *nos2* in the hippocampus as markers of inflammaging.

Expression of *tnfa* (two-way ANOVA age effect: $F_{2,28} = 7.804$, $p = 0.002$) (Figure 6A), *il6* (two-way ANOVA age effect: $F_{2,28} = 15.08$, $p < 0.0001$) (Figure 6B) and *ccl2* (two-way ANOVA interaction genotype $\times$ age effect: $F_{2,29} = 4.738$; $p = 0.0166$; age effect: $F_{2,29} = 12.29$; $p = 0.0001$) increased with age in the hippocampus. In WT mice, we observed a steady increase of *tnfa* expression. The expression of *il6* increased from 3- to 12- months but then decreased from 12- to 18- months, back to the *il6* expression levels at 3- months (Figure 6B).

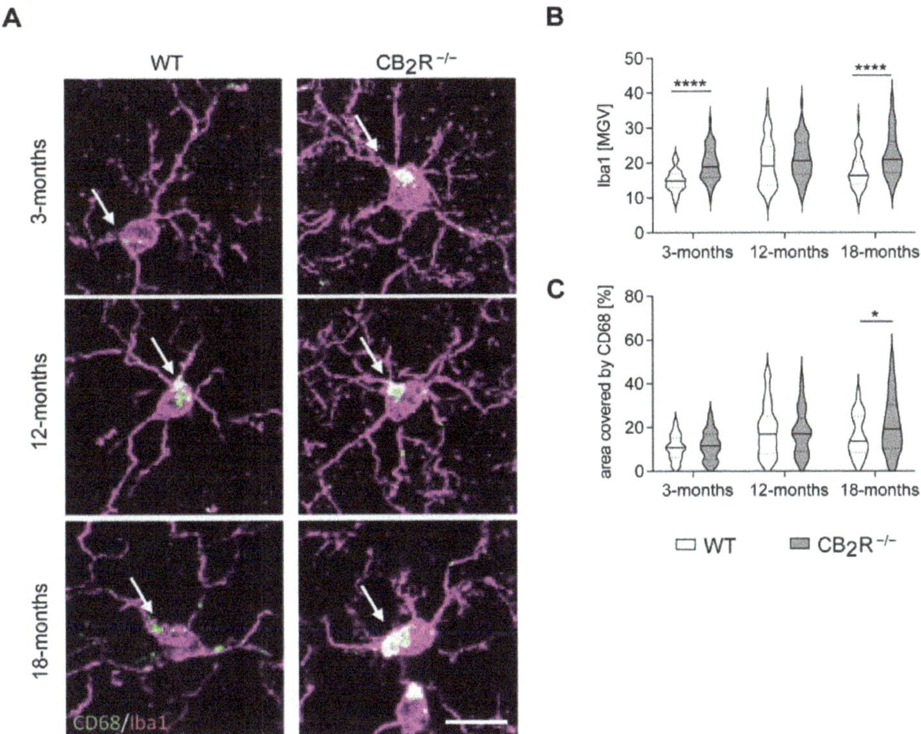

Figure 5. Iba1 and CD68 intensity is enhanced in CB$_2$R$^{-/-}$ microglia. Representative microscopy images from pyramidal microglia of 3-, 12- and 18- month old WT and CB$_2$R$^{-/-}$ mice with a scale bar of 10 μm (**A**). Iba1 intensity increased with age in both genotypes and was also enhanced in CB$_2$R$^{-/-}$ microglia when compared to WT microglia (**B**). CD68 expression was measured by area covered and increased significantly with age (**C**). Data displayed as median (full line) with 25 and 75 percentiles (dotted lines) (**B,C**). WT mice—white; CB$_2$R$^{-/-}$ mice—grey. N = 6 mice/genotype/age group. Two-way ANOVA followed by Sidak's multiple comparisons with * $p < 0.05$, **** $p < 0.0001$ significance between genotypes within the same age group.

The age-dependent increase of *tnfa* expression was more prominent in CB$_2$R$^{-/-}$ with significant increase between 3-months and 12-months and 3-months and 18-months (Figure 6A). Similarly, *il6* expression increased from 3- to 12- months, but in contrast to WT mice, it did not significantly decrease between 12-months and 18-months (Figure 6B). However, there was no significant difference between 3-months and 18-months old CB$_2$R$^{-/-}$ mice. Expression of *ccl2* increased significantly between 3- and 18-months and 12- and 18-months exclusively in WT mice (Figure 6C). This resulted in a lower expression of *ccl2* in 18-month old CB$_2$R$^{-/-}$ mice in comparison to WT controls.

The expression of *arg1* was not altered by aging or CB$_2$R deletion (Figure 6D), whereas for the expression of *nos2*, we detected an age effect, but no genotype effect (Figure 6E; two-way ANOVA age effect: F$_{2,26}$ = 4.275; p = 0.0248).

Thus, CB$_2$R deletion did not majorly alter cytokine expression in the hippocampus, but subsided an age-related increase in the *ccl2* expression.

To characterize the role of CB$_2$R microglial activation in the context of inflammaging, we analyzed 3D microglial morphology of WT and CB$_2$R$^{-/-}$ hippocampal microglia.

The soma size of microglia significantly increased with age in both WT and CB$_2$R$^{-/-}$ (soma size, age effect: F$_{2,1000}$ = 22.97; p < 0.0001) (Figure 7B). However, the increase was significantly less prominent in CB$_2$R$^{-/-}$ microglia as we detected a significant decrease in soma size in microglia from 18-months CB$_2$R$^{-/-}$ old mice.

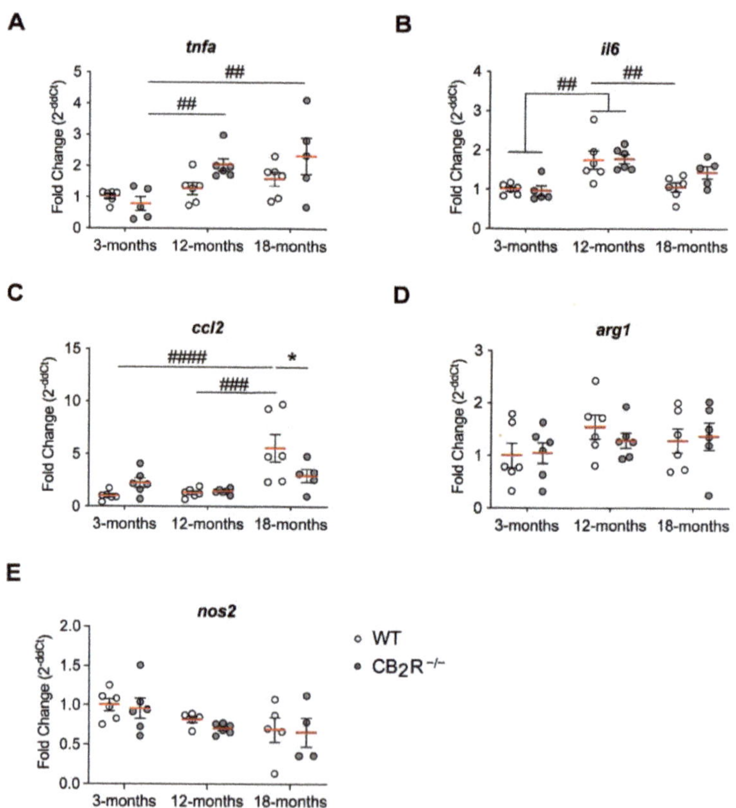

Figure 6. Age-dependent alteration in expression of *inflammatory mediators*. Expression of *tnfa* (**A**) and *il6* (**B**) in hippocampal tissue increases with age but does not differ between WT and CB$_2$R$^{-/-}$. Expression of *ccl2* (**C**) increases with age in WT, but not in CB$_2$R$^{-/-}$. Expression of *arg1* (**D**) did not differ between age groups and genotypes, whereas *nos2* expression (**E**) decreased with age. WT mice—white circle; CB$_2$R$^{-/-}$ mice—grey circles. N = 4–6 mice/genotype/age group. Each point represents a single mouse. Red line indicates the mean value ± SEM. Data were analyzed with two-way ANOVA followed by Sidak's multiple comparison test, with * p < 0.05, significance between genotypes within the same age group; ## p < 0.01, ### p < 0.001, #### p < 0.0001 significance between age groups within the same genotype.

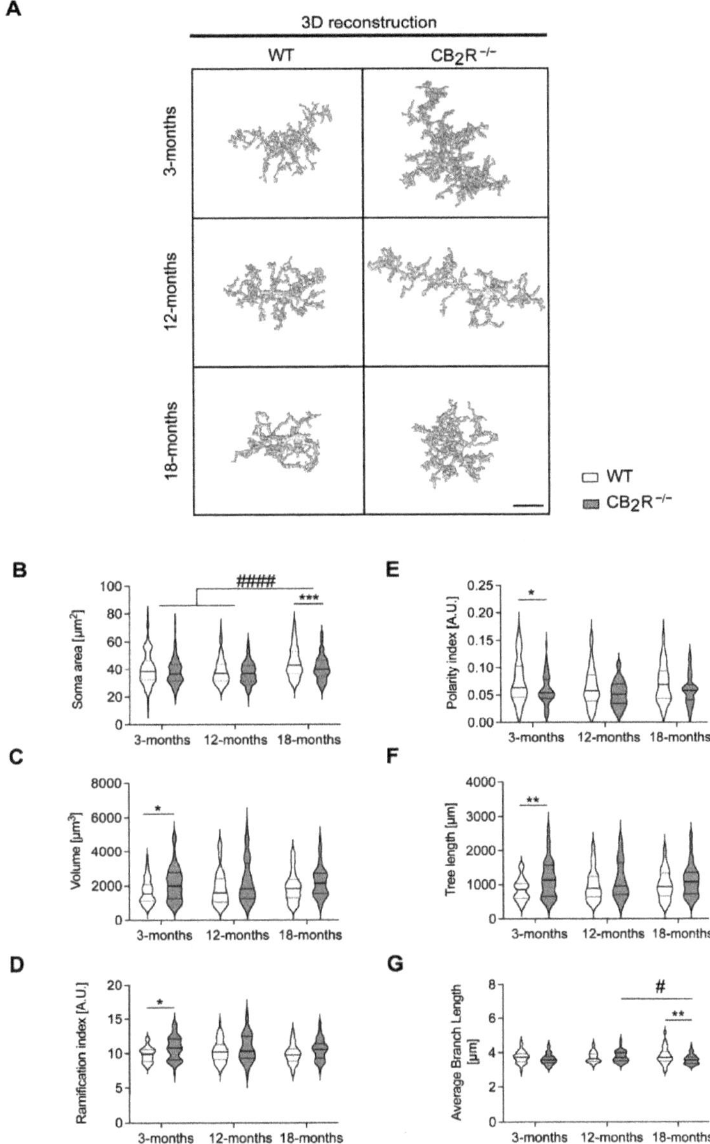

Figure 7. CB$_2$R deletion changes hippocampal microglial morphology. Representative reconstruction images with a scale bar of 20 μm (**A**). Microglia morphology was analyzed by measuring the soma size (**B**), volume (**C**), ramification index (**D**), polarity index (**E**), tree length (**F**), and average branch length (**G**). Microglia morphology differs between CB$_2$R$^{-/-}$ mice and WT mice. Soma size increased with age in both WT and CB$_2$R$^{-/-}$ microglia. Data displayed as median (full line) with 25 and 75 percentiles (dotted lines). WT mice—white circle; CB$_2$R$^{-/-}$ mice—grey circles. N = 6 mice/genotype/age group. Two-way ANOVA followed by Sidak's multiple comparisons with * $p < 0.05$, ** $p < 0.01$, *** $p < 0.001$, significance between genotypes within the same age group, # $p < 0.05$, #### $p < 0.0001$ significance between age groups within the same genotype.

Volume, ramification index, polarity index, tree length, and average branch length (Figure 7C–G) were not significantly altered with age (Supplementary Table S1). In contrast, microglia from $CB_2R^{-/-}$ mice showed an altered morphology with significant differences in volume, ramification index, polarity index, tree length, and average branch length (genotype effect, soma size: $F_{1,1000} = 15.99$; $p < 0.0001$; volume: $F_{1,316} = 11.6$; $p = 0.0007$; ramification index: $F_{1,317} = 15.18$; $p = 0.0001$; polarity: $F_{1,340} = 15.27$; $p = 0.0001$; tree length: $F_{1,329} = 13.02$; $p = 0.0004$; average branch length: $F_{1,326} = 5.847$; $p = 0.0161$) (Figure 7C–G), which was most prominent at the age of 3 months.

3. Discussion

We report that CB_2R deletion, in contrast to CB_1R deletion, has little to no effect on age-related changes in cognitive or anxiety-related behaviors. Nevertheless, we detected subtle genotype effects on inflammaging and microglial function. $CB_2R^{-/-}$ microglia exhibited an increased age-related lipofuscin accumulation and enhanced Iba1 and CD68 levels. Molecular changes were accompanied by altered microglial morphology, and moderately changed secretion of proinflammatory cytokines.

Several studies have shown that administration of THC to old animals was able to reverse many of the adverse consequences of aging on brain physiology and cognitive functions [30,34,35]. Whereas it has been established that CB1 receptors are the main target for these pro-cognitive effects of THC, the involvement of CB2 is less clear. Our results now suggest that CB2 receptors have little influence on the age-related decline of cognitive functions. However, they modulate age-related changes in the brain's inflammatory milieu and thus may be involved in the effects of THC on brain inflammaging.

Recent reports confirming the presence of functional CB_2R on neurons prompted us to investigate if CB_2R deletion also results in an accelerated aging phenotype with early cognitive impairment, similar to what has been observed in $CB_1R^{-/-}$ mice. We found no evidence for an accelerated age-dependent memory decline in $CB_2R^{-/-}$ mice. On the contrary, we observed that the changes of social memory were inversely correlated with age in $CB_2R^{-/-}$ mice. While young $CB_2R^{-/-}$ mice showed a slight decrease in social memory, old mice performed slightly better than age-matched controls. In agreement with the former, we have recently reported an impairment in social memory in $CB_2R^{-/-}$ mice of both sexes aged between 4 and 6-months [34]. In contrast, we did not observe any significant differences in long-term spatial memory, although we detected trends indicating a slightly better performance of 18-month $CB_2R^{-/-}$ mice. The impairment in social memory in young mice did not arise due to altered anxiety-like behaviors. Contrary to previous reports, we measured an age-independent decrease in anxiety-like behavior in $CB_2R^{-/-}$ mice [35].

In agreement with our findings, other studies done on young $CB_2R^{-/-}$ mice showed a decrease in hippocampus-dependent fear memory [36,37]. Synaptic changes might underlie the cognitive deficits that we and others observed in younger $CB_2R^{-/-}$ mice, as CB_2R deletion decreased dendritic spine density in the hippocampus [36,38].

Previous studies have investigated CB_2R age-related changes in the context of Alzheimer's disease (AD), where modulating CB_2R function impacted microglia activity, amyloid plaque load and cognitive abilities of AD-related model mice [39–41]. Whether cognitive changes were due to a direct modulation of the CB_2R on neurons or through microglia activity regulation remains unclear. Likewise, it is possible that either or both neuronal and microglial CB_2Rs contributed to the social memory phenotype that we observed in our study. Nevertheless, our study strengthens an important role of CB_2R in microglia activity in the context of inflammaging as we observed that the deletion of the CB_2R increased accumulation of lipofuscin and CD68 levels in aged microglia.

CB_2R expression is increased in microglia and macrophages in many diseases and acute inflammatory states, but also during aging, which may be due to inflammaging [41,42]. Aged microglia frequently have dystrophic morphology, with increased pro-inflammatory and decreased neuroprotective functions [33,43–45]. Therefore, we investigated age-related microglial activation by assessing Iba1 expression and cell morphology. Microglia from young $CB_2R^{-/-}$ mice had slightly increased cell surface, processes tree-length, number of branches, and ramification index, while their polarity index was decreased. These morphological changes were in line with a less-reactive microglial state and became less pronounced with aging. Microglial soma size increased with age, which is in line with previous studies [46]. However, the somas of $CB_2R^{-/-}$ microglia were smaller than those of WT microglia—especially in old mice—which could indicate reduced microglial activity. The aforementioned morphological changes were consistent with a less reactive state of microglia observed in young $CB_2R^{-/-}$ mice. It is in agreement with previous findings showing a dampened immune response to a pro-inflammatory stimulus [40], thus indicating that CB_2R signaling is required for an efficient microglia activation in young mice. Furthermore, a subtle decrease in an average branch length in addition to an increase in whole-cell Iba1 levels and CD68 levels, was detected in old mice. Age-related increases in Iba1 levels [47,48] and enhanced CD68 levels [49] were reported previously, supporting the idea that microglial reactivity in older $CB_2R^{-/-}$ mice was accompanied by increased phagocytic activity. However, we did not observe corresponding changes in microglial morphology that would indicate a more reactive microglial state. Thus, it is possible that in $CB_2R^{-/-}$ mice, the upregulation of Iba1 and CD68 did not result in a functional change of microglia.

One of the key characteristics of aging is a disturbed proteostasis, which can be observed by an intracellular accumulation of potentially damaging protein aggregates [4]. Among others, lipofuscin has been shown to accumulate both in neurons and microglia in an age-dependent manner and impair cognition [31,50–52]. One main hypothesis states that lipofuscin accumulation in microglia occurs due to an increased phagocytosis of neuronal debris [53,54]. This is supported by findings that accumulation of lipofuscin-like lysosomal particles in microglia is connected with increased phagocytosis of myelin and suggests microglial degradative pathways as a critical target [32]. This accumulation in turn possibly leads to impaired microglial functions [55]. A loss of CB_1R accelerated lipofuscin accumulation in the hippocampus [56]. Therefore, we hypothesized that the age-dependent effect of CB_2R deletion on cognition could be a result of altered lipofuscin accumulation in either neurons or microglia. We detected an age-related increase of lipofuscin in both WT and $CB_2R^{-/-}$ hippocampal microglia, as reported previously [32,33,56]. However, exclusively $CB_2R^{-/-}$ microglia showed enhanced lipofuscin accumulation, suggesting a deficit in lysosomal degradation. This idea is supported by a recent study that showed that CB_2R activation promotes the autophagy flux in macrophages and that the autophagy-lysosome pathway was involved in CB_2R-mediated HMGB1 (High mobility group box 1) degradation [57]. Nevertheless, the role of CB_2R in lysosomal pathways is still not well understood. Future studies should therefore include a more detailed analysis as to whether CB_2R deletion impacts microglial phagocytosis and autophagy and whether these processes are also modulated by enhanced lipofuscin-accumulation.

We also analyzed the expression of inflammatory mediators *tnfa*, *il6*, *ccl2*, *nos2*, and *arg1*. *Tnfa*, *ccl2*, and *il6* expression increased with age as reported previously [33,58], whereas *arg1* expression was not changed with age. In our study, *il6* expression increased up to the age of 12 months and then decreased again at the age of 18 months. Since astrocytes also produce *il6*, we cannot exclude the possibility that astrocytes might dilute the direct effects of microglia. Interestingly, the age-induced increase of *tnfa* expression was even more pronounced in $CB_2R^{-/-}$ mice when compared to WT mice. In contrast, *ccl2* expression was significantly lower in $CB_2R^{-/-}$ mice than in WT at the age of 18 months. We observed similar effects before in vitro and in an AD mouse [40], supporting our idea that CB_2R deletion alters inflammaging.

CB_2R mediated signaling in microglial response during aging was previously investigated only in the context of age-related neuroinflammatory diseases (including Alzheimer's disease). CB_2R activation decreased microglial activity [41,59,60]. Similar results were also observed in AD-related mouse models after CB_2R deletion [40,41]. However, one should consider that pharmacological CB_2R activation/inhibition represents acute effects, whereas CB_2R deletion represents chronic effects, which might be highly variable, especially during long-term processes such as inflammaging. We have recently shown that CB_2R is necessary for toll like receptor (TLR)-mediated microglial activation [61]. Stimulated microglia from $CB_2R^{-/-}$ mice had distinct gene expression patterns, disturbed downstream signaling, and failed to show morphological signs of reactivity [61]. The findings suggest that CB_2R activation is not only able to shift microglial activity from a pro- to an anti-inflammatory state but is also necessary to induce microglial activation in general. These recent data suggest that the role of CB_2R on microglial activation is crucial and significantly more complex than previously thought and therefore needs to be investigated more thoroughly.

Taken together, we report that CB_2R deletion has no effects on long-term spatial memory but has mild effects on short-term social memory during aging. Furthermore, we showed an age-dependent increase of lipofuscin in $CB_2R^{-/-}$ microglia but not in $CB_2R^{-/-}$ neurons. Enhanced lipofuscin accumulation in $CB_2R^{-/-}$ hippocampal microglia was accompanied by increased Iba1 and CD68 levels. Microglial morphology was not majorly altered with age as aging exclusively increased microglia soma size but did not alter other investigated parameters. However, $CB_2R^{-/-}$ microglia showed morphological differences independent of age with increased cell volume, ramification index, and process tree length, and decreased polarity and soma size. We conclude that CB_2R plays a role in cognition and microglial regulation in an age-dependent manner. Furthermore, our data suggest that CB_2R deletion contributes to microglial activity and might be crucial for microglial activation itself.

4. Materials and Methods

4.1. Animals

The generation of $CB_2R^{-/-}$ mice has been previously described [62]. C57BL/6J were originally obtained from a commercial breeder (Charles River) and bred in house. $CB_2R^{-/-}$ mice were bred homozygous and backcrossed to the C57BL/6J line every six generations to minimize the risk of genetic drift.

All animals were housed in specific-pathogen-free conditions in the main animal facility of the University of Bonn. After weaning, mice were housed grouped in standard laboratory cages, with an automatic ventilation system, and *ad libitum* water and food access, under 12 h light-dark cycle (lights on at 09:00 a.m.). Cages were monitored daily and bedding, water, and food were changed weekly. Experiments were carried out with male mice at the age of around 3, 12, and 18 months.

Care of the animals and conduct of the experiments followed the guidelines of the European Communities Directive 86/609/EEC and the German Animal Protection Law regulating animal research and were approved by the Landesamt für Natur-, Umwelt-, und Verbraucherschutz (LANUV NRW), Germany (AZ 84-02.04.2017.A231).

4.2. Behavioral Testing

A week before the first behavioral test, mice were single-housed and transferred to a room with a reversed light-dark cycle (lights off at 9:00 a.m.). Tests were interspersed with 7-day intervals. Groups with 3, 12, and 18 month-old mice were tested independently and analyzed using Ethiovision XT 8.5 and 13 (Noldus, RRID:SCR_000441). The experimenter was blind to the genotype.

4.3. Partner Recognition

Partner recognition (PR) paradigm was used to assess social memory. The test was performed in an open-field box (44 cm × 44 cm) containing a thin layer (about 1 cm) of sawdust. For three consecutive days, mice were allowed to explore the arena freely for 10 min, and habituate to the environment. On the test day, mice underwent two trials. In trial 1, mice were given 9 min to freely explore the arena containing an object (metal can) and a grid cage (diameter about 10 cm, height about 12 cm) with an unfamiliar C57BL6/J male partner mouse. The can and the cage were in opposite corners, each placed about 6–7 cm from the wall. Partner mice were approximately 10 weeks old. Interaction was noted when the mouse nose point was within 2 cm of the cage/object. The time spent on top of any of the objects was deducted from the interaction time. After trial 1, mice were returned to their home cages for 30 min (Figure 1) or 1 h (Supplementary Figure S1A). Sociability in trial 1 was calculated as follows: sociability (%) = Tp/(Tp + Tc) * 100, where Tp is the time of interaction with a partner mouse, and Tc is the time of interaction with the object.

The mean sociability value was tested with a one-sample *t*-test against the chance level. Values above 50% indicated that the mouse spent more time interacting with a partner than with an object. In trial 2 the metal can was replaced by a grid cage with a novel mouse and the test mouse was given 3 min to freely explore and interact with both caged mice. Preference for the novel mouse was calculated as: preference (%) = Tn/(Tf + Tn) * 100, where Tf is the time spent with the familiar mouse and Tn is the time spent with a novel mouse. A preference for the new partner was interpreted as evidence for social memory. To detect learning in each group, we analyzed if the preference for the novel partner deviated statistically from the chance level with a one-sample *t*-test against a hypothetical mean (50%). Mice with sociability ≤55% were excluded from the analysis. If partner mice showed any signs of aggression, they were excluded from the analysis.

4.4. 0-Maze (Elevated Zero Maze)

The zero maze consisted of a circular runway with a diameter of 47 cm and width of 5.6 cm, elevated 30 cm above the ground. It was divided into four equally sized compartments, two of which were enclosed by 24 cm high walls. Mice were allowed to explore the maze for 5 min. Light intensity was around 200 lx. Head-dipping in the open compartment and stretched-attend postures were counted manually as described previously [59].

4.5. Morris Water Maze

The Morris water maze (MWM) was used (Morris 1981) to assess spatial learning and memory. In this paradigm, mice learn to locate a submerged and invisible platform in a round basin filled with turbid water, based on spatial cues. The experiment consisted of three phases: acquisition (days 1–6), probe trial (day 7), and reversal phase (days 8–10). During the acquisition phase the hidden platform remained in a fixed location and animals swam four times per day from different entry points. Inter-trial interval time was 1 h. The cut-off time for each swim was 90 s. If the mouse located the platform within the time limit, then it remained on it for an additional 5 s before being taken out of the maze. Otherwise, after the time limit passed, the mouse was guided to swim to the platform and remained there for an additional 20 s. The decrease of the time required to find the hidden platform indicated spatial learning. In the probe trial, the platform was removed and the time spent in the platform-associated quadrant was measured for 90 s. For the reversal phase, the platform was placed into the opposing quadrant, thus necessitating a re-learning of the position. Similar to the acquisition phase, mice swam four times daily and the latency to the platform was recorded. One mouse that stayed close to the wall at all times was excluded from the analysis.

4.6. Organ Extraction

Mice were anesthetized and perfused transcardially with PBS. Brains were hemisected and one hemisphere was post-fixed in 4% *w/v* formaldehyde for 3.5–4 h on ice. Afterwards, left hemispheres were incubated overnight in 10% sucrose, followed by an overnight incubation in 30% sucrose. The hemispheres were then frozen in dry ice-cooled isopentane and stored at −80 °C. The hippocampus was dissected from the right hemisphere and snap-frozen in liquid nitrogen.

4.7. RNA Isolation and DNase I Digestion

Total RNA was isolated from PBS-perfused, right hemispheric hippocampi (n = 6 hemispheric hippocampi per genotype per age group) using the TRIzol® protocol. Briefly, frozen tissue was homogenized in 1 mL or 800 µL TRIzol (Invitrogen, Camarillo, CA, USA). Tissue homogenates were centrifuged and mixed with 160 µL chloroform. RNA was precipitated with 400 µL ice-cold isopropanol, washed twice with 75% ethanol, and the resulting pellet was dried. Subsequently, RNA was incubated with 2 µL DNase buffer, 10U DNase I and RNase free water in a total volume of 20 µL for 30 min at 30 °C, followed by a DNase inactivation at 75 °C for 5 min. RNA samples were stored at −80 °C.

4.8. cDNA Synthesis

For cDNA synthesis, 1080 ng RNA was incubated for 5 min at 65 °C and then reverse transcribed at 42 °C for 50 min. A total volume of 20 µL included 4 µL first-strand buffer (Invitrogen), 2 µL 0.1 mol/L DTT, 1 µL 10 mmol/1 dNTPs, 0,5 µL oligo(dT) 20 primer (Invitrogen), and 200 U Super-Script II reverse transcriptase (Invitrogen).

4.9. Quantitative Real Time PCR (qPCR)

Analysis by qPCR of cDNA samples was performed using a BioRad CXF384 Cycler and ThermoFisher TaqMan® Gene Expression system. A 30 ng quantity of cDNA was used per reaction. A standard program was applied as follows: step 1 (1×) 95 °C, 10 min; step 2 (40×) 95 °C, 15 s and 60 °C, 1 min. TaqMan primer (all Applied Biosystems, Foster City, CA, USA): *hprt* (Mm03024075_m1), *tnfa* (Mm00443258 _m1), *il6* (Mm00446190 _m1), *ccl2* (Mm00441242_m1), *arg1* (Mm00475988_m1), *nos2* (Mm00440502_m1).

4.10. Immunohistochemistry and Imaging

Mouse brains were sectioned coronally at a thickness of 50 µm using a cryostat. Five dorsal hippocampal sections per mouse were stained as free-floating sections. Sections were post-fixed in 4% PFA for 2 h at room temperature (RT). After three washing steps with PBS, slides were blocked overnight at 4 °C in 10% *w/v* bovine serum albumin (BSA), 2% normal goat serum, and 0.5% Triton X-100 in PBS. Blocked sections were incubated with primary antibodies for 48 h at RT in the dark and, after several washing steps, incubated with secondary antibodies for 4 h at RT. Finally, slices were incubated with 0.1 µg/mL DAPI for 15 min and mounted on a slide using Fluoromount-G™ Mounting Medium. The following antibodies were used: Iba1 (AB_839504), CD68 (AB_322219), goat-anti-rabbit AF647 (AB_2535813), and goat-anti-rat AF488 (AB_2534074). High-resolution images were acquired with a confocal laser scanning microscope (Leica TCS SP8) using a 63x water-immersion objective lens (NA = 1.2). In each experiment, two z-stacks (about 30 µm; step size 0.5 µm; 0.18 µm/px; resolution 1024 × 1024 px) were acquired per mouse of the strata radiatum and pyramidale in the CA1 hippocampal region. Lipofuscin accumulation was measured as autofluorescence (576–640 nm).

4.11. Image Analysis

Quantitative cellular parameters were determined using ImageJ (FIJI ver. 2.0, and higher). Two z-stacks with at least 5 microglia/stack were analyzed for each animal. Stacks from 6 mice/genotype were analyzed per age group.

4.12. CD68 Area Fraction

The CD68 content was determined within each microglial soma. Briefly, maximum intensity projections of the CD68 channel z-stacks were generated using the 'z project' command. Images were binarized with the 'threshold' command. Mean grey value threshold was kept constant among all groups. The threshold used was determined as an average individual threshold of all WT images. The area fraction of CD68 signal was measured within each microglial soma using the 'measure' command.

4.13. Microglial 3D Reconstruction and Analysis of Microglial Branching

Microglial morphology was quantified using a custom-written ImageJ toolbox designed to reconstruct and analyze microglial cells, similar to previous studies from Plescher et al. (2018) and Schmöle et al. (2018) [41,60]. The toolbox consists of three ImageJ plugins for single-cell image generation, image segmentation, and cell analysis. Per group in each genotype, at least 50 microglial cells were selected in all z-slices of confocal z-stacks using the single-cell selection plugin by an investigator who was blind to the experimental conditions. The resulting single-cell images were segmented using the image segmentation plugin. An intensity threshold (algorithm: "Huang") was calculated in an 8-bit converted, 0.5-fold scaled, and maximum-intensity projected copy of the original image. The threshold was applied to the unmodified original image. Segmented images were analyzed using the cell analysis plugin after applying a particle-filter (Length calibration = 0.3608 μm/pixel, Voxel Depth = 0.5 μm/voxel, minimum particle volume = 10,000 voxels). The microglial mean Iba1 intensity was determined as the mean intensity of all voxels in the original image that were positive in the particle-filtered, segmented image. The 3D microglial ramification index was defined as: cell surface area/$(4\pi \cdot [((3 \cdot \text{cell volume})/(4\pi))]^{(2/3)})$, which describes the ratio of cell surface to cell volume and serves as a sensitive measure for cell shape complexity. To determine the "Branch number" and "Tree length", the segmented images, after particle filtering, were Gauss-filtered (Sigma XY = 1.0 and Sigma Z = 0.0), skeletonized using the Fiji plugin "Skeletonize3D" [63], and analyzed using the Fiji plugin "Analyze Skeleton" [63]. The polarity index indicates how equally the process tree is distributed around the cell soma. It was defined as the length of the vector from the center of mass of the microglial cell to the center of the convex hull around the microglial cell, normalized to the size of the convex hull: polarity index = vector length/$(2 \cdot \sqrt[3]{3 \cdot spanned\ volume / (4\pi)})$.

4.14. Lipofuscin Analysis

Neuronal accumulation of lipofuscin was measured in the stratum pyramidale of the hippocampal CA1 region from a binarized max z-projection (7 image planes) with a defined start. Number of lipofuscin particles of a size >0.5 μm was counted using the 'particle analyzer' plugin in ImageJ. Density was calculated as the number of lipofuscin particles divided by area of selection. Lipofuscin levels were measured in the soma of single stratum radiale microglia using binarized maximum z-stack projections as described above for the CD68 area fraction with at least 12 cells per mouse.

4.15. Soma Size and Iba1 Intensity

The somas of microglia were manually delineated using the 'polygon selection' tool and saved as 'regions of interest' (ROI). Iba1 intensity and soma size were measured within each ROI with the 'measure' command. Soma size was measured in both Lipofuscin and CD68 area fraction experiments and both datasets were pooled together.

4.16. Statistical Analysis and Data Presentation

Microsoft Excel (v 16.43) was used for data analysis followed by statistical analysis and data visualization in GraphPad Prism version 7.0.0 and 9.1.2 for Mac, GraphPad Software, San Diego, CA, USA, www.graphpad.com. Figures were created in Adobe Illustrator (v 24.0.2). For presentation, representative images were post-processed in ImageJ (Fiji) to adjust brightness and contrast. All images within one experiment were adjusted the same way. Behavioral data were analyzed using Ethiovision XT 8.5 and 13 (Noldus, RRID:SCR_000441).

For datasets consisting of more than two groups with two independent variables (e.g., genotype and gender), two-way analysis of variance (ANOVA) was used followed by Sidak's multiple comparison test. For MWM repeated measurement (RM) ANOVA was used. For PR, the mean of the group was tested against a theoretical mean (50) with one-sample *t*-test. For single microglia analysis, an outlier test was performed with

ROUT = 5% prior to the analysis, with the detected outliers excluded. For expression analysis, an outlier test was performed with ROUT = 10% prior to the analysis, with the detected outliers excluded. Datasets with more than 20 points were depicted using a violin plot to precisely visualize the distribution of the data, whereas datasets with fewer than 20 points were depicted as scatter plots. Statistical significance was stated when p-value < 0.05 at a 95% confidence interval. Detailed results of statistical analysis are presented in Supplementary Table S1.

Supplementary Materials: The following are available online, Figure S1: Memory; Table S1: Statistical analysis; Figure S2: Motility.

Author Contributions: J.A.K.-M., A.Z. and A.-C.S. conceived the study; J.A.K.-M., T.R. and B.F.O. performed the experiments; J.A.K.-M., T.R., B.F.O. and A.-C.S. analyzed the data; J.A.K.-M., A.Z. and A.-C.S. wrote the original manuscript, which was edited by all authors; A.Z. and A.-C.S. supervised the project. All authors read and approved the manuscript.

Funding: This research leading to these results was funded by the Deutsche Forschungsgemeinschaft (DFG, German Research Foundation) accorded to A.Z. and A.-C.S. under Germany's Excellence Strategy—EXC2151—390873048 and BONFOR funding by the Medical Faculty, University of Bonn to A.-C.S.

Institutional Review Board Statement: Care of the animals and conduct of the experiments followed the guidelines of the European Communities Directive 86/609/EEC and the German Animal Protection Law regulating animal research and were approved by the Landesamt für Natur-, Umwelt-, und Verbraucherschutz (LANUV NRW), Germany (AZ 84-02.04.2017.A231).

Informed Consent Statement: Not applicable.

Data Availability Statement: Datasets are available on request. The raw data supporting the conclusions of this article will be made available by the authors, without undue reservation, to any qualified researcher.

Acknowledgments: The authors would like to thank Edda Erxlebe, Kerstin Nicolai, Hanna Schrage and Anne Zimmer for excellent technical support. We would like to thank Britta Schürmann for the introduction to the behavioral paradigms and critical discussion of the behavioral data as well as Andras Bilkei-Gorzo for critical discussion of the manuscript.

Conflicts of Interest: The authors declare no conflict of interest.

References

1. Kennedy, B.K.; Berger, S.L.; Brunet, A.; Campisi, J.; Cuervo, A.M.; Epel, E.S.; Franceschi, C.; Lithgow, G.J.; Morimoto, R.I.; Pessin, J.E.; et al. Geroscience: Linking Aging to Chronic Disease. *Cell* **2014**, *159*, 709–713. [CrossRef]

2. Franceschi, C.; Campisi, J. Chronic Inflammation (Inflammaging) and Its Potential Contribution to Age-Associated Diseases. *J. Gerontol. A Biol. Sci. Med. Sci.* **2014**, *69*, 4–9. [CrossRef] [PubMed]

3. Franceschi, C.; Bonafè, M.; Valensin, S.; Olivieri, F.; De Luca, M.; Ottaviani, E.; De Benedictis, G. Inflamm-aging. An evolutionary perspective on immunosenescence. *Ann. N. Y. Acad. Sci.* **2000**, *908*, 244–254. [CrossRef] [PubMed]

4. López-Otín, C.; Blasco, M.A.; Partridge, L.; Serrano, M.; Kroemer, G. The hallmarks of aging. *Cell* **2013**, *153*, 1194. [CrossRef] [PubMed]

5. Angelova, D.M.; Brown, D.R. Microglia and the aging brain: Are senescent microglia the key to neurodegeneration? *J. Neurochem.* **2019**, *151*, 676–688. [CrossRef] [PubMed]

6. Cunningham, C.; Wilcockson, D.C.; Campion, S.; Lunnon, K.; Perry, V.H. Central and systemic endotoxin challenges exacerbate the local inflammatory response and increase neuronal death during chronic neurodegeneration. *J. Neurosci.* **2005**, *25*, 9275–9284. [CrossRef] [PubMed]

7. Niraula, A.; Sheridan, J.F.; Godbout, J.P. Microglia Priming with Aging and Stress. *Neuropsychopharmacology* **2017**, *42*, 318–333. [CrossRef] [PubMed]

8. Correa, F.; Hernangómez, M.; Mestre, L.; Loría, F.; Spagnolo, A.; Docagne, F.; Di Marzo, V.; Guaza, C. Anandamide enhances IL-10 production in activated microglia by targeting CB2 receptors: Roles of ERK1/2, JNK, and NF-κB. *Glia* **2010**, *58*, 135–147. [CrossRef]

9. Mecha, M.; Feliú, A.; Carrillo-Salinas, F.J.; Rueda-Zubiaurre, A.; Ortega-Gutiérrez, S.; de Sola, R.G.; Guaza, C. Endocannabinoids drive the acquisition of an alternative phenotype in microglia. *Brain. Behav. Immun.* **2015**. [CrossRef]

10. Ma, L.; Jia, J.; Liu, X.; Bai, F.; Wang, Q.; Xiong, L. Activation of murine microglial N9 cells is attenuated through cannabinoid receptor CB2 signaling. *Biochem. Biophys. Res. Commun.* **2015**, *458*, 92–97. [CrossRef]

11. Castillo, P.E.; Younts, T.J.; Chávez, A.E.; Hashimotodani, Y. Endocannabinoid Signaling and Synaptic Function. *Neuron* **2012**, *76*, 70–81. [CrossRef]

12. Liu, Q.-R.; Pan, C.-H.; Hishimoto, A.; Li, C.-Y.; Xi, Z.-X.; Llorente-Berzal, A.; Viveros, M.-P.; Ishiguro, H.; Arinami, T.; Onaivi, E.S.; et al. Species differences in cannabinoid receptor 2 (CNR2) gene: Identification of novel human and rodent CB2 isoforms, differential tissue expression, and regulation by cannabinoid receptor ligands. *Genes, Brain Behav.* **2009**, *8*, 819–830. [CrossRef]

13. Li, Y.; Kim, J. Neuronal expression of CB2 cannabinoid receptor mRNAs in the mouse hippocampus. *Neuroscience* **2015**, *311*, 253–267. [CrossRef]

14. Onaivi, E.S. Neuropsychobiological evidence for the functional presence and expression of cannabinoid CB2 receptors in the brain. *Neuropsychobiology* **2007**, *54*, 231–246. [CrossRef]

15. Schmöle, A.C.; Lundt, R.; Gennequin, B.; Schrage, H.; Beins, E.; Krämer, A.; Zimmer, T.; Limmer, A.; Zimmer, A.; Otte, D.M. Expression analysis of CB2-GFP BAC transgenic mice. *PLoS ONE* **2015**, *10*, 1–16. [CrossRef]

16. López, A.; Aparicio, N.; Pazos, M.R.; Grande, M.T.; Barreda-Manso, M.A.; Benito-Cuesta, I.; Vázquez, C.; Amores, M.; Ruiz-Pérez, G.; García-García, E.; et al. Cannabinoid CB 2 receptors in the mouse brain: Relevance for Alzheimer's disease. *J. Neuroinflammation* **2018**, *15*, 1–11. [CrossRef] [PubMed]

17. Benito, C.; Tolón, R.M.; Pazos, M.R.; Núñez, E.; Castillo, A.I.; Romero, J. Cannabinoid CB 2 receptors in human brain inflammation. *Br. J. Pharmacol.* **2008**, *153*, 277–285. [CrossRef]

18. Stempel, A.V.; Stumpf, A.; Zhang, H.Y.; Özdoğan, T.; Pannasch, U.; Theis, A.K.; Otte, D.M.; Wojtalla, A.; Rácz, I.; Ponomarenko, A.; et al. Cannabinoid Type 2 Receptors Mediate a Cell Type-Specific Plasticity in the Hippocampus. *Neuron* **2016**, *90*, 795–809. [CrossRef]

19. Zhang, H.-Y.; Gao, M.; Liu, Q.-R.; Bi, G.-H.; Li, X.; Yang, H.-J.; Gardner, E.L.; Wu, J.; Xi, Z.-X. Cannabinoid CB 2 receptors modulate midbrain dopamine neuronal activity and dopamine-related behavior in mice. *Proc. Natl. Acad. Sci. USA* **2014**, *111*, E5007–E5015. [CrossRef] [PubMed]

20. Zhang, H.Y.; Bi, G.H.; Li, X.; Li, J.; Qu, H.; Zhang, S.J.; Li, C.Y.; Onaivi, E.S.; Gardner, E.L.; Xi, Z.X.; et al. Species differences in cannabinoid receptor 2 and receptor responses to cocaine self-administration in mice and rats. *Neuropsychopharmacology* **2015**, *40*, 1037–1051. [CrossRef] [PubMed]

21. Zhang, H.-Y.; Gao, M.; Shen, H.; Bi, G.-H.; Yang, H.-J.; Liu, Q.-R.; Wu, J.; Gardner, E.L.; Bonci, A.; Xi, Z.-X. Expression of functional cannabinoid CB 2 receptor in VTA dopamine neurons in rats. *Addict. Biol.* **2017**, *22*, 752–765. [CrossRef]

22. Liu, Q.R.; Canseco-Alba, A.; Zhang, H.Y.; Tagliaferro, P.; Chung, M.; Dennis, E.; Sanabria, B.; Schanz, N.; Escosteguy-Neto, J.C.; Ishiguro, H.; et al. Cannabinoid type 2 receptors in dopamine neurons inhibits psychomotor behaviors, alters anxiety, depression and alcohol preference. *Sci. Rep.* **2017**, *7*, 1–17. [CrossRef]

23. Albayram, O.; Alferink, J.; Pitsch, J.; Piyanova, A.; Neitzert, K.; Poppensieker, K.; Mauer, D.; Michel, K.; Legler, A.; Becker, A.; et al. Role of CB1 cannabinoid receptors on GABAergic neurons in brain aging. *Proc. Natl. Acad. Sci. USA* **2011**, *108*, 11256–11261. [CrossRef]

24. Albayram, O.; Bilkei-Gorzo, A.; Zimmer, A. Loss of CB1 receptors leads to differential age-related changes in reward-driven learning and memory. *Front. Aging Neurosci.* **2012**, *4*, 1–8. [CrossRef]

25. Bilkei-Gorzo, A.; Racz, I.; Valverde, O.; Otto, M.; Michel, K.; Sarstre, M.; Zimmer, A. Early age-related cognitive impairment in mice lacking cannabinoid CB1 receptors. *Proc. Natl. Acad. Sci. USA* **2005**, *102*, 15670–15675. [CrossRef]

26. Bilkei-Gorzo, A. The endocannabinoid system in normal and pathological brain ageing. *Philos. Trans. R. Soc. B Biol. Sci.* **2012**, *367*, 3326–3341. [CrossRef] [PubMed]

27. Marchalant, Y.; Cerbai, F.; Brothers, H.M.; Wenk, G.L. Cannabinoid receptor stimulation is anti-inflammatory and improves memory in old rats. *Neurobiol. Aging* **2008**, *29*, 1894–1901. [CrossRef] [PubMed]

28. Piyanova, A.; Lomazzo, E.; Bindila, L.; Lerner, R.; Albayram, O.; Ruhl, T.; Lutz, B.; Zimmer, A.; Bilkei-Gorzo, A. Age-related changes in the endocannabinoid system in the mouse hippocampus. *Mech. Ageing Dev.* **2015**, *150*, 55–64. [CrossRef] [PubMed]

29. Romero, J.; Berrendero, F.; Garcia-Gil, L.; De La Cruz, P.; Ramos, J.A.; Fernández-Ruiz, J.J. Loss of cannabinoid receptor binding and messenger RNA levels and cannabinoid agonist-stimulated [35S]guanylyl-5'-O-(thio)-triphosphate binding in the basal ganglia of aged rats. *Neuroscience* **1998**, *84*, 1075–1083. [CrossRef]

30. Bilkei-Gorzo, A.; Albayram, O.; Draffehn, A.; Michel, K.; Piyanova, A.; Oppenheimer, H.; Dvir-Ginzberg, M.; Rácz, I.; Ulas, T.; Imbeault, S.; et al. A chronic low dose of Δ9-tetrahydrocannabinol (THC) restores cognitive function in old mice. *Nat. Med.* **2017**, *23*, 782–787. [CrossRef]

31. Moreno-García, A.; Kun, A.; Calero, O.; Medina, M.; Calero, M. An overview of the role of lipofuscin in age-related neurodegeneration. *Front. Neurosci.* **2018**, *12*, 1–13. [CrossRef]

32. Safaiyan, S.; Kannaiyan, N.; Snaidero, N.; Brioschi, S.; Biber, K.; Yona, S.; Edinger, A.L.; Jung, S.; Rossner, M.J.; Simons, M. Age-related myelin degradation burdens the clearance function of microglia during aging. *Nat. Neurosci.* **2016**, *19*, 995–998. [CrossRef] [PubMed]

33. Sierra, A.; Gottfried-Blackmore, A.C.; McEwen, B.S.; Bulloch, K. Microglia derived from aging mice exhibit an altered inflammatory profile. *Glia* **2007**, *55*, 412–424. [CrossRef] [PubMed]

34. Komorowska-Müller, J.A.; Ravichandran, K.A.; Zimmer, A.; Schürmann, B. Cannabinoid receptor 2 deletion influences social memory and synaptic architecture in the hippocampus. *Sci. Rep.* **2021**, *11*, 1–10. [CrossRef] [PubMed]

35. Ortega-Alvaro, A.; Aracil-Fernández, A.; García-Gutiérrez, M.S.; Navarrete, F.; Manzanares, J. Deletion of CB2 cannabinoid receptor induces schizophrenia-related behaviors in mice. *Neuropsychopharmacology* **2011**, *36*, 1489–1504. [CrossRef]

36. García-Gutiérrez, M.S.; Ortega-Álvaro, A.; Busquets-García, A.; Pérez-Ortiz, J.M.; Caltana, L.; Ricatti, M.J.; Brusco, A.; Maldonado, R.; Manzanares, J. Synaptic plasticity alterations associated with memory impairment induced by deletion of CB2 cannabinoid receptors. *Neuropharmacology* **2013**, *73*, 388–396. [CrossRef]

37. Li, Y.; Kim, J. CB2 cannabinoid receptor knockout in mice impairs contextual long-term memory and enhances spatial working memory. *Neural Plast.* **2016**, *2016*. [CrossRef]

38. Li, Y.; Kim, J. Deletion of CB2 cannabinoid receptors reduces synaptic transmission and long-term potentiation in the mouse hippocampus. *Hippocampus* **2016**, *26*, 275–281. [CrossRef]

39. Aso, E.; Juvés, S.; Maldonado, R.; Ferrer, I. CB2cannabinoid receptor agonist ameliorates alzheimer-like phenotype in AβPP/PS1 mice. *J. Alzheimer's Dis.* **2013**, *35*, 847–858. [CrossRef]

40. Schmöle, A.; Lundt, R.; Ternes, S.; Albayram, Ö.; Ulas, T.; Schultze, J.L.; Bano, D.; Nicotera, P.; Alferink, J.; Zimmer, A. Cannabinoid receptor 2 deficiency results in reduced neuroinflammation in an Alzheimer's disease mouse model. *Neurobiol. Aging* **2015**, *36*, 710–719. [CrossRef]

41. Schmöle, A.; Lundt, R.; Toporowski, G.; Hansen, J.N.; Beins, E.; Halle, A.; Zimmer, A. Cannabinoid Receptor 2-Deficiency Ameliorates Disease Symptoms in a Mouse Model with Alzheimer's Disease-Like Pathology. *J. Alzheimer's Dis.* **2018**, *64*, 379–392. [CrossRef] [PubMed]

42. Komorowska-Müller, J.A.; Schmöle, A.C. CB2 receptor in microglia: The guardian of self-control. *Int. J. Mol. Sci.* **2021**, *22*, 19. [CrossRef]

43. Dipatre, P.L.; Gelman, B.B. Microglial Cell Activation in Aging and Alzheimer Disease. *J. Neuropathol. Exp. Neurol.* **1997**, *56*, 143–149. [CrossRef] [PubMed]

44. Sheng, J.G.; Mrak, R.E.; Griffin, W.S.T. Enlarged and phagocytic, but not primed, interleukin-1α-immunoreactive microglia increase with age in normal human brain. *Acta Neuropathol.* **1998**, *95*, 229–234. [CrossRef] [PubMed]

45. Shobin, E.; Bowley, M.P.; Estrada, L.I.; Heyworth, N.C.; Orczykowski, M.E.; Eldridge, S.A.; Calderazzo, S.M.; Mortazavi, F.; Moore, T.L.; Rosene, D.L. Microglia activation and phagocytosis: Relationship with aging and cognitive impairment in the rhesus monkey. *GeroScience* **2017**, *39*, 199–220. [CrossRef]

46. Hefendehl, J.K.; Neher, J.J.; Sühs, R.B.; Kohsaka, S.; Skodras, A.; Jucker, M. Homeostatic and injury-induced microglia behavior in the aging brain. *Aging Cell* **2014**, *13*, 60–69. [CrossRef] [PubMed]

47. Hashizume, T.; Son, B.K.; Taniguchi, S.; Ito, K.; Noda, Y.; Endo, T.; Nanao-Hamai, M.; Ogawa, S.; Akishita, M. Establishment of Novel Murine Model showing Vascular Inflammation-derived Cognitive Dysfunction. *Sci. Rep.* **2019**, *9*, 1–12. [CrossRef]

48. Lana, D.; Ugolini, F.; Wenk, G.L.; Giovannini, M.G.; Zecchi-Orlandini, S.; Nosi, D. Microglial distribution, branching, and clearance activity in aged rat hippocampus are affected by astrocyte meshwork integrity: Evidence of a novel cell-cell interglial interaction. *FASEB J.* **2019**, *33*, 4007–4020. [CrossRef]

49. Hart, A.D.; Wyttenbach, A.; Hugh Perry, V.; Teeling, J.L. Age related changes in microglial phenotype vary between CNS regions: Grey versus white matter differences. *Brain. Behav. Immun.* **2012**, *26*, 754–765. [CrossRef]

50. Wong, W.T. Microglial aging in the healthy CNS: Phenotypes, drivers, and rejuvenation. *Front. Cell. Neurosci.* **2013**, *7*, 1–13. [CrossRef]

51. Singh Kushwaha, S.; Patro, N.; Kumar Patro, I. A Sequential Study of Age-Related Lipofuscin Accumulation in Hippocampus and Striate Cortex of Rats. *Ann. Neurosci.* **2019**, *25*, 223–233. [CrossRef]

52. Flood, J.F.; Morley, P.M.K.; Morley, J.E. Age-related changes in learning, memory, and lipofuscin as a function of the percentage of SAMP8 genes. *Physiol. Behav.* **1995**, *58*, 819–822. [CrossRef]

53. Nakanishi, H.; Wu, Z. Microglia-aging: Roles of microglial lysosome- and mitochondria-derived reactive oxygen species in brain aging. *Behav. Brain Res.* **2009**, *201*, 1–7. [CrossRef]

54. Tremblay, M.-È.; Zettel, M.L.; Ison, J.R.; Allen, P.D.; Majewska, A.K. Effects of aging and sensory loss on glial cells in mouse visual and auditory cortices. *Glia* **2012**, *60*, 541–558. [CrossRef]

55. Burns, J.C.; Cotleur, B.; Walther, D.M.; Bajrami, B.; Rubino, S.J.; Wei, R.; Franchimont, N.; Cotman, S.L.; Ransohoff, R.M.; Mingueneau, M. Differential accumulation of storage bodies with aging defines discrete subsets of microglia in the healthy brain. *Elife* **2020**, *9*, 1–71. [CrossRef]

56. Piyanova, A.; Albayram, O.; Rossi, C.A.; Farwanah, H.; Michel, K.; Nicotera, P.; Sandhoff, K.; Bilkei-Gorzo, A. Loss of CB1 receptors leads to decreased cathepsin D levels and accelerated lipofuscin accumulation in the hippocampus. *Mech. Ageing Dev.* **2013**, *134*, 391–399. [CrossRef] [PubMed]

57. Zhou, H.; Du, R.; Li, G.; Bai, Z.; Ma, J.; Mao, C.; Wang, J.; Gui, H. Cannabinoid receptor 2 promotes the intracellular degradation of HMGB1 via the autophagy-lysosome pathway in macrophage. *Int. Immunopharmacol.* **2020**, *78*, 106007. [CrossRef]

58. Gavilán, M.P.; Revilla, E.; Pintado, C.; Castaño, A.; Vizuete, M.L.; Moreno-González, I.; Baglietto-Vargas, D.; Sánchez-Varo, R.; Vitorica, J.; Gutiérrez, A.; et al. Molecular and cellular characterization of the age-related neuroinflammatory processes occurring in normal rat hippocampus: Potential relation with the loss of somatostatin GABAergic neurons. *J. Neurochem.* **2007**, *103*, 984–996. [CrossRef] [PubMed]

59. Shepherd, J.K.; Grewal, S.S.; Fletcher, A.; Bill, D.J.; Dourish, C.T. Behavioural and pharmacological characterisation of the elevated "zero-maze" as an animal model of anxiety. *Psychopharmacology* **1994**, *116*, 56–64. [CrossRef] [PubMed]
60. Plescher, M.; Seifert, G.; Hansen, J.N.; Bedner, P.; Steinhäuser, C.; Halle, A. Plaque-dependent morphological and electrophysiological heterogeneity of microglia in an Alzheimer's disease mouse model. *Glia* **2018**. [CrossRef]
61. Reusch, N.; Ravichandran, K.A.; Olabiyi, B.F.; Komorowska-Müller, J.A.; Hansen, J.N.; Ulas, T.; Beyer, M.; Zimmer, A.; Schmöle, A.C. Cannabinoid receptor 2 is necessary to induce toll-like receptor-mediated microglial activation. *Glia* **2021**, 1–18. [CrossRef]
62. Buckley, N.E.; McCoy, K.L.; Mezey, É.; Bonner, T.; Zimmer, A.; Felder, C.C.; Glass, M.; Zimmer, A. Immunomodulation by cannabinoids is absent in mice deficient for the cannabinoid CB2receptor. *Eur. J. Pharmacol.* **2000**, *396*, 141–149. [CrossRef]
63. Arganda-Carreras, I.; Fernández-González, R.; Muñoz-Barrutia, A.; Ortiz-De-Solorzano, C. 3D reconstruction of histological sections: Application to mammary gland tissue. *Microsc. Res. Tech.* **2010**, *73*, 1019–1029. [CrossRef] [PubMed]

molecules

MDPI

Article

Involvement of the γ Isoform of cPLA$_2$ in the Biosynthesis of Bioactive *N*-Acylethanolamines

Yiman Guo [1,2], Toru Uyama [1], S. M. Khaledur Rahman [1], Mohammad Mamun Sikder [1], Zahir Hussain [1], Kazuhito Tsuboi [3], Minoru Miyake [2] and Natsuo Ueda [1,*,†]

[1] Department of Biochemistry, Kagawa University School of Medicine, 1750-1 Ikenobe, Miki 761-0793, Kagawa, Japan; s18d710@stu.kagawa-u.ac.jp (Y.G.); uyama.toru@kagawa-u.ac.jp (T.U.); smk.rahman@just.edu.bd (S.M.K.R.); s20d710@stu.kagawa-u.ac.jp (M.M.S.); MDH106@pitt.edu (Z.H.)

[2] Department of Oral and Maxillofacial Surgery, Kagawa University School of Medicine, 1750-1 Ikenobe, Miki 761-0793, Kagawa, Japan; dentmm@med.kagawa-u.ac.jp

[3] Department of Pharmacology, Kawasaki Medical School, 577 Matsushima, Kurashiki 701-0192, Okayama, Japan; ktsuboi@med.kawasaki-m.ac.jp

*   Correspondence: nueda@med.kagawa-u.ac.jp; Tel.: +81-(87)-891-2104; Fax: +81-(87)-891-2105

†   Recipient of Mechoulam Award in 2020.

Citation: Guo, Y.; Uyama, T.; Rahman, S.M.K.; Sikder, M.M.; Hussain, Z.; Tsuboi, K.; Miyake, M.; Ueda, N. Involvement of the γ Isoform of cPLA$_2$ in the Biosynthesis of Bioactive *N*-Acylethanolamines. *Molecules* **2021**, *26*, 5213. https://doi.org/10.3390/molecules26175213

Academic Editor: Mauro Maccarrone

Received: 28 July 2021
Accepted: 23 August 2021
Published: 27 August 2021

Publisher's Note: MDPI stays neutral with regard to jurisdictional claims in published maps and institutional affiliations.

Abstract: Arachidonylethanolamide (anandamide) acts as an endogenous ligand of cannabinoid receptors, while other *N*-acylethanolamines (NAEs), such as palmitylethanolamide and oleylethanolamide, show analgesic, anti-inflammatory, and appetite-suppressing effects through other receptors. In mammalian tissues, NAEs, including anandamide, are produced from glycerophospholipid via *N*-acyl-phosphatidylethanolamine (NAPE). The ϵ isoform of cytosolic phospholipase A$_2$ (cPLA$_2$) functions as an *N*-acyltransferase to form NAPE. Since the cPLA$_2$ family consists of six isoforms (α, β, γ, δ, ϵ, and ζ), the present study investigated a possible involvement of isoforms other than ϵ in the NAE biosynthesis. Firstly, when the cells overexpressing one of the cPLA$_2$ isoforms were labeled with [^{14}C]ethanolamine, the increase in the production of [^{14}C]NAPE was observed only with the ϵ-expressing cells. Secondly, when the cells co-expressing ϵ and one of the other isoforms were analyzed, the increase in [^{14}C]*N*-acyl-lysophosphatidylethanolamine (lysoNAPE) and [^{14}C]NAE was seen with the combination of ϵ and γ isoforms. Furthermore, the purified cPLA$_2\gamma$ hydrolyzed not only NAPE to lysoNAPE, but also lysoNAPE to glycerophospho-*N*-acylethanolamine (GP-NAE). Thus, the produced GP-NAE was further hydrolyzed to NAE by glycerophosphodiesterase 1. These results suggested that cPLA$_2\gamma$ is involved in the biosynthesis of NAE by its phospholipase A$_1$/A$_2$ and lysophospholipase activities.

Keywords: *N*-acyltransferase; anandamide; endocannabinoid; phospholipase A$_2$

1. Introduction

N-Acylethanolamines (NAEs) are a class of bioactive lipids consisting of long-chain fatty acids and ethanolamine, and are widely present in animal and plant tissues [1]. They exhibit different biological activities depending on the type of constituent fatty acids. For example, *N*-arachidonoylethanolamine, also called arachidonylethanolamide or anandamide, functions as an endocannabinoid that binds to cannabinoid receptors CB1 and CB2 [2]. On the other hand, *N*-palmitoylethanolamine (palmitylethanolamine) [3] and *N*-oleoylethanolamine (oleylethanolamide) [4] act on the peroxisome proliferator-activated receptor (PPAR)-α as well as other receptors to show anti-inflammatory/analgesic and appetite-suppressing effects, respectively.

NAEs are biosynthesized from membrane phospholipids mainly in two-step enzyme reactions (Figure 1) [5,6]. The first reaction is the transfer of a fatty acyl chain from the *sn*-1 position of a glycerophospholipid molecule such as phosphatidylcholine (PC) to the amino group of a diacyl-type or plasmalogen-type phosphatidylethanolamine

187

(PE), resulting in the formation of *N*-acyl-phosphatidylethanolamine (NAPE), a unique phospholipid molecule with three fatty acyl chains. The enzymes catalyzing this reaction are collectively called *N*-acyltransferases, which are classified into two groups by Ca^{2+}-dependency [5–8]. The second reaction is the release of NAE from NAPE. The phospholipase D (PLD)-type enzyme NAPE-PLD directly produces NAE [9], while the alternative pathway does not involve NAPE-PLD, but consists of consecutive hydrolytic reactions via *N*-acyl-lysophosphatidylethanolamine (lysoNAPE) and glycerophospho-*N*-acylethanolamine (GP-NAE) (Figure 1) [10,11]. This pathway involves several hydrolases such as group IB, IIA, and V of secretory phospholipase A_2s (sPLA$_2$s) [12], α/β-hydrolase domain containing 4 (ABHD4) [13], and glycerophosphodiesterase (GDE) 1 [14]. The analysis of NAPE-PLD-deficient mice demonstrated the presence of the alternative pathway in the brain [11,15] and peripheral tissues such as heart, kidney, liver, and jejunum [16].

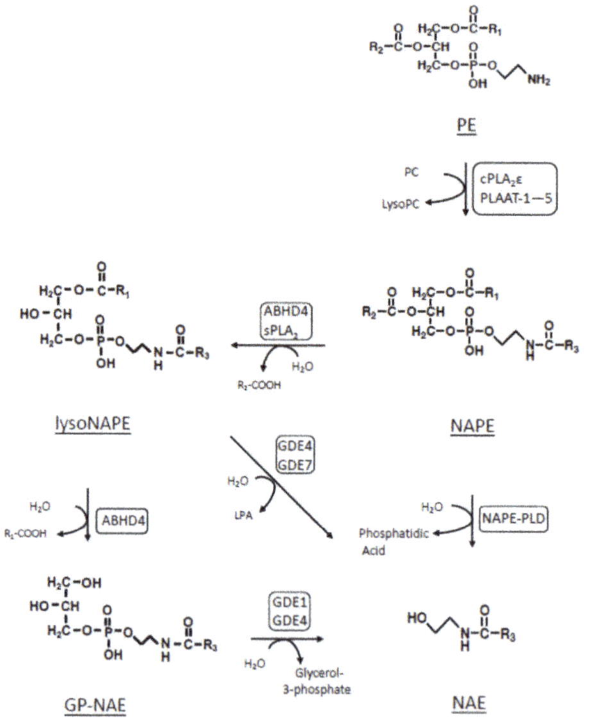

Figure 1. Biosynthetic pathways of NAE in mammals. PLAAT: phospholipase A and acyltransferase.

The cytosolic phospholipase A_2 (cPLA$_2$) family, also referred to as the group IV PLA$_2$ family, belongs to the PLA$_2$ superfamily and consists of six isoforms (α, β, γ, δ, ϵ, and ζ) [17]. Ogura et al. revealed that the ϵ isoform of the cPLA$_2$ family (cPLA$_2\epsilon$), also known as group IVE PLA$_2$ (PLA2G4E), functions as a Ca^{2+}-dependent *N*-acyltransferase to form NAPE [8]. On the other hand, the involvement of isoforms other than ϵ in the biosynthetic pathway of NAE has not yet been reported. In the present study, we examined the facilitatory effects of the isoforms of the cPLA$_2$ family on the NAE formation in living mammalian cells, as well as the reactivity of purified cPLA$_2\gamma$ with NAPE and lysoNAPE. The results suggested that the γ isoform deacylates NAPE to GP-NAE via the formation of lysoNAPE and constitutes the alternative pathway for the formation of NAE.

2. Results and Discussion

2.1. N-Acyltransferase Activity of cPLA₂ Isoforms in Living Cells

To examine whether mouse cPLA$_2$ isoforms have NAPE-producing N-acyltransferase activity, we transiently expressed each isoform (α, β, γ, δ, ϵ, or ζ) in human embryonic kidney (HEK)293 cells. Since these recombinant proteins were tagged with FLAG, their successful expression was confirmed by Western blotting using anti-FLAG antibody (Figure 2A). These FLAG-tagged proteins exhibited immunopositive bands at the position of the deduced molecular mass of each isoform (α, 85 kDa; β, 88 kDa; γ, 68 kDa; δ, 93 kDa; ϵ, 100 kDa; and ζ, 96 kDa), respectively. The extra bands were presumed to be degradative or modified proteins of each isoform.

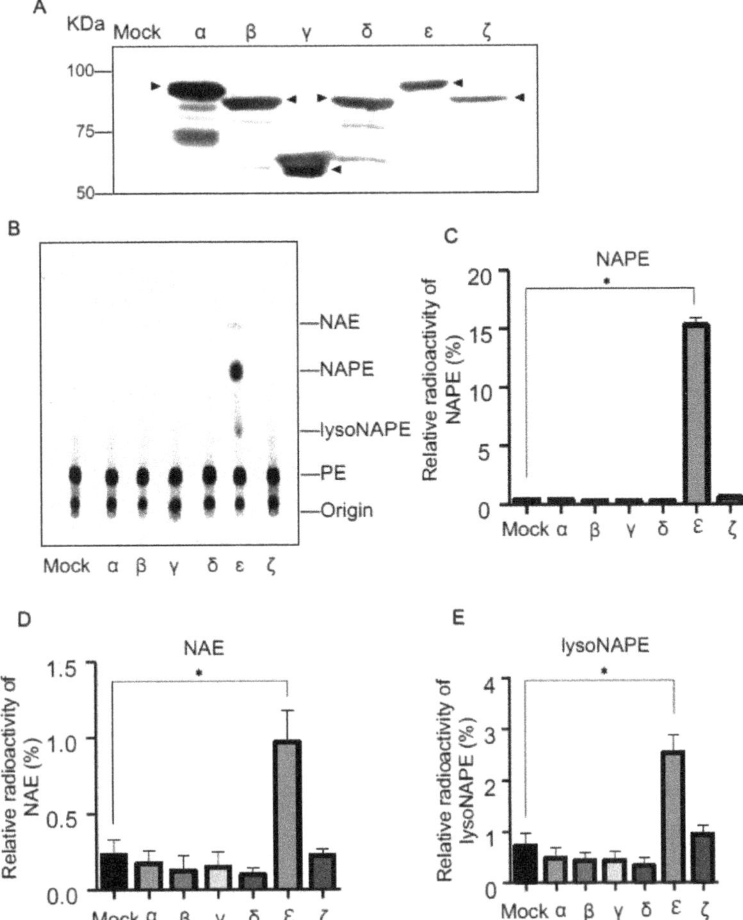

Figure 2. Metabolic labeling of cPLA$_2$-expressing cells with [^{14}C]ethanolamine. HEK293 cells were transfected with the insert-free vector (Mock) or the expression vector harboring cDNA for the indicated cPLA$_2$ isoforms tagged with FLAG. Their expressions were confirmed by Western blotting using anti-FLAG antibody (**A**). Arrowheads indicate the positions of the deduced molecular mass of each isoform. The cells were metabolically labelled with [^{14}C]ethanolamine, followed by the treatment with ionomycin. Total lipids were then analyzed by TLC (**B**). The positions of the origin and authentic compounds are shown. The relative radioactivities of NAPE (**C**), NAE (**D**), and lysoNAPE (**E**) are shown (mean values $\pm$ S.D., n = 3). * $p < 0.05$.

We recently reported that cPLA$_2\epsilon$-expressing cells pretreated with [^{14}C]ethanolamine produce a large amount of [^{14}C]NAPE in response to the Ca^{2+} ionophore ionomycin [18]. Thus, in the present study, we used this metabolic labeling as a simple method to detect Ca^{2+}-dependent N-acyltransferase activity in living cells. We cultured the HEK293 cells expressing each isoform in the presence of [^{14}C]ethanolamine for 18 h and further treated the cells with ionomycin for 30 min. Total lipids were then extracted from the cells and separated by thin-layer chromatography (TLC). The distribution of radioactivity on the thin-layer plate was visualized (Figure 2B) and quantified (Figure 2C–E). In comparison with the control cells transfected with an insert-free vector, the cPLA$_2\epsilon$-expressing cells exhibited remarkable increases in the intensities of the radioactive bands corresponding to authentic NAPE, NAE, and lysoNAPE. However, such increases were not seen with the cells expressing cPLA$_2$ isoforms other than ϵ.

The δ isoform (PLA2G4D) was previously reported not to exhibit N-acyltransferase activity [8]. Although our results suggested that the ϵ isoform is the sole enzyme functioning in living cells as NAPE-forming N-acyltransferase, we could not rule out the possibility that the other isoforms show N-acyltransferase activity under different assay conditions.

2.2. NAPE-PLA$_1$/A$_2$ Activity of cPLA$_2$ Isoforms in Living Cells

Since all the cPLA$_2$ isoforms were previously reported to exhibit PLA$_1$/A$_2$ activity for glycerophospholipids such as PC [17], it was likely that these isoforms also showed PLA$_1$/A$_2$ activity for NAPE in living cells. For this purpose, we used cPLA$_2\epsilon$/Tet-on cells, which stably expressed FLAG-tagged cPLA$_2\epsilon$ in the presence of doxycycline (DOX) and produced a large amount of NAPE when stimulated by a Ca^{2+} ionophore [18]. We transiently expressed one of α, β, γ, δ, and ζ isoforms with a FLAG tag in cPLA$_2\epsilon$/Tet-on cells in the presence of DOX. Successful co-expression of ϵ isoform and one of the other isoforms was confirmed by Western blotting using anti-FLAG antibody (Figure 3A). cPLA$_2\epsilon$ was stably expressed in cPLA$_2\epsilon$/Tet-on cells in the presence of doxycyclin, while other isoforms of cPLA$_2$ were transiently and potently expressed by the introduction of each cDNA using Lipofectamine 2000. The differences in the expression levels between ϵ and other isoforms were presumably attributed to the differences in the expression methods. Metabolic labeling of cPLA$_2\epsilon$/Tet-on cells with [^{14}C]ethanolamine, followed by the ionomycin treatment, exhibited the production of large amounts of radioactive NAPE, NAE, and lysoNAPE due to the activation of cPLA$_2\epsilon$ (Figure 3B) as reported previously [18]. Interestingly, as compared with the sole expression of ϵ, the co-expression of ϵ with γ, but not with α, β, δ, or ζ, showed a lower level of NAPE (Figure 3C) and higher levels of NAE (Figure 3D) and lysoNAPE (Figure 3E). These results suggested that NAPE, produced by ϵ, was hydrolyzed to lysoNAPE by the PLA$_1$/A$_2$ activity of γ. The increase in NAE levels by the expression of γ was presumably due to further hydrolysis of the increased lysoNAPE.

Earlier, mouse ABHD4 was reported to have the ability to hydrolyze NAPE to lysoNAPE, and then lysoNAPE to GP-NAE (Figure 1) [13]. In fact, substantial reductions in GP-NAE and plasmalogen-type lysoNAPE were observed in the brain of ABHD4-deficient mice [19]. Thus, we also transiently expressed mouse ABHD4 in cPLA$_2\epsilon$/Tet-on cells. Western blotting revealed the expression of FLAG-tagged ABHD4 with a molecular mass of 39 kDa (Figure 3A). However, in the metabolic labeling with [^{14}C]ethanolamine, the expression of ABHD4 did not significantly affect the levels of radioactive NAPE, NAE, or lysoNAPE (Figure 3B–E), despite the fact that the purified ABHD4 successfully hydrolyzed N-[^{14}C]palmitoyl-PE to N-[^{14}C]palmitoyl-lysoPE (Figure 4A,B). The reason for this discrepancy remained unclear.

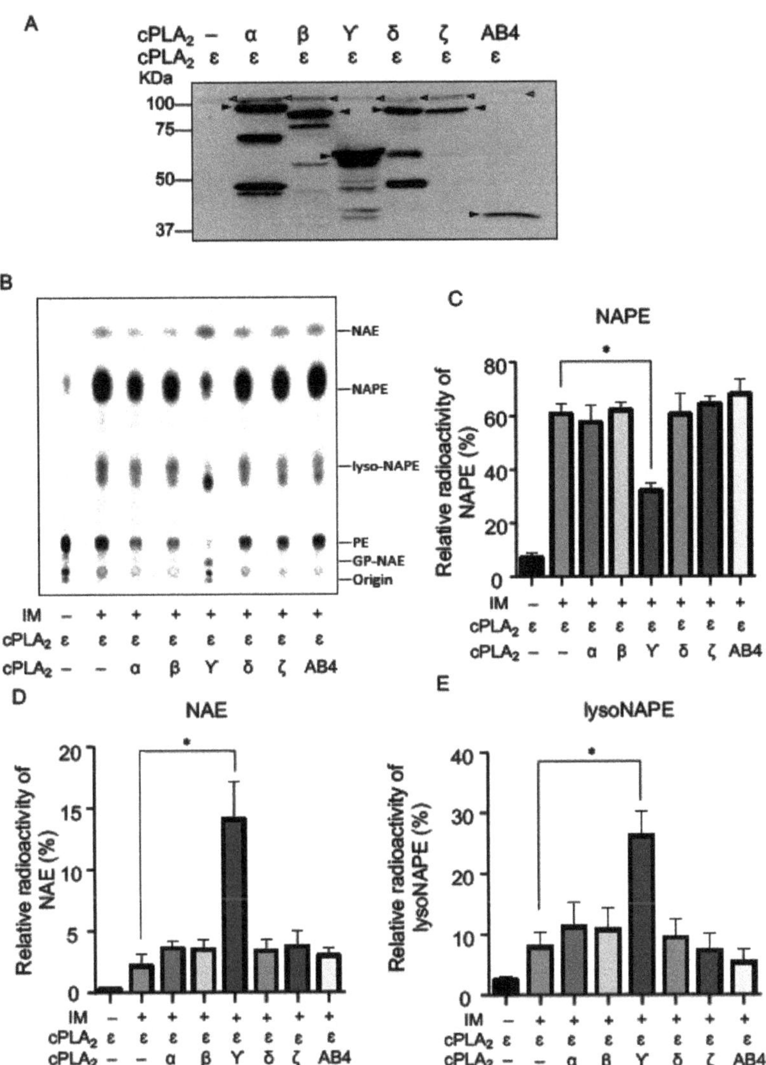

Figure 3. Metabolic labeling with [^{14}C]ethanolamine of the cells co-expressing cPLA$_2\varepsilon$ and one of the other isoforms. cPLA$_2\varepsilon$/Tet-on cells were transfected with the insert-free vector or the expression vector harboring cDNA for the indicated FLAG-tagged cPLA$_2$ isoforms or ABHD4 (AB4). Their expressions were confirmed by Western blotting using anti-FLAG antibody (**A**). Arrowheads indicate the positions of the deduced molecular mass of each isoform and ABHD4. The cells were labelled with [^{14}C]ethanolamine, followed by the treatment with ionomycin (IM). Total lipids were then analyzed by TLC (**B**). The positions of the origin and authentic compounds are shown. The relative radioactivities of NAPE (**C**), NAE (**D**), and lysoNAPE (**E**) are shown (mean values ± S.D., n = 3). * $p < 0.05$.

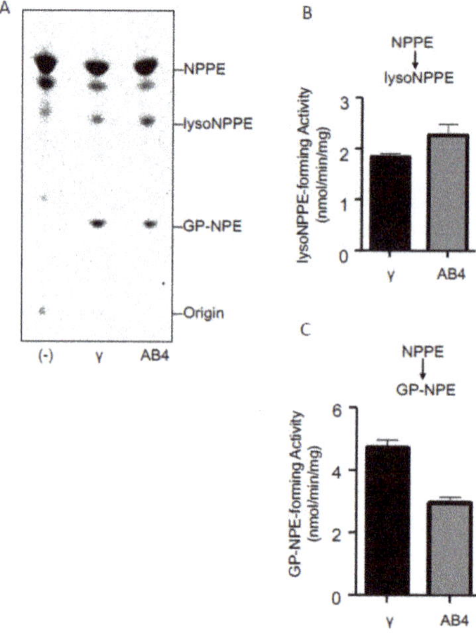

Figure 4. Reactivity of cPLA$_2$γ and ABHD4 with NAPE. The purified cPLA$_2$γ and ABHD4 (AB4) as well as buffer alone (-) were allowed to react with N-[^{14}C]palmitoyl-PE (NPPE), and the products were analyzed by TLC (**A**). The positions of the origin and authentic compounds are shown. The N-[^{14}C]palmitoyl-lysoPE (lysoNPPE)-forming activity (**B**) and GP-N-[^{14}C]palmitoylethanolamine (GP-NPE)-forming activity (**C**) are shown (mean values ± S.D., n = 3).

2.3. Activities of Purified cPLA$_2$γ and ABHD4

We expressed FLAG-tagged cPLA$_2$γ and ABHD4 in HEK293 cells and purified these enzymes by anti-FLAG affinity chromatography. The purified enzymes were allowed to react with N-[^{14}C]palmitoyl-PE, and the radioactive products were separated by TLC (Figure 4A). The results showed that both enzymes produced two radioactive bands corresponding to N-[^{14}C]palmitoyl-lysoPE (Figure 4B) and GP-N-[^{14}C]palmitoylethanolamine (Figure 4C). Moreover, when the purified enzymes were incubated with N-[^{14}C]palmitoyl-lysoPE, the production of GP-N-[^{14}C]palmitoylethanolamine was observed (Figure 5A,B). These results showed that the purified cPLA$_2$γ, as well as the purified ABHD4 catalyzed two sequential hydrolytic reactions to convert NAPE to GP-NAE via lysoNAPE. Notably, cPLA$_2$γ catalyzed the latter reaction at a higher rate than the former reaction, suggesting the efficient formation of GP-NAE from NAPE.

cPLA$_2$γ was earlier cloned from a human [20] and characterized as a novel membrane-bound, Ca^{2+}-independent PLA$_2$ [17,20]. Our preliminary assay also showed that the purified cPLA$_2$γ can hydrolyze [^{14}C]PC (data not shown). In addition to PLA$_2$ activity, cPLA$_2$γ exhibited PLA$_1$, lysophospholipase, and acyltransferase activity [21]. The substrates used were PC, PE, lysoPC, and lysoPE. Thus, cPLA$_2$γ sequentially hydrolyzed two acyl chains from *sn*-1 and -2 positions of the glycerol backbone of PC and PE, resulting in the formation of glycerophosphocholine or glycerophosphoethanolamine, respectively. The ability to convert NAPE to GP-NAE via lysoNAPE (Figures 4 and 5) may be explained by this multi-function of cPLA$_2$γ.

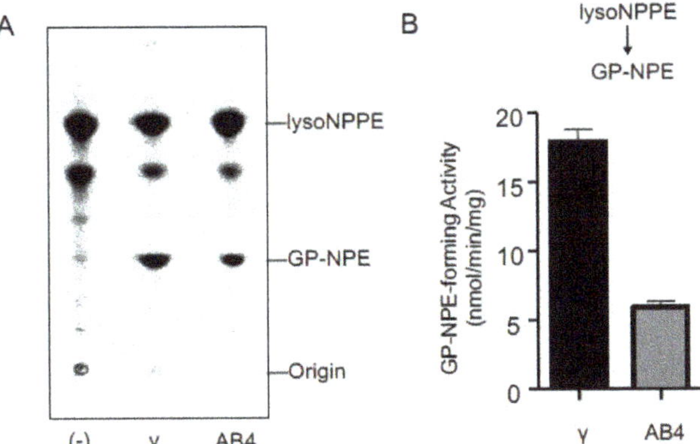

Figure 5. Reactivity of cPLA$_2$γ and ABHD4 with lysoNAPE. The purified cPLA$_2$γ and ABHD4 (AB4), as well as buffer alone (-) were allowed to react with N-[^{14}C]palmitoyl-lysoPE (lysoNPPE), and the products were then analyzed by TLC (**A**). The positions of the origin and authentic compounds are shown. The GP-N-[^{14}C]palmitoylethanolamine (GP-NPE)-forming activity is shown (mean values ± S.D., n = 3) (**B**).

To identify one of the cPLA$_2$γ products as GP-N-[^{14}C]palmitoylethanolamine, we extracted this radioactive product from silica gel with organic solvent and allowed the substance to react with purified recombinant mouse GDE1, which is known to hydrolyze GP-NAE to NAE and glycerol 3-phoshate [14]. As shown in Figure 6A, GDE1 converted the substance to a radioactive band corresponding to authentic N-[^{14}C]palmitoylethanolamine. Furthermore, we incubated the purified cPLA$_2$γ with N-[^{14}C]palmitoyl-PE for 30 min and then added the purified GDE1 to the reaction mix, followed by further incubation for 15 min (Figure 6B–E). This sequential reaction led to the production of the radioactive band corresponding to N-[^{14}C]palmitoylethanolamine. In contrast, GDE1 was inactive with N-[^{14}C]palmitoyl-PE. These results suggest that GP-NAE produced by cPLA$_2$γ is converted to NAE by GDE1.

2.4. Tissue Distributions of cPLA$_2$γ and ABHD4

We examined the distribution of mRNAs of cPLA$_2$γ and ABHD4 in mouse tissues by reverse transcription-PCR (Figure 7A). cPLA$_2$γ mRNA was widely distributed in various tissues with higher expression levels in the liver, colon, and testis, followed by many other tissues (Figure 7B). On the other hand, ABHD4 mRNA was widely distributed with higher levels in the brain, heart, lung, ileum, kidney, testis, and skeletal muscle (Figure 7C). Previously, the tissue distribution of mouse ABHD4 mRNA was reported with the highest expression in the central nervous system and testis, followed by the liver and kidney, with negligible signals in the heart [13].

In the nervous system, ABHD4 did not appear to be the sole enzyme that hydrolyzed NAPE and lysoNAPE [19]. Moreover, the enzyme(s) responsible in peripheral tissues have not fully been understood. Considering the wide distribution of cPLA$_2$γ in mouse tissues, cPLA$_2$γ may function as an alternative of ABHD4 in the NAE biosynthesis. Specific inhibitors for these enzymes will be useful to quantitatively estimate the contribution of each enzyme, particularly if used in primary culture.

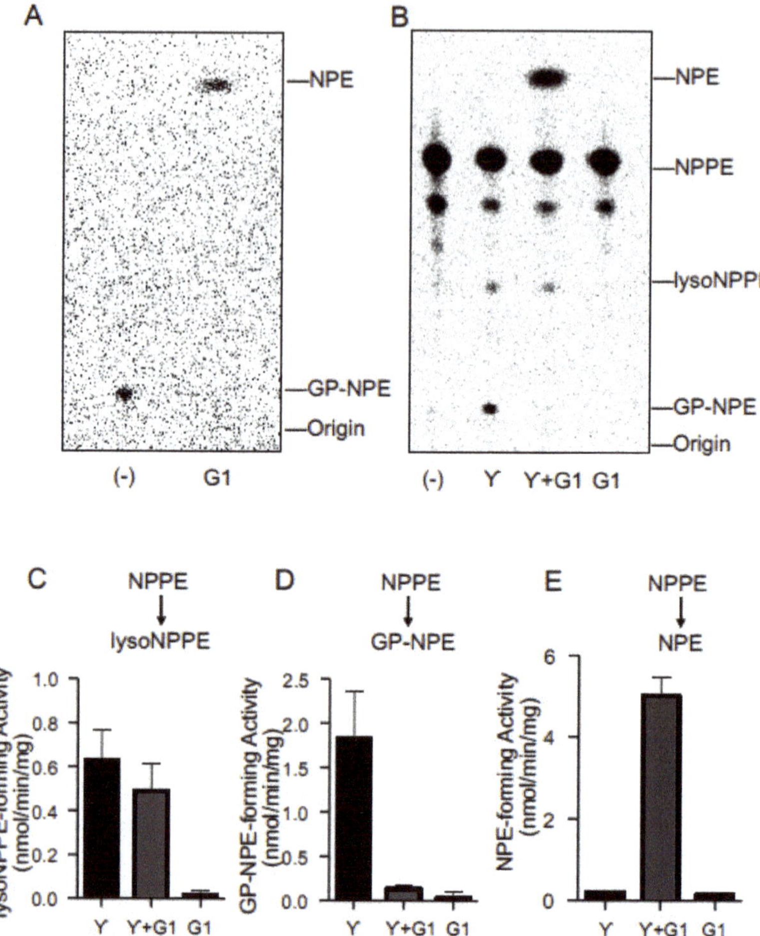

Figure 6. The formation of NAE from GP-NAE by GDE1. The radioactive band corresponding to GP-N-[^{14}C]palmitoylethanolamine (GP-NPE), which was produced by purified cPLA$_2$γ, was scraped from the TLC plate. The radioactive compound was then extracted by the Bligh and Dyer protocol, and incubated with purified GDE1 (G1) or water (-) (**A**). N-[^{14}C]Palmitoyl-PE (NPPE) was incubated with the purified cPLA$_2$γ for 30 min and additionally with the purified GDE1 for 15 min (γ+G1) (**B**). GDE1 or cPLA$_2$γ was omitted in γ and G1, respectively. The products were analyzed by TLC. The positions of the origin and authentic compounds are shown. The N-[^{14}C]palmitoyl-lysoPE (lysoNPPE)-forming activity (**C**), GP-NPE-forming activity (**D**), and N-[^{14}C]palmitoylethanolamine (NPE)-forming activity (**E**) are shown (mean values ± S.D., n = 3).

As for human cPLA$_2$γ, Northern blot analysis indicated that cPLA$_2$γ mRNA is most abundant in the skeletal muscle and heart, with lower levels in the spleen, brain, placenta, and pancreas [20]. Human cPLA$_2$γ (accession number, NP_003697) [20] and mouse cPLA$_2$γ (NM_001004762) were deduced to comprise 541 and 597 amino acids, respectively. Arg-54, Ser-82, and Asp-385, forming the catalytic center of human cPLA$_2$γ, were conserved as Arg-55, Ser-83, and Asp-417 in mouse cPLA$_2$γ [17]. Human cPLA$_2$γ [20] and mouse cPLA$_2$γ (Guo et al., unpublished observation) showed the PLA$_1$/A$_2$ activity for PC. Thus, human cPLA$_2$γ was considered to be the ortholog of the mouse enzyme.

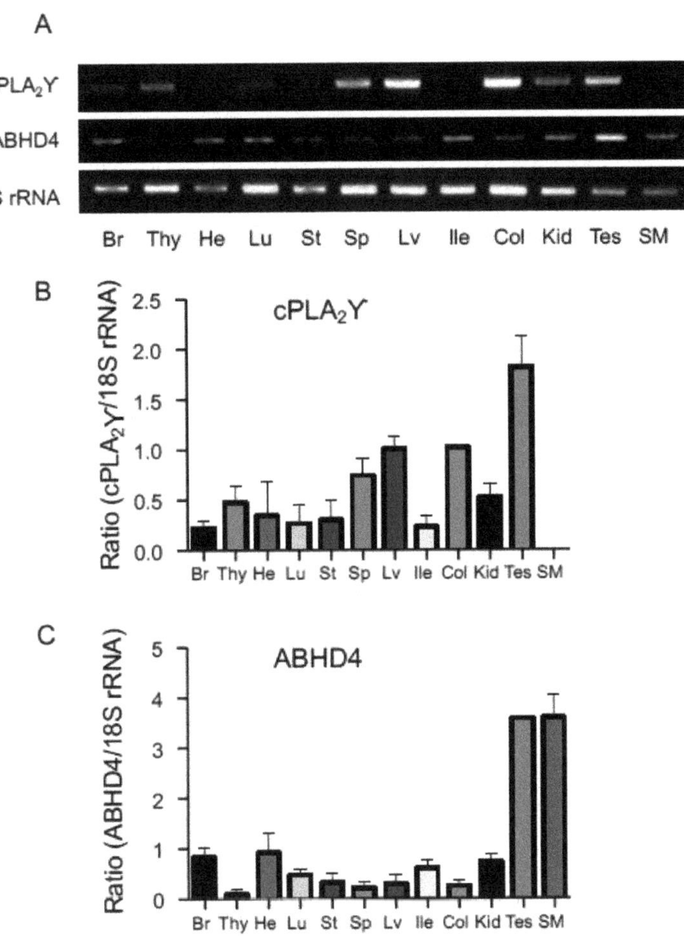

Figure 7. Tissue distribution of cPLA$_2$γ and ABHD4 in mice. mRNAs from the indicated mouse tissues were analyzed by reverse transcription-PCR using primers specific for cPLA$_2$γ, ABHD4, and 18S rRNA (a control) (**A**). The semi-quantitative results are also shown (mean values ± S.D., n = 3) (**B,C**). Br: brain; Thy: thymus; He: heart; Lu: lung; St: stomach; Sp: spleen; Lv: liver; Ile: ileum; Col: colon; Kid: kidney; Tes: testis; SM: skeletal muscle.

3. Materials and Methods

3.1. Materials

[1,2-^{14}C]Ethanolamine-HCl ([^{14}C]ethanolamine) was purchased from Moravek Biochemicals (Brea, CA, USA); anti-FLAG M2-conjugated agarose affinity gel and FLAG peptide were from Sigma-Aldrich (St. Louis, MO, USA); rabbit anti-FLAG (DYKDDDDK) monoclonal antibody was from Cell Signaling Technology (Danvers, MA, USA); horseradish peroxidase-linked anti-rabbit IgG was from GE Healthcare (Piscataway, NJ, USA); protein assay dye reagent concentrate was from Bio-Rad (Hercules, CA, USA); PrimeScript RT reagent kit was from Takara Bio (Kusatsu, Japan); precoated silica gel 60 F254 aluminum sheets for TLC (20 × 20 cm, 0.2 mm thickness) and Immobilon-P were from Merck Millipore (Darmstadt, Germany); fetal bovine serum, Lipofectamine 2000, TRIzol, pEF6/Myc-His vector, and Pierce Western Blotting Substrate Plus were from Invitrogen/Thermo Fisher Scientific (Carlsbad, CA, USA); Nonidet P-40 was from Nacalai Tesque (Kyoto, Japan); Dulbecco's modified Eagle's medium (DMEM), dithiothreitol (DTT), 3(2)-*t*-butyl-4-hydroxyanisole (BHA), Tween

20, and ionomycin were from FUJIFILM Wako Pure Chemical (Osaka, Japan); KOD-Plus-Neo polymerase and Quick taq DNA polymerase were from TOYOBO (Osaka, Japan); *n*-octyl-β-D-glucoside and 3-[(3-cholamidopropyl)dimethylammonio]-propanesulfonate (CHAPS) were from Dojindo (Kumamoto, Japan); DOX was from Clontech (Mountain View, CA, USA); HEK293 cells were from Health Science Research Resources Bank (Osaka, Japan). 1,2-Dioleoyl-*sn*-glycero-3-phospho(N-[1'-^{14}C]palmitoyl)ethanolamine (N-[^{14}C]palmitoyl-PE), 1-oleoyl-2-hydroxy-*sn*-glycero-3-phospho(N-[1'-^{14}C]palmitoyl)ethanolamine (N-[^{14}C] palmitoyl-lysoPE), *sn*-glycero-3-phospho(N-[1'-^{14}C]palmitoyl)ethanolamine (glycerophospho-N-[^{14}C]palmitoylethanolamine or GP-N-[^{14}C]palmitoylethanolamine) and N-[^{14}C] palmitoylethanolamine were enzymatically prepared as described previously [22]. The products were purified by TLC with a mixture of chloroform/methanol/28% ammonium hydroxide (80:20:2, by vol.) or chloroform/methanol/acetic acid (9:1:1, by vol.).

3.2. Construction of Expression Vectors

C57BL/6 mice (male, 8 weeks old) (Japan SLC, Inc., Hamamatsu, Japan) were anesthetized and sacrificed by decapitation according to the guidelines for care and use of animals established by Kagawa University (Kagawa, Japan). Total RNAs were then isolated using TRIzol from the mouse tissues indicated in Table 1. First-strand cDNA was prepared from 5 µg of total RNA using a PrimeScript RT reagent kit. The cDNA encoding N-terminally FLAG-tagged mouse cPLA$_2$α, cPLA$_2$β, cPLA$_2$γ, cPLA$_2$δ, cPLA$_2$ζ, and ABHD4 was amplified by PCR with KOD-Plus-Neo DNA polymerase. The primers used are shown in Table 1. PCR was carried out for 35 cycles at 94 °C for 20 s, 56 °C for 20 s, and 68 °C for 3 min. The obtained DNA fragments were subcloned into the corresponding sites of pEF6/Myc-His. All the constructs were sequenced in both directions using an ABI 3130 Genetic Analyzer (Invitrogen/Life Technologies, Carlsbad, CA, USA). The expression vectors harboring N-terminally FLAG-tagged mouse cPLA$_2$ε [23] and C-terminally FLAG-tagged mouse GDE1 [22] were constructed as described previously.

Table 1. Primers and mouse tissues used for the construction of expression vectors.

cDNA (Accession Number)	Direction	Sequence (Restriction Sites and a Tag Sequence)	Tissue
cPLA$_2$α (NM_008869)	forward	5'-cgcactagtggaaaatggattacaaggatgacgacgataagtctttcatagatccttatcagcac-3' (*Spe* I site, in-frame FLAG sequence)	Thymus
	reverse	5'-cgcgcggccgcttacacagtgggtttacttagaaa-3' (*Not* I site)	
cPLA$_2$β (NM_145378)	forward	5'-cgcactagtggaaaatggattacaaggatgacgacgataaggctctgcaaacctgcccagtctac-3' (*Spe* I site, in-frame FLAG sequence)	Brain
	reverse	5'-cgcgcggccgctcactccggcctaaactgtttgcg-3' (*Not* I site)	
cPLA$_2$γ (NM_001004762)	forward	5'-cgcactagtggaaaatggattacaaggatgacgacgataaggaactaagctctggggtctgccct-3' (*Spe* I site, in-frame FLAG sequence)	Brain
	reverse	5'-cgcgcggccgcttaatccttagatatgttgtggga-3' (*Not* I site)	
cPLA$_2$δ (NM_001024137)	forward	5'-cgcactagtggaaaatggattacaaggatgacgacgataagtggagtggagatagaagagtaggc-3' (*Spe* I site, in-frame FLAG sequence)	Testis
	reverse	5'-cgcgcggccgctcacgtcttcactcccaatggcct-3' (*Not* I site)	
cPLA$_2$ζ (NM_001024145)	forward	5'-cgcactagtggaaaatggattacaaggatgacgacgataagccctggactctccagccaaagtgg-3' (*Spe* I site, in-frame FLAG sequence)	Large intestine
	reverse	5'-cgcgcggccgctcagcccccaacccttcccccagc-3' (*Not* I site)	
ABHD4 (NM_134076)	forward	5'-cgcactagtggaaaatggattacaaggatgacgacgataaggctgatgatctggagcagcagcctcag-3' (*Spe* I site, in-frame FLAG sequence)	Brain
	reverse	5'-cgcgcggccgctcagtcaactgagttgcagatctcttc-3' (*Not* I site)	

3.3. Metabolic Labeling

A Tet-on cell line (FL-cPLA$_2$ε/Tet-on), which DOX-dependently expresses FLAG-tagged cPLA$_2$ε, was established by the transfection of HEK293 cells with pcDNA5/TO vector harboring FLAG-tagged cPLA$_2$ε as reported previously [18]. The cells were maintained for at least four days in the presence of 1 µg/mL DOX.

HEK293 cells and FL-cPLA$_2\varepsilon$/Tet-on cells were grown at 37 °C to 80% confluency in 6-well plastic plates containing DMEM with 10% fetal bovine serum in a humidified 5% CO$_2$ and 95% air incubator. For the transient expression of FLAG-tagged enzymes, the expression vectors harboring cDNA of each enzyme were introduced into the cells using Lipofectamine 2000 according to the manufacturer's instructions. Twenty-four hours after transfection, the cells were labeled with [^{14}C]ethanolamine (0.16 µCi/well) for 18 h. [^{14}C]Ethanolamine was then removed and serum-free fresh medium with or without 2 µM ionomycin was added to the wells. After further incubation at 37 °C for 30 min, total lipids were extracted by the method of Bligh and Dyer [24], spotted on a silica gel thin-layer plate (20 cm height), and developed at 4 °C for 90 min with a mixture of chloroform/methanol/28% ammonium hydroxide (80:20:2, by vol.). The distribution of radioactivity on the plate was visualized and quantified using an image reader FLA-7000 (FUJIFILM, Tokyo, Japan). All assays were performed in triplicate.

3.4. Expression and Purification of Recombinant Proteins

HEK293 cells were grown at 37 °C to 90% confluency in 150 mm plastic dishes containing DMEM with 10% fetal bovine serum in a humidified 5% CO$_2$ and 95% air incubator. For the expression of recombinant FLAG-tagged cPLA$_2\gamma$, ABHD4, or GDE1, their expression vectors were introduced into HEK293 cells using Lipofectamine 2000 according to the manufacturer's instructions. Forty-eight hours after transfection, cells were harvested from 2 to 3 dishes and sonicated twice each for 5 s in 20 mM Tris-HCl (pH 7.4).

For the purification of cPLA$_2\gamma$ and ABHD4, soluble fractions were prepared from the cell homogenates by centrifugation in the presence of 0.1% Nonidet P-40 at 105,000× *g* for 30 min at 4 °C; they were then mixed with 1 mL of a 50% slurry of anti-FLAG M2 affinity gel pre-equilibrated with 50 mM Tris-HCl (pH 7.4) containing 150 mM NaCl and 0.05% Nonidet P-40 (buffer A). After overnight incubation at 4 °C under gentle mixing, the gel was packed into a column and washed three times each with 12 mL of buffer A. The FLAG-tagged protein was eluted with buffer A containing 0.1 mg/mL of FLAG peptide, and every 0.25 mL fraction was collected.

For the purification of GDE1, particulate fractions were prepared from the homogenates of the GDE1-expressing cells by centrifugation in the presence of 0.1% octyl glucoside at 105,000× *g* for 30 min at 4 °C. GDE1 was then solubilized and purified as described previously [22].

The protein concentration was determined by the method of Bradford with bovine serum albumin as a standard.

3.5. Enzyme Assay

Purified cPLA$_2\gamma$ (2 µg protein) and ABHD4 (2 µg protein) were allowed to react with 25 µM *N*-[^{14}C]palmitoyl-PE (25,000 cpm, dissolved in 5 µL ethanol) or 25 µM *N*-[^{14}C]palmitoyl-lysoPE (25,000 cpm, dissolved in 5 µL ethanol) in 100 µL of 50 mM Tris-HCl (pH 7.4), 0.1% CHAPS, and 5 mM EDTA at 37 °C for 30 min. The reaction was terminated by adding 0.32 mL of chloroform/methanol/1 M citric acid (8:4:1, *v/v*) containing 5 mM BHA. After centrifugation, 100 µL of the organic phase was spotted on a thin-layer silica gel plate (20 cm height), and developed in chloroform/methanol/water (65:25:4, by vol.) at 4 °C for 90 min.

Purified GDE1 (2 µg protein) was incubated with ^{14}C-labeled compounds in 100 µL of 50 mM Tris-HCl (pH 7.4), 3 mM DTT, and 2 mM MgCl$_2$ at 37 °C for 30 min. The reaction was terminated by adding 0.32 mL of chloroform/methanol/1 M citric acid (8:4:1, by vol.) containing 5 mM BHA. After centrifugation, 100 µL of the organic phase was spotted on a thin-layer silica gel plate (20 cm height), and developed in chloroform/methanol/28% ammonium hydroxide (80:20:2, by vol.) at 4 °C for 90 min.

After the development by TLC, the radioactive substances on the plate were quantified by an FLA7000 image analyzer. All enzyme assays were performed in triplicate.

3.6. Statistical Analysis

Statistical significance was assessed using one-way ANOVA followed by Tukey's multiple comparison test using GraphPad Prism 8 (GraphPad Software Inc., La Jolla, CA, USA), with $p < 0.05$ considered statistically significant. Data are presented as mean $\pm$ standard deviation (S.D.) of the mean.

3.7. Western Blotting

The homogenates (30 µg of protein) of the cells expressing FLAG-tagged enzymes were separated by SDS-PAGE on 10% gel and electrotransferred to a hydrophobic polyvinylidene difluoride membrane (Immobilon-P). The membrane was blocked with PBS containing 5% dried skimmed milk and 0.1% Tween 20 (buffer B) and then incubated with anti-FLAG antibody (1:2000 dilution) in buffer B at room temperature for 1 h, followed by incubation with horseradish peroxidase-labeled anti-rabbit IgG antibody (1:4000 dilution) in buffer B at room temperature for 1 h. The membrane was finally treated with Pierce Western Blotting Substrate Plus, and the labeled proteins were visualized with the aid of a LAS1000plus luminoimaging analyzer (FUJIX Ltd., Tokyo, Japan).

3.8. Reverse Transcription-PCR

First-strand cDNA was prepared as described in "*4.2. Construction of Expression Vectors*" and subjected to PCR amplification by Quick taq DNA polymerase. The primer sequences used are shown in Table 2. The PCR conditions were as follows: 30 cycles with denaturation at 98 °C for 10 s, annealing and extension at 68 °C for 60 s for $cPLA_2\gamma$; 30 cycles with denaturation at 98 °C for 10 s, annealing and extension at 68 °C for 60 s for ABHD4; 18 cycles with denaturation at 98 °C for 10 s, annealing at 55 °C for 30 s, and extension at 68 °C for 30 s for 18S rRNA as a control.

Table 2. Primers used for reverse transcription-PCR.

Gene	Direction	Sequence
$cPLA_2\gamma$	forward	5′-tgaggtgagcgaggatcagctgaag-3′
	reverse	5′-atgagtcagatagttttactgtccc-3′
ABHD4	forward	5′-ggcacagtttgggaggattcctggc-3′
	reverse	5′-gaggtgcacggatctcactagggtc-3′
18S rRNA	forward	5′-gtaacccgttgaaccccatt-3′
	reverse	5′-ccatccaatcggtagtagcg-3′

4. Conclusions

In the present study, we first suggested that in living cells, the γ isoform of $cPLA_2$ has PLA_1/A_2 activity to generate lysoNAPE from NAPE, which was produced by the ε isoform of $cPLA_2$. We then showed that the purified $cPLA_2\gamma$ hydrolyzes not only NAPE to lysoNAPE, but also lysoNAPE to GP-NAE. These consecutive hydrolytic reactions starting from NAPE were previously reported with ABHD4. Considering the wide distribution of $cPLA_2\gamma$ in mouse tissues, $cPLA_2\gamma$ may function as an alternative of ABHD4, constituting the NAPE-PLD-independent pathway for the biosynthesis of bioactive NAEs.

Author Contributions: Conceptualization, T.U. and N.U.; investigation and formal analysis, Y.G., T.U., S.M.K.R., M.M.S. and Z.H.; resources, T.U. and K.T.; writing—original draft, Y.G.; writing—review and editing, Y.G., T.U., Z.H., K.T. and N.U.; visualization, Y.G. and T.U.; supervision, T.U., M.M. and N.U.; funding acquisition, T.U., Z.H. and N.U. All authors have read and agreed with the published version of the manuscript.

Funding: The authors are supported by Grants-in-Aid for Scientific Research from the Japan Society for the Promotion of Science (T.U., grant number JP 18K06915; Z.H., grant number JP 19K23828; N.U., grant number JP 19K07353) as well as a research grant from Charitable Trust MIU Foundation Memorial Fund (to T.U.).

Institutional Review Board Statement: The study was conducted according to the guidelines of the Declaration of Helsinki, and approved by the Animal Care and Use Committee for Kagawa University (Approval No. 21618).

Informed Consent Statement: Not applicable.

Data Availability Statement: Not applicable.

Acknowledgments: We acknowledge Divisions of Research Instrument and Equipment and Radioisotope Research, Life Science Research Center, Kagawa University.

Conflicts of Interest: The authors declare no conflict of interest.

References

1. Schmid, H.H.O.; Schmid, P.C.; Natarajan, V. *N*-acylated glycerophospholipids and their derivatives. *Prog. Lipid Res.* **1990**, *29*, 1–43. [CrossRef]
2. Devane, W.A.; Hanus, L.; Breuer, A.; Pertwee, R.G.; Stevenson, L.A.; Griffin, G.; Gibson, D.; Mandelbaum, A.; Etinger, A.; Mechoulam, R. Isolation and structure of a brain constituent that binds to the cannabinoid receptor. *Science* **1992**, *258*, 1946–1949. [CrossRef]
3. Lo Verme, J.; Fu, J.; Astarita, G.; la Rana, G.; Russo, R.; Calignano, A.; Piomelli, D. The nuclear receptor peroxisome proliferator-activated receptor-alpha mediates the anti-inflammatory actions of palmitoylethanolamide. *Mol. Pharmacol.* **2005**, *67*, 15–19. [CrossRef]
4. Rodríguez de Fonseca, F.; Navarro, M.; Gómez, R.; Escuredo, L.; Nava, F.; Fu, J.; Murillo-Rodríguez, E.; Giuffrida, A.; loVerme, J.; Gaetani, S.; et al. An anorexic lipid mediator regulated by feeding. *Nature* **2001**, *414*, 209–212. [CrossRef]
5. Ueda, N.; Tsuboi, K.; Uyama, T. *N*-acylethanolamine metabolism with special reference to *N*-acylethanolamine-hydrolyzing acid amidase (NAAA). *Prog. Lipid Res.* **2010**, *49*, 299–315. [CrossRef]
6. Hussain, Z.; Uyama, T.; Tsuboi, K.; Ueda, N. Mammalian enzymes responsible for the biosynthesis of *N*-acylethanolamines. *Biochim. Biophys. Acta* **2017**, *1862*, 1546–1561. [CrossRef]
7. Jin, X.H.; Okamoto, Y.; Morishita, J.; Tsuboi, K.; Tonai, T.; Ueda, N. Discovery and characterization of a Ca^{2+}-independent phosphatidylethanolamine *N*-acyltransferase generating the anandamide precursor and its congeners. *J. Biol. Chem.* **2007**, *282*, 3614–3623. [CrossRef]
8. Ogura, Y.; Parsons, W.H.; Kamat, S.S.; Cravatt, B.F. A calcium-dependent acyltransferase that produces *N*-acyl phosphatidylethanolamines. *Nat. Chem. Biol.* **2016**, *12*, 669–671. [CrossRef]
9. Okamoto, Y.; Morishita, J.; Tsuboi, K.; Tonai, T.; Ueda, N. Molecular characterization of a phospholipase D generating anandamide and its congeners. *J. Biol. Chem.* **2004**, *279*, 5298–5305. [CrossRef]
10. Natarajan, V.; Schmid, P.C.; Reddy, P.V.; Schmid, H.H.O. Catabolism of *N*-acylethanolamine phospholipids by dog brain preparations. *J. Neurochem.* **1984**, *42*, 1613–1619. [CrossRef] [PubMed]
11. Tsuboi, K.; Okamoto, Y.; Ikematsu, N.; Inoue, M.; Shimizu, Y.; Uyama, T.; Wang, J.; Deutsch, D.G.; Burns, M.P.; Ulloa, N.M.; et al. Enzymatic formation of *N*-acylethanolamines from *N*-acylethanolamine plasmalogen through *N*-acylphosphatidylethanolamine-hydrolyzing phospholipase D-dependent and -independent pathways. *Biochim. Biophys. Acta* **2011**, *1811*, 565–577. [CrossRef]
12. Sun, Y.X.; Tsuboi, K.; Okamoto, Y.; Tonai, T.; Murakami, M.; Kudo, I.; Ueda, N. Biosynthesis of anandamide and *N*-palmitoylethanolamine by sequential actions of phospholipase A_2 and lysophospholipase D. *Biochem. J.* **2004**, *380*, 749–756. [CrossRef]
13. Simon, G.M.; Cravatt, B.F. Endocannabinoid biosynthesis proceeding through glycerophospho-*N*-acyl ethanolamine and a role for alpha/beta-hydrolase 4 in this pathway. *J. Biol. Chem.* **2006**, *281*, 26465–26472. [CrossRef]
14. Simon, G.M.; Cravatt, B.F. Anandamide biosynthesis catalyzed by the phosphodiesterase GDE1 and detection of glycerophospho-*N*-acyl ethanolamine precursors in mouse brain. *J. Biol. Chem.* **2008**, *283*, 9341–9349. [CrossRef]
15. Leung, D.; Saghatelian, A.; Simon, G.M.; Cravatt, B.F. Inactivation of *N*-acyl phosphatidylethanolamine phospholipase D reveals multiple mechanisms for the biosynthesis of endocannabinoids. *Biochemistry* **2006**, *45*, 4720–4726. [CrossRef]
16. Inoue, M.; Tsuboi, K.; Okamoto, Y.; Hidaka, M.; Uyama, T.; Tsutsumi, T.; Tanaka, T.; Ueda, N.; Tokumura, A. Peripheral tissue levels and molecular species compositions of *N*-acyl-phosphatidylethanolamine and its metabolites in mice lacking *N*-acyl-phosphatidylethanolamine-specific phospholipase D. *J. Biochem.* **2017**, *162*, 449–458. [CrossRef]
17. Ghosh, M.; Tucker, D.E.; Burchett, S.A.; Leslie, C.C. Properties of the Group IV phospholipase A_2 family. *Prog. Lipid Res.* **2006**, *45*, 487–510. [CrossRef]
18. Binte Mustafiz, S.S.; Uyama, T.; Morito, K.; Takahashi, N.; Kawai, K.; Hussain, Z.; Tsuboi, K.; Araki, N.; Yamamoto, K.; Tanaka, T.; et al. Intracellular Ca^{2+}-dependent formation of *N*-acyl-phosphatidylethanolamines by human cytosolic phospholipase $A_2\varepsilon$. *Biochim. Biophys. Acta* **2019**, *1864*, 158515. [CrossRef]
19. Lee, H.C.; Simon, G.M.; Cravatt, B.F. ABHD4 regulates multiple classes of *N*-acyl phospholipids in the mammalian central nervous system. *Biochemistry* **2015**, *54*, 2539–2549. [CrossRef] [PubMed]
20. Underwood, K.W.; Song, C.; Kriz, R.W.; Chang, X.J.; Knopf, J.L.; Lin, L.L. A novel calcium-independent phospholipase A_2, $cPLA_2$-gamma, that is prenylated and contains homology to $cPLA_2$. *J. Biol. Chem.* **1998**, *273*, 21926–21932. [CrossRef] [PubMed]

21. Yamashita, A.; Tanaka, K.; Kamata, R.; Kumazawa, T.; Suzuki, N.; Koga, H.; Waku, K.; Sugiura, T. Subcellular localization and lysophospholipase/transacylation activities of human group IVC phospholipase A_2 (cPLA$_2$gamma). *Biochim. Biophys. Acta* **2009**, *1791*, 1011–1022. [CrossRef] [PubMed]
22. Tsuboi, K.; Okamoto, Y.; Rahman, I.A.; Uyama, T.; Inoue, T.; Tokumura, A.; Ueda, N. Glycerophosphodiesterase GDE4 as a novel lysophospholipase D: A possible involvement in bioactive *N*-acylethanolamine biosynthesis. *Biochim. Biophys. Acta* **2015**, *1851*, 537–548. [CrossRef] [PubMed]
23. Hussain, Z.; Uyama, T.; Kawai, K.; Binte Mustafiz, S.S.; Tsuboi, K.; Araki, N.; Ueda, N. Phosphatidylserine-stimulated production of *N*-acyl-phosphatidylethanolamines by Ca^{2+}-dependent *N*-acyltransferase. *Biochim. Biophys. Acta* **2018**, *1863*, 493–502. [CrossRef] [PubMed]
24. Bligh, E.G.; Dyer, W.J. A rapid method of total lipid extraction and purification. *Can. J. Biochem. Physiol.* **1959**, *37*, 911–917. [CrossRef] [PubMed]

Article

Preclinical Investigation in Neuroprotective Effects of the GPR55 Ligand VCE-006.1 in Experimental Models of Parkinson's Disease and Amyotrophic Lateral Sclerosis

Sonia Burgaz [1,2,3], Concepción García [1,2,3], Claudia Gonzalo-Consuegra [1,2,3], Marta Gómez-Almería [1,2,3], Francisco Ruiz-Pino [4], Juan Diego Unciti [4], María Gómez-Cañas [1,2,3], Juan Alcalde [1], Paula Morales [5], Nadine Jagerovic [5], Carmen Rodríguez-Cueto [1,2,3], Eva de Lago [1,2,3], Eduardo Muñoz [4,6,7,8] and Javier Fernández-Ruiz [1,2,3,*]

[1] Instituto Universitario de Investigación en Neuroquímica, Departamento de Bioquímica y Biología Molecular, Facultad de Medicina, Universidad Complutense, 28040 Madrid, Spain; soniabur@ucm.es (S.B.); conchig@med.ucm.es (C.G.); clagon11@ucm.es (C.G.-C.); margom27@ucm.es (M.G.-A.); mgc@med.ucm.es (M.G.-C.); jualcald@ucm.es (J.A.); carc@med.ucm.es (C.R.-C.); elagofem@med.ucm.es (E.d.L.)
[2] Centro de Investigación Biomédica en Red de Enfermedades Neurodegenerativas (CIBERNED), 28040 Madrid, Spain
[3] Instituto Ramón y Cajal de Investigación Sanitaria (IRYCIS), 28040 Madrid, Spain
[4] Emerald Health Biotechnology España, 14014 Córdoba, Spain; b62rupif@uco.es (F.R.-P.); jdunciti@gmail.com (J.D.U.); fi1muble@uco.es (E.M.)
[5] Instituto de Química Médica, CSIC, 28006 Madrid, Spain; paula.morales@iqm.csic.es (P.M.); nadine@iqm.csic.es (N.J.)
[6] Instituto Maimónides de Investigación Biomédica de Córdoba (IMIBIC), 14004 Córdoba, Spain
[7] Department of Cellular Biology, Physiology and Immunology, University of Córdoba, 14071 Córdoba, Spain
[8] Hospital Universitario Reina Sofía, 14004 Córdoba, Spain
* Correspondence: jjfr@med.ucm.es; Tel.: +34–913941450

Citation: Burgaz, S.; García, C.; Gonzalo-Consuegra, C.; Gómez-Almería, M.; Ruiz-Pino, F.; Unciti, J.D.; Gómez-Cañas, M.; Alcalde, J.; Morales, P.; Jagerovic, N.; et al. Preclinical Investigation in Neuroprotective Effects of the GPR55 Ligand VCE-006.1 in Experimental Models of Parkinson's Disease and Amyotrophic Lateral Sclerosis. *Molecules* **2021**, *26*, 7643. https://doi.org/10.3390/molecules26247643

Academic Editors: Mauro Maccarrone and Simona Rapposelli

Received: 13 November 2021
Accepted: 13 December 2021
Published: 16 December 2021

Publisher's Note: MDPI stays neutral with regard to jurisdictional claims in published maps and institutional affiliations.

Abstract: Cannabinoids act as pleiotropic compounds exerting, among others, a broad-spectrum of neuroprotective effects. These effects have been investigated in the last years in different preclinical models of neurodegeneration, with the cannabinoid type-1 (CB_1) and type-2 (CB_2) receptors concentrating an important part of this research. However, the issue has also been extended to additional targets that are also active for cannabinoids, such as the orphan G-protein receptor 55 (GPR55). In the present study, we investigated the neuroprotective potential of VCE-006.1, a chromenopyrazole derivative with biased orthosteric and positive allosteric modulator activity at GPR55, in murine models of two neurodegenerative diseases. First, we proved that VCE-006.1 alone could induce ERK1/2 activation and calcium mobilization, as well as increase cAMP response but only in the presence of lysophosphatidyl inositol. Next, we investigated this compound administered chronically in two neurotoxin-based models of Parkinson's disease (PD), as well as in some cell-based models. VCE-006.1 was active in reversing the motor defects caused by 6-hydroxydopamine (6-OHDA) in the pole and the cylinder rearing tests, as well as the losses in tyrosine hydroxylase-containing neurons and the elevated glial reactivity detected in the substantia nigra. Similar cytoprotective effects were found in vitro in SH-SY5Y cells exposed to 6-OHDA. We also investigated VCE-006.1 in LPS-lesioned mice with similar beneficial effects, except against glial reactivity and associated inflammatory events, which remained unaltered, a fact confirmed in BV2 cells treated with LPS and VCE-006.1. We also analyzed GPR55 in these in vivo models with no changes in its gene expression, although GPR55 was down-regulated in BV2 cells treated with LPS, which may explain the lack of efficacy of VCE-006.1 in such an assay. Furthermore, we investigated VCE-006.1 in two genetic models of amyotrophic lateral sclerosis (ALS), mutant SOD1, or TDP-43 transgenic mice. Neither the neurological decline nor the deteriorated rotarod performance were prevented with this compound, and the same happened with the elevated microglial and astroglial reactivities, albeit modest spinal motor neuron preservation was achieved in both models. We also analyzed GPR55 in these in vivo models and found no changes in both TDP-43 transgenic and mSOD1 mice. Therefore, our findings support the view that targeting the GPR55 may afford neuroprotection in experimental PD, but not in ALS, thus stressing the specificities for the development of cannabinoid-based therapies in the different neurodegenerative disorders.

Keywords: cannabinoids; GPR55 receptors; VCE-006.1; chromenopyrazole; Parkinson's disease; 6-hydroxydopamine; lipopolysaccharide; amyotrophic lateral sclerosis; mSOD1 mice; TDP-43 transgenic mice

1. Introduction

Phytocannabinoids, the active constituents of the *Cannabis* plant, as well as endocannabinoids and synthetic cannabinoids, have been proposed as promising neuroprotective agents in accidental brain damage (e.g., stroke, brain trauma, spinal injury) and in chronic progressive disorders (e.g., Alzheimer's disease, amyotrophic lateral sclerosis (ALS), Parkinson's disease (PD), Huntington's disease, and others) [1–3]. This potential derives from their pleiotropism and ability to activate numerous cytoprotective targets within the endocannabinoid system, but also outside this signaling system [3]. An important part of these neuroprotective properties described for cannabinoids have been related to the activation of the type-1 cannabinoid (CB_1) receptor [1,2]. This receptor is predominantly located in neurons in the CNS, which facilitates its role in the control of excitotoxic damage in glutamatergic synapses [4], as well as a possible contribution in the autophagy-mediated elimination of protein aggregates [5]. Data supporting CB_1 receptor-mediated neuroprotective effects have been collected in experimental models of Alzheimer's disease [6–8], PD [9,10], ALS [11–13], Huntington's disease [4,14–16], and multiple sclerosis [17,18].

Important neuroprotective effects have also been described for the activation of the type-2 cannabinoid (CB_2) receptor [1–3,19]. This receptor is predominantly located in activated astrocytes and reactive microglial cells in the CNS of neuroinflammatory/neurodegenerative conditions, in which it becomes significantly up-regulated with the purpose to control glial toxicity for neurons as well as other beneficial effects [1,19]. Data supporting CB_2 receptor-mediated neuroprotective effects have been collected in experimental models of Alzheimer's disease and related dementias [7,20–23], PD [7,24–27], ALS [28–32], Huntington's disease [33–35], and multiple sclerosis [36–38].

These broadly-demonstrated neuroprotective effects of cannabinoids have also been extended to additional targets, within or outside the endocannabinoid system, which are also active for cannabinoids [3]. This includes, for example, the nuclear receptors of the peroxisome proliferator-activating receptor (PPAR) family, which have been investigated for their role in the control of inflammatory/neurodegenerative events [39,40] in experimental PD [41–44], and, to a lower extent, in experimental ALS [45] and Alzheimer's disease [46,47]. More recent data have indicated the orphan G-protein receptor 55 (GPR55) as an additional neuroprotective and anti-inflammatory target [48–50]. This has been investigated mainly in PD given the abundant presence of GPR55 receptors in the basal ganglia [51,52] and the important motor impairment found in mice lacking GPR55 [53].

GPR55 receptor was considered for years as an orphan receptor, but some recent evidence has positioned this receptor as a possible new cannabinoid receptor type [54]. However, such an assumption has been controversial due to the important differences in homology, conformational structure, pharmacology, signaling, and functional relevance shown by GPR55 compared to classic CB_1 and CB_2 receptors [55–57]. The human GPR55 protein has 319 amino acids and is also a member of the rhodopsin-like 7TM/GPCR family [55,57]. It was isolated and cloned in 1999, when it was found to be located in chromosome 2 (2q37) in humans [58]. Its naturally-occurring ligand is lysophosphatidyl inositol (LPI) [59]. Its pharmacology is complex and still remains to be clarified, including some non-cannabinoid compounds that do not bind CB_1/CB_2 receptors (e.g., GSK-494,581, CID-16020046 [60]), but also certain phytocannabinoids (e.g., cannabidiol), endocannabinoids (e.g., anandamide, 2-arachidonoylglycerol) and synthetic cannabinoids (e.g., WIN

55,212-2, HU-210, SR141716, AM251, methanandamide), which may also be active at other cannabinoid receptors [61,62]. GPR55 is widely distributed in the CNS, in particular in the basal ganglia, hippocampus, thalamus, and cerebellum [63], and is also present in the periphery (e.g., vasculature, gastrointestinal tract, bones, lung, spleen, liver, kidney, uterus) [64]. This distribution has prompted research on this receptor in relation to pathogenesis and/or development of novel therapies against different central and peripheral pathologies, including, as mentioned above, neurodegenerative disorders for which targeting GPR55 has been proposed as a promising anti-inflammatory and neuroprotective strategy [48–51].

In the present study, we have further investigated the neuroprotective potential of this new target for cannabinoids, using VCE-006.1, a chromenopyrazole derivative designed, synthesized, and investigated as GPR55 ligand in a previous study of our group [65]. VCE-006.1 is the compound 2-[2-(4-cyclohexylcarbonylpiperazinyl)ethyl]-2,4-dihydro-7-methoxy-4,4-dimethylchromeno[4,3-c]pyrazole (compound 23 in [65]), which showed affinity at the GPR55 receptor analyzed in a label-free cell-impedance-based assay in hGPR55-HEK293 cells, whereas having negligible or poor affinity for the CB_1 and CB_2 receptor (as measured in competitive radioligand assays), respectively [65]. The patent generated with this and other similar compounds [66] was acquired by the company Emerald Health Biotechnology-Spain in 2018, and the compound was renamed as VCE-006.1. In this study, we have extended the analysis of its activity at the GPR55 receptor, using several cell-based assays, which has situated this compound as a potential biased positive allosteric modulator (PAM) for the GPR55 receptor. Next, we have investigated its neuroprotective profile in vitro (cell-based assays) and in vivo (neurotoxin-based models or genetically-modified mice) models of two neurodegenerative diseases, PD and ALS, in which the potential of GPR55 as a neuroprotective target has been claimed [32,51].

2. Results

2.1. Studies on PAM Activity of VCE-006.1

Our first objective was to further explore the activity of VCE-006.1 (see chemical structure in Figure 1A) at the GPR55. Previous studies [65] have indicated VCE-006.1 to be a selective ligand of this receptor with activity as a partial agonist and having no relevant affinity at the classic CB_1 and CB_2 receptors tested in competitive radioligand binding assays. Here, we have explored canonical GPR55 signalling pathways in cells expressing the native receptor (DU145 and U937 cells) and in cells overexpressing the receptor (HEK-293-GPR55 cells). We found that both LPI and VCE-006.1 induced ERK1/2 phosphorylation in DU145 cells and that a combination of both further increased this phosphorylation (Figure 1B). Ca^{2+} mobilization in response to VCE-006.1 and LPI was studied in U937 cells, and as depicted in Figure 1C, both compounds were able to induce Ca^{2+} mobilization, with LPI being more potent than VCE-006.1, suggesting a different mode of action for each compound. Next, we stimulated HEK293-GPR55-CRE-Luc cells with either VCE-006.1 or LPI, separately or in combination, and the luciferase activity was measured as indicative of cAMP induction. VCE-006.1 did not induce CRE-Luc activity but significantly enhanced the effect of LPI as a potential orthosteric ligand (F(7,40) = 17.36, $p < 0.0001$; Figure 1D). Altogether, our results showed that VCE-006.1 activated GPR55 in a biased manner compared to LPI, showing characteristics of both partial orthosteric agonist and PAM depending on the specific cell assay used.

2.2. Studies in Experimental PD

Our second objective was to investigate this compound when administered chronically in two neurotoxin-based models of PD, as well as in some cell-based models of this disease. We first used a classic PD model of mitochondrial damage, 6-OHDA-lesioned mice, which proved the expected hemiparesis in the cylinder rearing test (Figure 2A) and an elevated latency to descend in the pole test (Figure 2B). VCE-006.1 was active in reversing these motor defects caused by 6-OHDA in the cylinder rearing test (F(3,29) = 17.49, $p < 0.0001$;

Figure 2A) and in the pole test (F(3,27) = 8.803, $p < 0.0005$; Figure 2B), effects evident in 6-OHDA-lesioned mice, but absent in sham-operated mice.

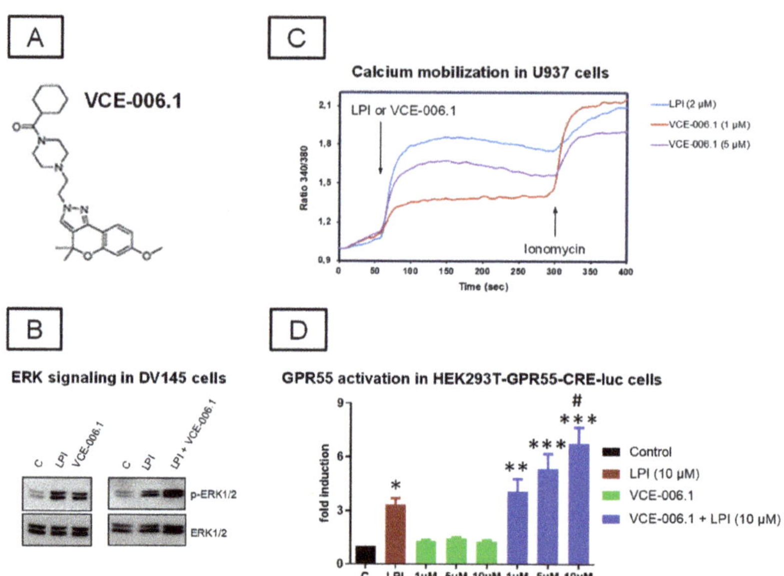

Figure 1. (**A**) Chemical structure of VCE-006.1. (**B**) VCE-006.1 and LPI induces ERK1/2 activation in DU145 cells. The cells were stimulated as indicated and the expression of phospho-ERK1/2 and total ERK1/2 determined by immunoblots. (**C**) VCE-006.1 and LPI induces [Ca^{2+}] immobilization in U937 cells. U937 cells were loaded with Indo1-AM, treated with the compounds, and the calcium mobilization was measured by ratiometric fluorescence as indicated under Materials and Methods. (**D**) GPR55 activity of VCE-006.1 at different concentrations (1, 5, and 10 µM) in the absence or the presence of 10 µM LPI on HEK293T-GPR55-CRE-luc cells. Results are expressed as the fold induction of GPR55 activity and represent means ± SEM of data generated in 6 independent experiments, each conducted in triplicates. Statistical significance was determined by one-way ANOVA followed by the Tukey test (* $p < 0.05$, ** $p < 0.01$, *** $p < 0.005$ vs. control (basal) and VCE-006.1 alone; [#] $p < 0.05$ vs. LPI and VCE-006.1 (1 µM) + LPI).

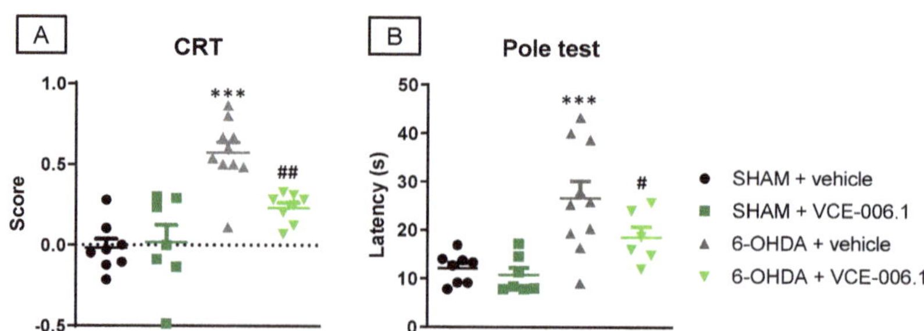

Figure 2. Response in the cylinder rearing test (**A**) and in the pole test (**B**) of male mice subjected to unilateral 6-OHDA lesions or sham-operated and daily treated with VCE-006.1 (20 mg/kg, i.p.) for 2 weeks. Values are means ± SEM of more than 6 animals per group. Data were assessed by one-way ANOVA followed by the Tukey test (*** $p < 0.005$ vs. the two sham-operated groups; [#] $p < 0.05$, [##] $p < 0.01$ vs. the vehicle-treated 6-OHDA lesioned mice).

These benefits with VCE-006.1 were associated with a reduction in the loss of TH-containing neurons caused by a 6-OHDA lesion in the substantia nigra ($F_{(3,27)} = 25.57$, $p < 0.0001$; Figure 3A,B). The 6-OHDA lesion also caused a modest elevation of LAMP-1 immunostaining, a marker of autophagy, which was attenuated by the treatment with VCE-006.1 ($F_{(3,29)} = 4.77$, $p < 0.01$; Figure 3C,D).

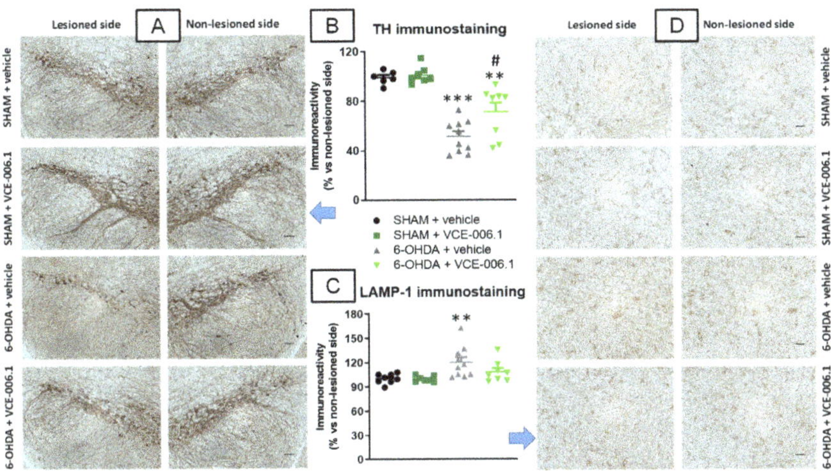

Figure 3. Quantification of TH (**B**) and LAMP-1 (**C**) immunoreactivities, including representative images (**A**) (TH; scale bar = 100 μm) and (**D**) (LAMP-1; scale bar = 50 μm)), measured in a selected area of the substantia nigra pars compacta of male mice subjected to unilateral 6-OHDA lesions or sham-operated and daily treated with VCE-006.1 (20 mg/kg, i.p.) for 2 weeks. Values correspond to % of the ipsilateral lesioned side vs. contralateral non-lesioned side and are expressed as means ± SEM of more than 6 animals per group. Data were assessed by one-way ANOVA followed by the Tukey test (** $p < 0.01$, *** $p < 0.005$ vs. the two sham-operated groups; # $p < 0.05$ vs. the vehicle-treated 6-OHDA lesioned mice).

Our histological analysis of the substantia nigra also proved an elevated glial reactivity detected in this structure when lesioned with 6-OHDA, visible for Cd68 immunolabelling (reflecting reactive microgliosis) and with GFAP immunostaining. Both responses were notably attenuated by the treatment with VCE-006.1 (Cd68: $F_{(3,29)} = 15.43$, $p < 0.0001$; Figure 4A,B; GFAP: $F_{(3,29)} = 22.72$, $p < 0.0001$; Figure 4C,D). VCE-006.1 had no effect on these markers in sham-operated mice.

In a second experiment, we investigated whether VCE-006.1 also exerts similar cytoprotective effects in vitro in SH-SY5Y cells, which express GPR55 [67], exposed to 6-OHDA. Our data revealed that 6-OHDA reduced cell viability up to close to 50% in these cells, which was attenuated by VCE-006.1 in a concentration-related manner with a maximum at 1 μM ($F_{(6,40)} = 40.80$, $p < 0.0001$), lower effects at higher concentrations (5 and 10 μM), and no effect at 20 μM (Figure 5).

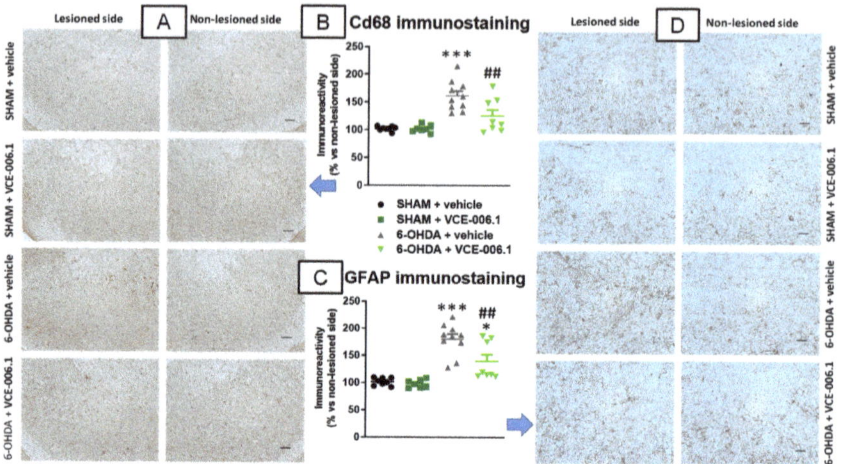

Figure 4. Quantification of Cd68 (**B**) and GFAP (**C**) immunoreactivities, including representative images (**A**) (Cd68; scale bar = 100 μm) and (**D**) (GFAP; scale bar = 50 μm)), measured in a selected area of the substantia nigra pars compacta of male mice subjected to unilateral 6-OHDA lesions or sham-operated and daily treated with VCE-006.1 (20 mg/kg, i.p.) for 2 weeks. Values correspond to % of the ipsilateral lesioned side vs. contralateral non-lesioned side and are expressed as means ± SEM of more than 6 animals per group. Data were assessed by one-way ANOVA followed by the Tukey test (* $p < 0.05$, *** $p < 0.005$ vs. the two sham-operated groups; ## $p < 0.01$ vs. the vehicle-treated 6-OHDA lesioned mice).

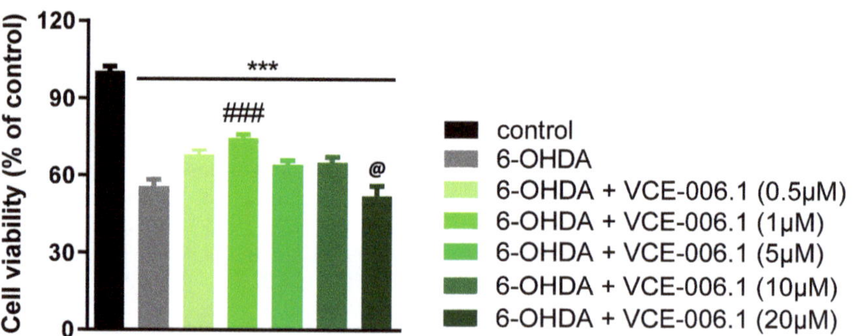

Figure 5. Cell viability measured with the MTT assay in cultured SH-SY5Y cells at 24 h to be treated with different concentrations of VCE-006.1 (0.5, 1, 2, 5, 10, and 20 μM) against 6-OHDA (200 μM). In all cases, a group with cells exposed to vehicle was also included to determine the 100% of cell viability. Values are means ± SEM of at least 4 independent experiments, each performed in triplicate. Data were assessed by the one-way ANOVA followed by the Tukey (*** $p < 0.005$ vs. control cells; ### $p < 0.005$ vs. cells exposed to 6-OHDA + vehicle; @ $p < 0.05$ vs. cells treated with the other VCE-006.1 concentrations).

Next, we also investigated VCE-006.1 in an inflammatory model of PD, LPS-lesioned mice, having relatively similar beneficial effects. Again, LPS-lesioned mice exhibited motor defects in the cylinder rearing test (hemiparesis) and in the pole test (elevated latency to descend the pole), which were attenuated by the treatment with VCE-006.1 (CRT: $F_{(2,17)} = 9.34$, $p < 0.005$; Figure 6A; pole test: $F_{(2,19)} = 11.75$, $p < 0.0005$; Figure 6B).

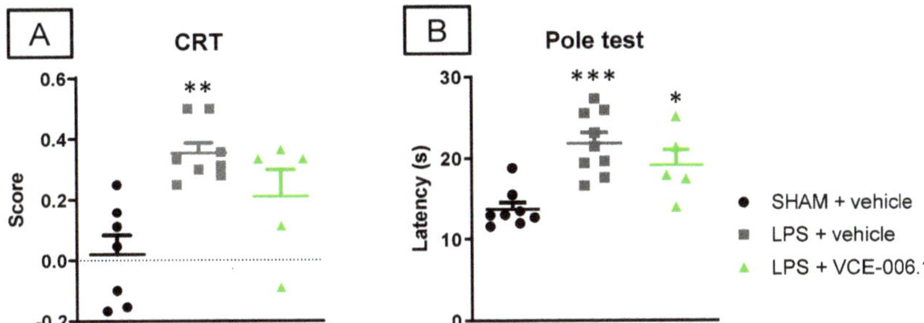

Figure 6. Response in the cylinder rearing test (**A**) and in the pole test (**B**) of male mice subjected to unilateral LPS lesions or sham-operated and daily treated with VCE-006.1 (20 mg/kg, i.p.) for 2 weeks. Values are means ± SEM of more than 6 animals per group. Data were assessed by one-way ANOVA followed by the Tukey test (* $p < 0.05$, ** $p < 0.01$, *** $p < 0.005$ vs. the two sham-operated groups).

These benefits of VCE-006.1 on the neurological state of LPS-lesioned mice were again accompanied by higher survival or TH-positive neurons in the substantia nigra ($F(2,19) = 3.45$, $p < 0.05$; Figure 7A,B), an effect that was modest and reflected in the loss of statistically significant differences vs. sham-operated animals. However, this effect, surprisingly, was not accompanied by a reduction in the LPS-induced elevation of the autophagy marker LAMP-1 ($F(2,19) = 42.56$, $p < 0.0001$; Figure 7C). The same happened with the reactive microgliosis (elevated Cd68 immunoreactivity; $F(2,19) = 45.80$, $p < 0.0001$; Figure 7D) and astroglial reactivity (elevated GFAP immunolabelling; $F(2,19) = 69.94$, $p < 0.0001$; Figure 7E), which remained elevated in LPS-lesioned mice irrespective of VCE-006.1 treatment.

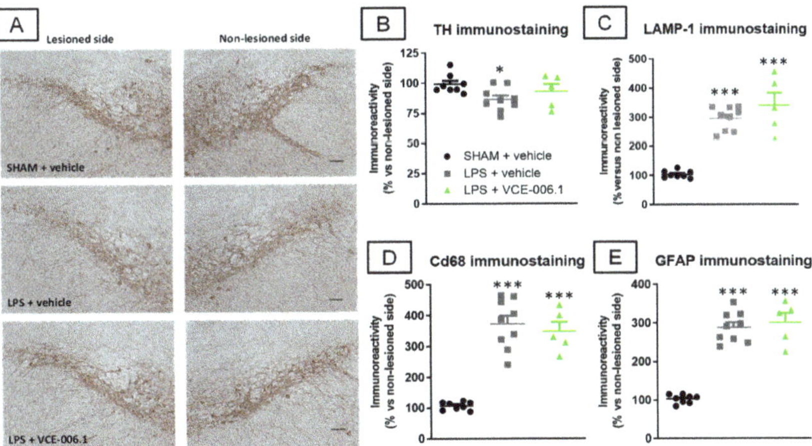

Figure 7. Quantification of TH (**B**), LAMP-1 (**C**), Cd68 (**D**), and GFAP (**E**) immunoreactivities, including representative images for TH immunostaining ((**A**); scale bar = 100 μm), measured in a selected area of the substantia nigra pars compacta of male mice subjected to unilateral LPS lesions or sham-operated and daily treated with VCE-006.1 (20 mg/kg, i.p.) for 2 weeks. Values correspond to % of the ipsilateral lesioned side vs. contralateral non-lesioned side and were expressed as means ± SEM of more than 5 animals per group. Data were assessed by one-way ANOVA followed by the Tukey test (* $p < 0.05$, *** $p < 0.005$ vs. the two sham-operated groups).

Such absence of VCE-006.1 effects against glial reactivity was also evident against some associated inflammatory events elicited by LPS lesion, for example the elevated gene expression detected in the striatum in proinflammatory cytokines TNF-α ($F_{(2,18)}$ = 33.34, $p < 0.0001$; Figure 8A) and IL-1β ($F_{(2,18)}$ = 9.41, $p < 0.005$; Figure 8B), as well as in proinflammatory enzymes iNOS ($F_{(2,16)}$ = 4.24, $p < 0.05$; Figure 8C) and COX-2 ($F_{(2,17)}$ = 9.13, $p < 0.005$; Figure 8D), which remained unaltered after VCE-006.1 treatment. This was also evident for the LPS-induced reduction in the CB$_1$ receptor ($F_{(2,18)}$ = 28.63, $p < 0.0001$; Figure 8E), elevation of the CB$_2$ receptor ($F_{(2,18)}$ = 31.31, $p < 0.0001$; Figure 8F), and no effect in PPAR-γ ($F_{(2,18)}$ = 1.14, ns; Figure 8G)

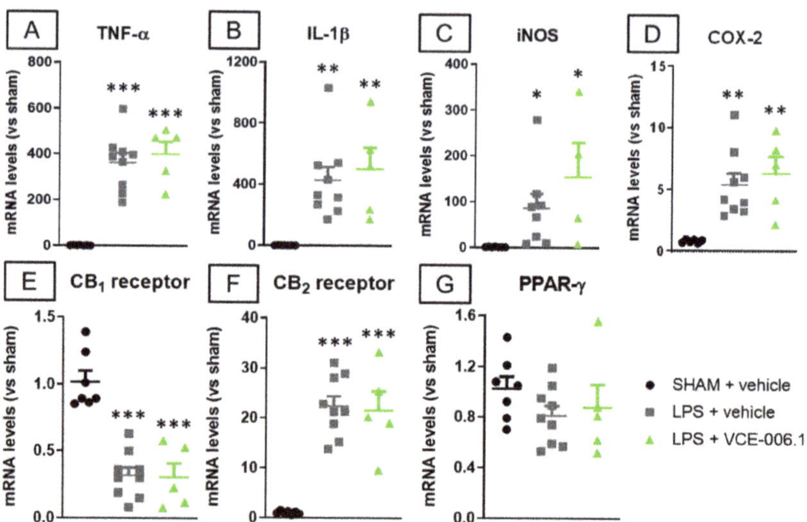

Figure 8. mRNA levels for TNF-α (**A**), IL-1β (**B**), iNOS (**C**), COX-2 (**D**), CB$_1$ receptor (**E**), CB$_2$ receptor (**F**), and PPAR-γ (**G**) measured by qPCR in the striatum of male mice subjected to unilateral LPS lesions or sham-operated and daily treated with VCE-006.1 (20 mg/kg, i.p.) for 2 weeks. GAPDH was used as an endogenous reference gene for data normalization. Values correspond to fold of change vs. sham-operated controls and are expressed as means ± SEM of more than 5 animals per group. Data were assessed by one-way ANOVA followed by the Tukey test (* $p < 0.05$, ** $p < 0.01$, *** $p < 0.005$ vs. the two sham-operated groups).

Lastly, the absence of VCE-006.1 effects against glial reactivity and associated inflammatory events detected in LPS-lesioned mice was also confirmed in BV2 cells (which also express GPR55 [68]) treated with LPS and VCE-006.1, as the elevated levels of gene expression detected for TNF-α ($F_{(2,15)}$ = 15.14, $p < 0.0005$; Figure 9A) and IL-1β ($F_{(2,15)}$ = 12.21, $p < 0.001$; Figure 9B) after LPS again remained unaltered by the treatment with VCE-006.1. This may be related to the strong reduction in GPR55 mRNA levels found in BV2 cells treated with LPS in the absence or presence of VCE-006.1 in comparison with control cells ($F_{(2,15)}$ = 10.53, $p < 0.005$; Figure 9C). However, the analysis of gene expression for GPR55 in the striatum of LPS-lesioned mice proved no changes in this receptor ($F_{(2,18)}$ = 0.57, ns; Figure 9D), and the same happened in 6-OHDA-lesioned mice ($F_{(3,22)}$ = 0.65, ns; Figure 9E).

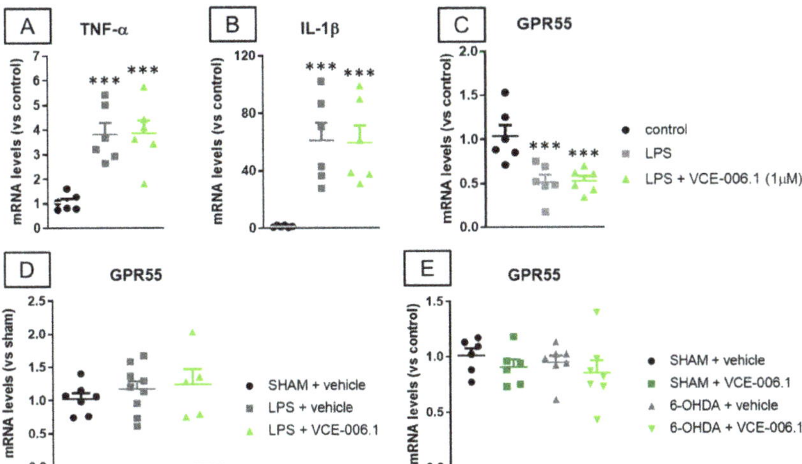

Figure 9. mRNA levels for TNF-α (**A**), IL-1β (**B**), and GPR55 (**C**) measured by qPCR in BV2 cells exposed to LPS and/or VCE-006.1 (1 μM) for 20 h, and mRNA levels for GPR55 measured by qPCR in the striatum of male mice subjected to unilateral 6-OHDA (**D**) or LPS (**E**) lesions or sham-operated and daily treated with VCE-006.1 (20 mg/kg, i.p.) for 2 weeks. In all cases, GAPDH was used as an endogenous reference gene for data normalization, and values correspond to fold change vs. controls and are expressed as means ± SEM of more than 5 animals per group. Data were assessed by one-way ANOVA followed by the Tukey test (*** $p < 0.005$ vs. the control group).

2.3. Studies in Experimental ALS

Our third objective was to investigate VCE-006.1 when administered chronically in two genetic murine models of ALS. We first used the classic mSOD-1 model which showed several motor abnormalities such as: (i) a progressive reduction in the time on wire (2-way interaction: $F_{(10,155)} = 13.25$, $p < 0.0001$; Figure 10A) visible in the hanging wire test; (ii) a progressively marked deterioration in the rotarod performance (2-way interaction: $F_{(18,270)} = 15.43$, $p < 0.0001$; Figure 10B) detected in the rotarod test; and (iii) a rapid elevation in a specific neurological score for ALS signs recapitulated in mice (2-way interaction: $F_{(18,288)} = 10.23$, $p < 0.0001$; Figure 10C). VCE-006.1 was not active against any of these neurological decline signs, then indicating no effects at the functional level. However, the strong loss of Nissl-stained motor neurons visible in the ventral horn of the spinal cord (lumbar levels) in mSOD-1 mice was partially attenuated by the chronic treatment with VCE-006.1 ($F_{(2,31)} = 98.79$, $p < 0.0001$; Figure 10D,E), although this does not have any influence on possible neurological recoveries as seen in the above behavioral data. This may be in part related to the persistence of higher levels of glial reactivity in the ventral horn of the spinal cord (lumbar levels) in mSOD-1 mice after the treatment with VCE-006.1 (GFAP immunolabelling: $F_{(2,30)} = 53.34$, $p < 0.0001$; Figure 11A,B); Iba-1 immunolabelling: $F_{(2,31)} = 62.56$, $p < 0.0001$; Figure 11C,D), which were similar to mSOD-1 mice treated with vehicle.

Next, we investigated the same issue in an alternative and more recent ALS model based on the RNA-binding protein TDP-43. Again, TDP-43 transgenic mice showed several motor abnormalities such as: (i) a progressively higher clasping response (2-way interaction: $F_{(8,88)} = 4.50$, $p < 0.0001$; Figure 12A); and (ii) a progressively marked deterioration in the rotarod performance (2-way interaction: $F_{(8,88)} = 4.46$, $p < 0.0001$; Figure 12B) detected in the rotarod test. Again, VCE-006.1 was not active against any of these motor signs, then indicating no effects at the functional level, despite the strong loss of Nissl-stained motor neurons visible in the ventral horn of the spinal cord (lumbar levels) in TDP-43 transgenic mice was partially attenuated by the chronic treatment with VCE-006.1 ($F_{(2,21)} = 82.28$, $p < 0.0001$; Figure 12C,D).

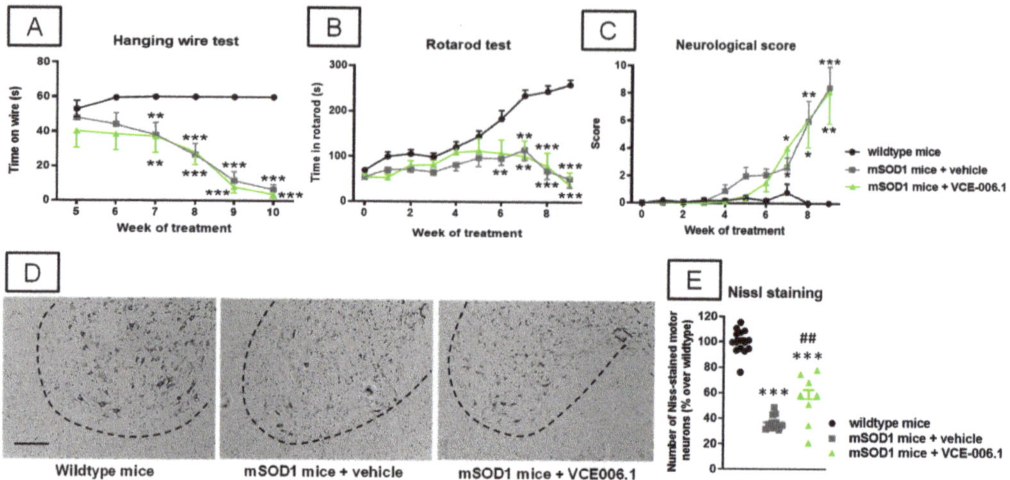

Figure 10. Hanging wire response (**A**), rotarod performance (**B**), and neurological score (**C**), analyzed mSOD1 transgenic and wild-type male mice at specific weeks during a chronic treatment from 63 day-old to 125 day-old with VCE-006.1 (20 mg/kg, daily and i.p.) or vehicle, and quantification of the number of Nissl-stained motor neurons (**E**), including representative images ((**D**); scale bar = 100 μm), in the lumbar ventral horn (marked with a dotted line) of the spinal cord in all experimental groups after the chronic treatment. Values are means ± SEM of more than 6 animals per group. Behavioral data were assessed by two-way ANOVA (with repeated measures), whereas Nissl staining data were assessed by one-way ANOVA, in both cases followed by the Tukey test (* $p < 0.05$, ** $p < 0.01$, *** $p < 0.005$ vs. wild-type mice; ## $p < 0.01$ vs. mSOD1 mice treated with vehicle).

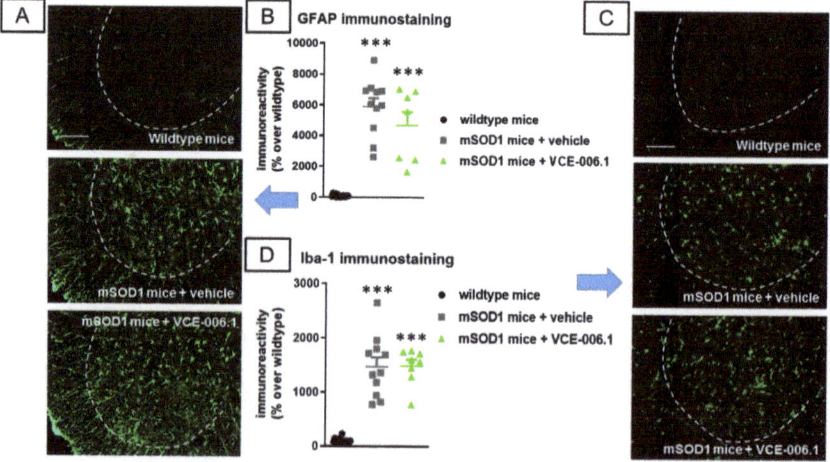

Figure 11. Quantification of GFAP (**B**) and Iba-1 (**D**) immunoreactivities, including representative images ((**A**) and (**C**), respectively; scale bar = 100 μm), in the lumbar ventral horn (marked with a dotted line) of the spinal cord in wild-type and mSOD1 transgenic mice after a chronic treatment from 63 day-old to 125 day-old with VCE-006.1 (20 mg/kg, daily and i.p.) or vehicle. Values are means ± SEM of more than 6 animals per group. Data were assessed by one-way ANOVA followed by the Tukey test (*** $p < 0.005$ vs. wild-type mice).

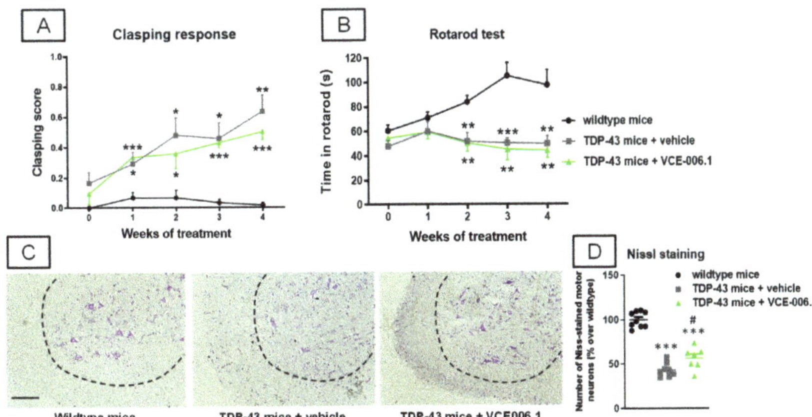

Figure 12. Clasping response (**A**) and rotarod performance (**B**) analyzed TDP-43 transgenic and wild-type male mice at specific weeks during a chronic treatment of 30 days with VCE-006.1 (20 mg/kg, daily and i.p.) or vehicle, and quantification of the number of Nissl-stained motor neurons (**D**), including representative images ((**C**); scale bar = 100 μm), in the lumbar ventral horn (marked with a dotted line) of the spinal cord in all experimental groups after the chronic treatment. Values are means ± SEM of more than 6 animals per group. Behavioral data were assessed by two-way ANOVA (with repeated measures), whereas Nissl staining data were assessed by one-way ANOVA, in both cases followed by the Tukey test (* $p < 0.05$, ** $p < 0.01$, *** $p < 0.005$ vs. wild-type mice; # $p < 0.05$ vs. TDP-43 mice treated with vehicle).

Again, we may attribute this effect in part to the persistence of higher levels of glial reactivity in the ventral horn of the spinal cord (lumbar levels) in TDP-43 transgenic mice after the treatment with VCE-006.1 (GFAP immunolabelling: $F_{(2,21)} = 21.08$, $p < 0.0001$; Figure 13A,B); Iba-1 immunolabelling: $F_{(2,20)} = 8.82$, $p < 0.005$; Figure 13C,D), which were similar to TDP-43 transgenic mice.

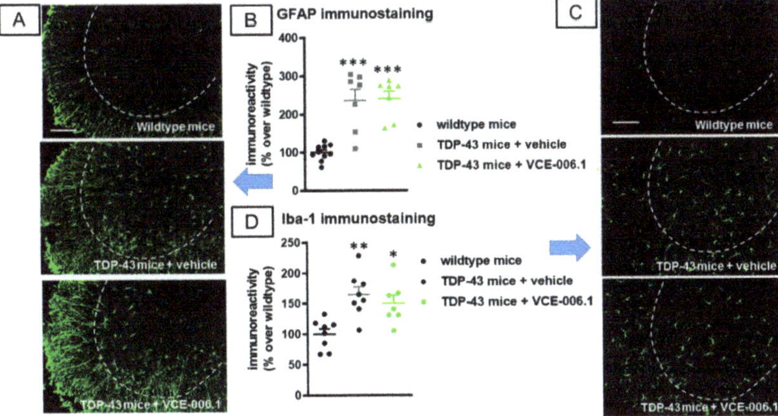

Figure 13. Quantification of GFAP (**B**) and Iba-1 (**D**) immunoreactivity, including representative images ((**A,C**), respectively; scale bar = 100 μm), in the lumbar ventral horn (marked with a dotted line) of the spinal cord in wild-type and TDP-43 transgenic mice after chronic treatment of 30 days with VCE-006.1 (20 mg/kg, daily and i.p.) or vehicle. Values are means ± SEM of more than 6 animals per group. Data were assessed by one-way ANOVA followed by the Tukey test (* $p < 0.05$, ** $p < 0.01$, *** $p < 0.005$ vs. wildtype mice).

Lastly, as in the experimental models of PD, we also analyzed GPR55 gene expression in these in vivo ALS models. Our data indicated that GPR55-mRNA levels did not experience any changes in the case of mSOD1 mice compared to wild-type animals when analyzed at a late symptomatic phase (123 days; Figure 14A), and the same happened with TDP-43 transgenic mice at two specific ages: 65 (early symptomatic stage; Figure 14B) and 105 days (advanced symptomatic phase; Figure 14C).

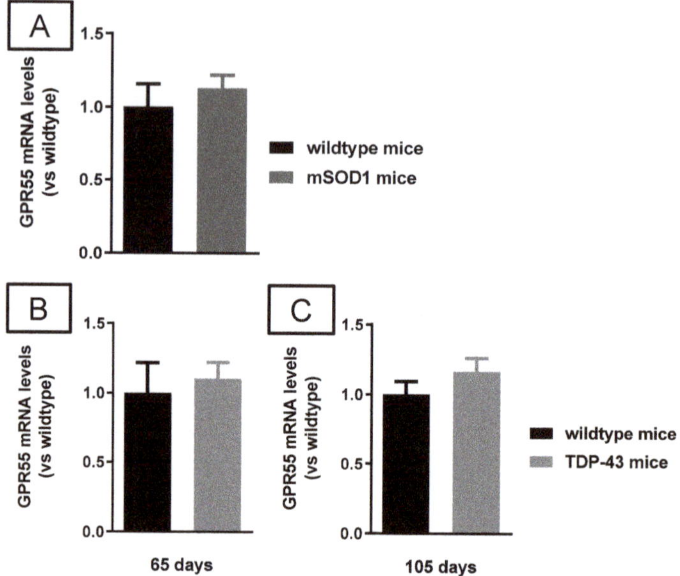

Figure 14. mRNA levels for GPR55 measured by qPCR in the spinal cord of male mSOD1 (at 123 days of age; (**A**)) or TDP-43 (at 65 (**B**) and 105 (**C**) days of age) transgenic mice, and their corresponding wild-type mice. GAPDH was used as an endogenous reference gene for data normalization. Values correspond to fold change vs. controls and are expressed as means ± SEM of more than 5 animals per group. Data were assessed by the unpaired Student's *t*-test.

3. Discussion

The orphan receptor GPR55 has emerged in the last years as a potential new component of the endocannabinoid signaling system [54], despite its differences with the classic CB_1 and CB_2 receptors [55–57], as well as a promising neuroprotective target for the development of novel therapies for neurodegenerative conditions [48–52]. One of the key areas, involving GPR55 activity in the CNS, is the control of movement and motor coordination, which is supported by the fact that motor-related areas (e.g., basal ganglia, cerebellum) are within the CNS structures with higher GPR55 expression [63]. In addition, GPR55-deficient mice develop, among others, important impairments in motor control and coordination [53]. This possibly explains that neurodegenerative disorders such as Alzheimer's disease and related dementias have been explored for determining the neuroprotective potential of GPR55-targeting compounds only recently [69,70], whereas movement-related disorders, in particular PD, are within those neurodegenerative pathologies investigated earlier and more extensively in relation with the GPR55 ligands [51,52,71,72]. Our present study has been designed to pursue the objective of developing a GPR55-based neuroprotective therapy for PD and also by other motor-related pathologies, for example, ALS. To do that, we used a chromenopyrazole derivative, VCE-006.1, which a priori showed selective properties as a partial agonist at the GPR55 receptors [65]. Our first objective was to extend the characterization of this compound to its activity at the GPR55 receptor, using specific cell assays that revealed a biased activity of VCE-006.1 on this receptor as a partial orthosteric agonist or PAM, depending on the specific cell assay used.

Once we confirmed this activity of VCE-006.1 at the GPR55 receptor, we wanted to explore whether this enables the compound to afford neuroprotection in cells and murine models of the two neurodegenerative diseases indicated before, i.e., PD and ALS. Our experiments in PD demonstrated that VCE-006.1 was highly active in the preservation of TH-containing nigral neurons damaged in this disease, and that this has an important reflect in the improvement of motor defects associated with this damage. In our study, this neuroprotective effect was evident in two in vivo models of PD generated by 6-OHDA or, to a lower extent, LPS lesions in mice, and was also confirmed in an in vitro cell-based model (SH-SY5Y cells exposed to 6-OHDA). Similar benefits have been observed with other GPR55-acting compounds using additional experimental models, such as MPTP-lesioned mice and a murine model of haloperidol-induced catalepsy [51], and the same happens with more recent studies conducted by Martínez-Pinilla and coworkers [52,71]. However, whereas the neuroprotection seen in 6-OHDA-lesioned mice with VCE-006.1 in our study was accompanied by an attenuation of the reactive gliosis elicited by the neurotoxin, this did not occur in the LPS-lesioned mice, in which the inflammatory response caused by LPS has been proposed to be the primary cause of further neuropathological events (e.g., loss of TH-positive neurons, motor defects). These paradoxical effects remain to be investigated, but, in support of this in vivo effect, the lack of VCE-006.1 effect against glial reactivity and associated inflammatory events (elevated generation of proinflammatory cytokines) was also evident in BV2 cells treated with LPS and VCE-006.1. This could be related to an LPS-induced down-regulation of GPR55 receptors in the BV2 cells, although such down-regulation was not found in LPS-lesioned mice, and the same was seen in 6-OHDA-lesioned mice. In addition, in preliminary studies carried out with post mortem tissues from PD patients and control subjects, we detected apparently similar levels of GPR55 and an equivalent cell distribution, although this will require further confirmation (García, Burgaz and Fernández-Ruiz, unpublished results). To make the issue more complicated and justify the need for additional studies, a previous experiment also conducted in BV2 cells, and in part in rat microglial cell primary cultures, showed activity of LPI against LPS-induced nitric oxide production and iNOS expression [50]. By contrast, a similar study was carried out with anandamide, which also binds GPR55; instead, LPI resulted in inactivity [73].

As indicated before, we also investigated VCE-006.1 in another motor-related neurodegenerative disorder, ALS, using two genetic models of this pathology, the classic mSOD-1 model and the more recent TDP-43 transgenic mice. In both cases, our results confirmed that VCE-006.1 was poorly active, exerting only partial preservation of spinal motor neurons, which was not sufficient to reverse the intense neurological decline and muscle strength deterioration seen in these animals during the progression of the pathological phenotype. This may be related to the lack of effect of VCE-006.1 on the elevated microglial and astroglial reactivities seen in both models, a fact that, in this case, was not associated with a reduction in the levels of GPR55 receptors, which resulted in being similar to those found in the corresponding wild-type mice for both TDP-43 transgenic and mSOD-1 mice. Combining neuroprotection (preservation of motor neurons) and anti-inflammatory (attenuation of glial reactivity) effects appear to be an important determinant for disease-modifying effects of cannabinoids in experimental ALS. For example, cannabinoids targeting the CB_2 (e.g., HU-308) or the PPAR-γ receptors (e.g., VCE-003.2) afforded important levels of neuroprotection, being able to preserve motor neurons and to attenuate glial reactivity, which results in an improvement against the neurological (motor) deterioration [30,32,45]. However, such neurological improvement was not observed in studies that used cannabinoids that were not active at the same time against both the loss of motor neurons and the elevated glial reactivity [74]. Therefore, we assume that the potential of VCE-006.1 for ALS would require its combination with other cannabinoids also active at other endocannabinoid-related targets (e.g., CB_2 receptors, PPAR-γ receptors).

We also have evidence that VCE-006.1 does not activate PPAR-γ receptors (Muñoz et al., unpublished results).

4. Materials and Methods

4.1. Synthesis and Characterization as PAM of VCE-006.1 in Cell-Based Assays

VCE-006.1 (2-[2-(4-cyclohexylcarbonylpiperazinyl)ethyl]-2,4-dihydro-7-methoxy-4,4-dimethylchromeno[4,3-c]pyrazole) was designed, synthesized, and characterized for the first time as a partial agonist at the GPR55 receptor by Morales and coworkers (compound 23 in [65]). In this new study, we have further characterized its biological activity profile both in HEK-293 cells overexpressing GPR55 and in cell lines expressing the native receptor.

4.1.1. Determination of ERK 1/2 Activation

DU145 cells expressing endogenous GPR55 were stimulated with either VCE-006.1 (5 µM), LPI (2 µM), or a combination of both for 30 min. Then, cells were washed with phosphate-buffered saline (PBS) and proteins extracted in lysis buffer (50 mM Tris–HCl pH 7.5, 150 mM NaCl, 10% glycerol, and 1% NP-40) supplemented with 10 mM NaF, 1 mM Na_3VO_4, 10 µg/mL leupeptin, 1 µg/mL pepstatin and aprotinin, and 1 µL/mL saturated PMSF. Thirty µg of proteins were boiled at 95 °C in Laemmli buffer and electrophoresed in 10% SDS/PAGE gels. Total ERK was used as a loading control. Separated proteins were transferred to PVDF membranes, and after blocking with non-fat milk in TBST buffer, primary antibodies were added. The washed membranes were incubated with appropriate secondary antibodies coupled to horseradish peroxidase that were detected by an enhanced chemiluminescence system (USB). Antibodies against total and phospho-ERK1/2 were purchased from Sigma-Aldrich (Madrid, Spain).

4.1.2. Ca^{2+} Mobilization Assay

U937 cells expressing endogenous GPR55 receptor were incubated for 1 h at 37 °C in Tyrode's salt solution (137 mM NaCl, 2.7 mM KCl, 1.8 mM $CaCl_2$, 1.0 mM $MgCl_2$, 0.4 mM NaH_2PO_4, 12.0 mM $NaHCO_3$, and 5.6 mM D-glucose) containing 5 µM Indo1-AM (Invitrogen, Waltham, MA, USA) for 30 min at 37 °C in the dark. Cells were then harvested, washed three times with buffer to remove extracellular Indo1 dye, readjusted to 10^6 cells/mL in the appropriate buffer, and analyzed in a spectrofluorometer operated in the ratio mode (model F-2500; Hitachi Ltd., Tokyo, Japan) under continuous stirring and at a constant temperature of 37 °C using a water-jacketed device. After a 5-min accommodation to equilibrate temperatures, samples were excited at 338 nm, and emission was collected at 405 and 485 nm, corresponding to the fluorescence emitted by Ca^{2+} bound and -free Indo1, respectively. The cells were stimulated with either LPI or VCE-006.1, and maximal ratio values for calculations were determined by the addition at the end of the measurements of 10 µM ionomycin. [Ca^{2+}]i changes are presented as changes in the ratio of bound to free calcium (340 nm/380 nm).

4.1.3. cAMP Signaling Induced by GPR55 Activation

The determination of GPR55 activity was carried out using the HEK293T-GPR55 cells stably transfected with the human GPR55 cDNA. Briefly, HEK293T-GPR55 cells were transiently transfected with 0.2 µg of the reporter plasmid CRE-Luc that contains six consensus cAMP-responsive elements (CRE) linked to the firefly luciferase reporter gene using Roti©-Fect (Carl Roth, Karlsruhe, Germany). Transfected cells were treated with either VCE-006.1, LPI, or a combination of both. After 6 h of stimulation, cells were washed twice with PBS 1× and lysed in 100 µL lysis buffer containing 25 mM Tris-phosphate (pH 7.8), 8 mM $MgCl_2$, 1 mM DTT, 1% Triton X-100, and 7% glycerol for 15 min at room temperature in a horizontal shaker. Luciferase activity was measured using a TriStar2 Berthold/LB942 multimode reader (Berthold Technologies, Bad Wildbad, Germany) following the instructions of the luciferase assay kit (Promega, Madison, WI,

USA). The RLUs (relative light units) were calculated, and the results were expressed as fold induction over unstimulated cells. The experiment was performed 5–6 times.

4.2. Animals and Cell Experiments

4.2.1. PD Experiments

Male C57BL/6 mice were housed in a room with a controlled photoperiod (08:00–20:00 light) and temperature (22 ± 1 °C). They had free access to standard food and water and were used at adult age (3–4 month-old; 25–30 g weight). All experiments were conducted according to national and European guidelines (directive 2010/63/EU), as well as conformed to ARRIVE guidelines and approved by the "Comité de Experimentación Animal" of our university (PROEX: 056/19).

In a first experiment, male C57BL/6 mice were subjected to stereotaxic unilateral application of 6-hydroxydopamine (6-OHDA) or saline [24,75]. To do that, mice were anesthetized (ketamine 40 mg/kg + xylazine 4 mg/kg, i.p.) 30 min after pretreatment with desipramine (25 mg/kg, i.p.), and then 6-OHDA free base (2 μL at a concentration of 2 μg/μL saline in 0.2% ascorbate to avoid oxidation) or saline (for control mice) were injected stereotaxically into the right striatum at a rate of 0.5 μL/min, using the following coordinates: + 0.4 mm AP, −1.8 mm ML and −3.5 mm DV, as described in [75]. Once injected, the needle was left in place for 5 min before being slowly withdrawn, thus avoiding reflux and a rapid increase in intracranial pressure. Control animals were sham-operated and injected with 2 μL of saline using the same coordinates. The lesions were generated using unilateral injection, the advantage of which is that contralateral structures serve as controls for the different analyses. After the application of 6-OHDA or saline, animals were subjected to a daily treatment with VCE-006.1 (20 mg/kg, i.p.) or vehicle (cremophor-saline, 1:18) for two weeks, at the end of which (24 h after the last injection), they were analyzed in the pole test and the cylinder rearing test just before being killed by rapid and careful decapitation and their brains rapidly removed. Brains were divided coronally into two parts, following the procedure described by Palkovits and Brownstein [76]. The anterior halves were used to dissect the striatum (both ipsilateral and contralateral sides separately), and tissues were rapidly frozen by immersion in cold 2-methylbutane and stored at −80 °C for qPCR analysis. The posterior halves containing the midbrains were fixed for one day at 4 °C in fresh 4% paraformaldehyde prepared in 0.1 M PBS, pH 7.4. Samples were cryoprotected by immersion in a 30% sucrose solution for a further day, and finally stored at −80 °C for immunohistochemical analysis in the substantia nigra.

In a second experiment, mice were anesthetized (ketamine 40 mg/kg + xylazine 4 mg/kg, i.p.) and subjected to unilateral injections of *S. Minnesota* LPS (Sigma-Aldrich, Madrid, Spain) into two points of the right striatum following the procedure developed by Hunter et al. [77]. We used the following stereotaxic coordinates from bregma: + 1.1 mm AP, −1.8 mm ML, and −3.5 mm DV, as well as −0.3 mm AP, −2.5 mm ML, and −3.2 mm DV (see details in [77]). At each intrastriatal coordinate, 5 μg of LPS in a volume of 1 μL of saline was injected slowly (0.5 μL/30 s), and the needle was again left in place for 5 min before being slowly withdrawn. This avoids generating reflux and a rapid increase in intracranial pressure. Control animals were sham-operated and injected with 1 μL of saline using the same coordinates. Again, the lesions were generated using unilateral administration, the advantage of which is that contralateral structures serve as controls for the different analyses. After the application of LPS or saline, animals were subjected to a daily treatment with VCE-006.1 (20 mg/kg, i.p.) or vehicle (cremophor-saline, 1:18) for two weeks, at the end of which (24 h after the last injection), they were analyzed in the pole test and the cylinder rearing test just before being killed by rapid and careful decapitation and their brains rapidly removed and processed as described before 6-OHDA-lesioned mice.

In a third experiment, cultures of SH-SY5Y neuronal cell line (kindly provided by Dr. Ana Martínez, CIB-CSIC, Madrid, Spain) were used to induce cell death with 6-OHDA and to investigate in vitro the possible cytoprotective effects of VCE-006.1, following a procedure described previously [78]. To this end, SH-SY5Y cells were maintained in Dulbecco's Modified Eagle's Medium (DMEM; Lonza, Verviers, Belgium) supplemented with 10% fetal bovine serum (FBS), 2 mM Ultraglutamine, and 1% antibiotics (Lonza, Verviers, Belgium) under a humidified 5% CO_2 atmosphere at 37 °C. For cytotoxicity experiments, cells were seeded at 60,000 cells/well in 96-well plates and maintained under a humidified atmosphere (5% CO_2) at 37 °C overnight. For experiments, 24 h after seeding, cells were treated with the vehicle (DMEM + 0.1% DMSO) or with five different concentrations of VCE-006.1 (0.5, 1, 2, 5, 10, and 20 μM; selected according to [65]), 60 min before being exposed to 200 μM 6-OHDA (or saline) following our previously published studies with different concentrations of 6-OHDA in these cells [43,44]. Cells were incubated 24 h before the neuronal death was analyzed with the MTT assay (Panreac AppliChem., Barcelona, Spain). Data of cell viability were normalized in relation to the corresponding control group (cells exposed to vehicles for 6-OHDA and VCE-006.1).

In a fourth experiment, cultured BV-2 cells were maintained in DMEM (Lonza, Verviers, Belgium) supplemented with 10% FBS (Sigma-Aldrich, Madrid, Spain), 2 mM Ultraglutamine, and antibiotics (Lonza, Verviers, Belgium) in a humidified atmosphere of 5% CO_2 at 37 °C. Cells were plated at a density of 45×10^4 cells per well in 12-well culture plates and incubated in DMEM with a reduction of FBS to 1%. Three hours later, cells were treated with 0.5 μg/mL LPS (from *Escherichia coli* 055:B5, Sigma-Aldrich, Madrid, Spain), alone or in combination with VCE-006.1, used at a concentration of 1 μM, and added 1 h before LPS. Twenty hours after the addition of LPS, media were removed, and cell pellets were collected for analyzing mRNA levels of GPR55, tumor necrosis factor-α (TNF-α), and interleukin-1β (IL-1β) using qPCR analysis.

4.2.2. ALS Experiments

Experiments were conducted with two mouse colonies: (i) B6SJL-Tg(SOD1*G93A) 1Gur/J transgenic (mSOD1 mice) and non-transgenic littermate sibling mice bred in our animal facilities from initial breeders provided by Dr. Rosario Osta (LagenBio-Ingen, University of Zaragoza, Spain), and (ii) Prp-hTDP-43(A315T) transgenic and non-transgenic littermate sibling mice bred in our animal facilities from initial breeders purchased from Jackson Laboratories (Bar Harbor, ME, USA). In both cases, animals were subjected to genotyping for identifying the presence or absence of the transgene containing the SOD-1 or the TDP-43 mutation (see details in [30,45], respectively). As in PD experiments, all animals were housed in a room with controlled photoperiod (08:00–20:00 light) and temperature (22 ± 1 °C) with free access to standard food or, in the case of TDP-43 transgenic mice, to a high-fat jelly diet (DietGel Boost, ClearH20, Portland, ME, USA) [79], and water. All experiments were conducted according to local and European rules (directive 2010/63/EU), as well as conformed to ARRIVE guidelines. They were approved by the ethical committees of our university and the regulatory institution (PROEX: 056/19).

In a first experiment, wild-type and mSOD-1 transgenic mice were identified by numbered ear marks, and prior to the start of the different experiments, they were randomly allocated to the different treatment groups. We treated B6SJL-Tg(SOD-1*G93A)1Gur/J transgenic male mice with VCE-006.1, synthesized as previously described [65], and administered i.p. to mice at the dose of 20 mg/kg. Additional transgenic mice, as well as wild-type animals, were treated with vehicle (cremophor-saline, 1:18). The treatment was initiated when animals were 63 days old and prolonged daily up to the age of 18 weeks (125 days of age). During this period, animals were weighed every day and subjected to several neurological analyses and behavioral tests at specific time points. Twenty-four hours after the last injection, animals were euthanized by rapid decapitation, and their spinal cords were dissected and removed. The spinal cords (lumbar level) to be used for histology were fixed for one day at 4 °C in 4% formaldehyde solution in PBS. Samples were then cryoprotected by immersion in a 30% sucrose solution for a further day, and finally stored at −80 °C for Nissl staining and immunohistochemical analysis. The spinal samples (also lumbar area) to be used for qPCR analyses were collected and rapidly frozen by immersion in cold 2-methylbutane and stored at −80 °C for qPCR analysis.

In a second experiment, we treated non-transgenic and Prp-hTDP-43(A315T) transgenic male mice with VCE-006.1, again synthesized as previously described [65] and administered i.p. to mice at the dose of 20 mg/kg. Additional transgenic mice, as well as wild-type animals, were treated with vehicle (cremophor-saline, 1:18). The treatment was initiated when animals were 65 days old and prolonged daily up to the age of 95 days, the same treatment window used in our previous study [30], which extends from early symptomatic phases (around the 9th week of age) up to an advanced stage (around the 13th week of age). Animal weight was logged daily. Weight loss of 20% was established as the human end-point. Rotarod performance and clasping reflex to detect dystonia were recorded weekly during the 4 weeks of the treatment period (including a recording just before the first injection). All animals were euthanized by rapid decapitation at the age of 95 days, at least 24 h after the last administration. Their spinal cords were rapidly removed and processed as described for mSOD-1 mice.

4.3. Behavioral Recording

4.3.1. Pole Test

Mice were placed head-upward on the top of a vertical rough-surfaced pole (diameter 8 mm; height 55 cm), and the time until animals descended to the floor was recorded with a maximum duration of 120 s. When the mouse was not able to turn downward and instead dropped from the pole, the time was taken as 120 s (default value) (see details in [44]).

4.3.2. Cylinder Rearing Test

Given that the lesion was unilateral in the experiment with 6-OHDA or LPS, this test attempted to quantify the degree of forepaw (ipsilateral, contralateral, or both) preference for wall contacts after placing the mouse in a methacrylate transparent cylinder (diameter: 15.5 cm; height: 12.7 cm [80]). Each score was made out of a 3 min trial with a minimum of 4 wall contacts.

4.3.3. Neurological Score

Mice were evaluated for neurological decline using a numerical scale published previously [45]. The scale ranged from 0 to 15 distributed in three sub-scales (0–5) concentrated on ambulation, strength analysis, and hind-foot reflex test. A final score of 0 corresponds to animals that are not symptomatic, whereas a score of 15 reflects a state of total functional loss in hindlimbs and postural control. The assessment of ambulation was carried out by placing the animal inside a corridor ($10 \times 10 \times 80$ cm) while evaluating postural control and the way in which hindlimbs were leaned during motion. The strength test evaluated the animal's ability to drag and offer resistance when the tail was pulled softly to the opposite direction in which the animal moves. Lastly, the hind-foot reflex test evaluated the stiffness of the limbs and their coordination when the mouse was suspended by the tail 10 cm over the surface. The final score was calculated from the sum of values reached in each sub-scale.

4.3.4. Rotarod Test

Mice were evaluated for possible motor weakness using the rotarod test, using an LE8200 device (Panlab, Barcelona, Spain). Mice were exposed to a period of acclimation and training (first session: 0 r.p.m. for 30 s; second and third sessions: 4 r.p.m. for 60 s, with periods of 10 min between sessions), followed 30 min later by the assay. Mice were placed into the apparatus, and the rotational speed was increased from 4 to 40 r.p.m. over a period of 300 s to measure the time to fall off. Mice were tested for 3 consecutive trials with a rest period of approximately 15 min between trials, and the mean of the 3 trials was calculated.

4.3.5. Clasping Response

Dystonia was evaluated by picking up the mouse by the base of the tail for 30 s so that the mouse was facing downwards away from any object. The position of the hindlimbs was observed and scored following the scale reported by Guyenet et al. [81]. Animals were scored as follows: 0 if the hindlimbs were consistently extended away from the abdomen; 1 if one hindlimb was retracted toward the abdomen; 2 if both hindlimbs were partially retracted toward the abdomen; 3 if both hindlimbs were entirely retracted and touching the abdomen. Mice were tested for three consecutive trials, and the mean clasping score of the three trials was calculated.

4.3.6. Hanging Wire Test

The latency of mice to fall from a wire cage top, which was slowly inverted and suspended at approximately 30 cm to the floor, was also used as an index of motor weakness. The test was repeated three times to obtain the mean value of the three trials.

4.4. Histological Procedures

4.4.1. Tissue Slicing

In the PD experiment, brains were sliced in coronal sections (containing the substantia nigra) in a cryostat (30 µm thick) and collected on antifreeze solution (glycerol/ethylene glycol/PBS; 2:3:5) and stored at -20 °C until used for immunostaining. In the ALS experiment, fixed spinal cords were sliced with a cryostat at the lumbar level (L4-L6) to obtain coronal sections (20 µm thick) that were collected on gelatin-coated slides. Sections were used for procedures of Nissl-staining and immunostaining.

4.4.2. Immunohistochemistry Analysis in the PD Experiment

Brain sections containing the substantia nigra were mounted on gelatin-coated slides and, once adhered, washed in 0.1 M potassium PBS (KPBS) at pH 7.4. Endogenous peroxidase was blocked by 30 min incubation at room temperature in peroxidase blocking solution (Dako Cytomation, Glostrup, Denmark). After several washes with KPBS, sections were incubated overnight at room temperature with the following polyclonal antibodies: (i) rabbit anti-tyrosine hydroxylase (TH) (Chemicon-Millipore, Temecula, CA, USA) used at 1/200; (ii) rat anti-mouse Cd68 antibody (AbD Serotec, Oxford, UK) used at 1/200; or (iii) rabbit anti-mouse GFAP antibody (Dako Cytomation, Glostrup, Denmark) used at 1/200. In the case of LAMP-1 immunostaining, we used the hybridoma monoclonal rat anti-mouse LAMP-1 antibody 1D4B, which was deposited by Dr. J. Thomas in the Developmental Studies Hybridoma Bank (DSHB; Hybridoma Product 1D4B), created by the NICHD (NIH, Bethesda, MD, USA) and maintained at The University of Iowa, Department of Biology, Iowa City, IA, USA. Dilutions were carried out in KPBS containing 2% bovine serum albumin and 0.1% Triton X-100 (Sigma Chem., Madrid, Spain). After incubation, sections were washed in KPBS, followed by incubation with the corresponding biotinylated secondary antibody (1/200) (Vector Laboratories, Burlingame, CA, USA) for 1 h at room temperature. Avidin-biotin complex (Vector Laboratories, Burlingame, CA, USA) and 3,3'-diaminobenzidine substrate–chromogen system (Dako Cytomation, Glostrup, Denmark) were used to obtain a visible reaction product. Negative control sections were obtained using the same protocol with omission of the primary antibody. A Leica DMRB microscope and a DFC300FX camera (Leica, Wetzlar, Germany) were used for the observation and photography of the slides, respectively. For quantification of TH, LAMP-1, GFAP, or Cd68 immunostaining in the substantia nigra, we used the NIH Image Processing and Analysis software (ImageJ; NIH, Bethesda, MD, USA) using 4–5 sections, separated approximately by 200 µm, and observed with 5x-20x objectives depending on the method and the brain area under quantification. In all sections, the same area of the substantia nigra pars compacta was analyzed. Analyses were always conducted by experimenters who were blinded to all animal characteristics. Data were expressed as a percentage of immunostaining intensity in the ipsilateral (lesioned) side over the contralateral (non-lesioned) side.

4.4.3. Nissl Staining

Slices were used for Nissl staining using cresyl violet, as previously described [82], which permitted us to determine the effects of particular treatments on cell numbers. A Leica DMRB microscope (Leica, Wetzlar, Germany) and a DFC300Fx camera (Leica) were used to study and photograph the tissue, respectively. To count the number of Nissl-stained motor neurons ($>400 \ \mu m^2$) in the ventral horn, high-resolution photomicrographs were taken with a $10\times$ objective under the same conditions of light, brightness, and contrast. Counting was carried out with ImageJ software (U.S. National Institutes of Health, Bethesda, MD, USA, http://imagej.nih.gov/ij/, 1997–2012). At least 6 images per animal were analyzed to establish the mean of all animals studied in each group. Analyses were always conducted by experimenters who were blinded to all animal characteristics. In all analyses, data were transformed to the percentage over the mean obtained in the wild-type group for each parameter.

4.4.4. Immunofluorescence Analysis in the ALS Experiment

Spinal slices were used for the detection and quantification of GFAP or Iba-1 immunofluorescence. After preincubation for 1 h with Tris-buffered saline with 0.1% Triton X-100 (pH 7.5), sections were sequentially incubated overnight at 4 °C with the following polyclonal antibodies: (i) anti-Iba-1 (Wako Chemicals, Richmond, VI, USA) used at 1:500; or (ii) anti-GFAP (Dako Cytomation, Glostrup, Denmark) used at 1:200, followed by washing in Tris-buffered saline and a new incubation (at 37 °C for 2 h) with an anti-rabbit secondary antibody conjugated with Alexa 488 (Invitrogen, Carlsbad, CA, USA). A DMRB microscope and a DFC300Fx camera (Leica, Wetzlar, Germany) were used for slide observation and photography. The mean density of immunolabelling was measured in the selected areas. Again, all data were transformed to the percentage over the mean obtained in the wild-type group for each parameter.

4.5. Real Time qRT-PCR Analysis

Tissues (striatum and spinal cord) from in vivo experiments and cell pellets from the in vitro experiments were also used for qRT-PCR analysis. Total RNA was isolated from the different samples using Trizol reagent (Sigma-Aldrich, Madrid, Spain). The total amount of RNA extracted was quantitated by spectrometry at 260 nm and its purity from the ratio between the absorbance values at 260 and 280 nm. After genomic DNA was removed (to eliminate DNA contamination), single-stranded complementary DNA was synthesized from up to 1 μg of total RNA using the commercial kits Rneasy Mini Quantitect Reverse Transcription (Qiagen, Hilgen, Germany) and iScript^{TM} cDNA Synthesis Kit (Bio-Rad, Hercules, CA, USA). The reaction mixture was kept frozen at -20 °C until enzymatic amplification. Quantitative RT-PCR assays were performed using TaqMan Gene Expression Assays (Applied Biosystems, Foster City, CA, USA) to quantify mRNA levels for TNF-α (ref. Mm99999068_m1), IL-1β (ref. Mm00434228_m1), iNOS (ref. Mm01309902_m1), COX-2 (ref. Mm00478372_m1), CB$_1$ receptor (ref. Mm00432621_s1), CB$_2$ receptor (ref. Mm00438286_m1), GPR55 (ref. Mm03978245_m1), and PPARγ (ref. Mm01184322_m1), using GAPDH expression (ref. Mm99999915_g1) as an endogenous control gene for normalization. The PCR assay was performed using the 7300 Fast Real-Time PCR System (Applied Biosystems, Foster City, CA, USA), and the threshold cycle (Ct) was calculated by the instrument's software (7300 Fast System, Applied Biosystems, Foster City, CA, USA). Expression levels were calculated using the $2^{-\Delta\Delta Ct}$ method.

4.6. Statistics

Data were assessed using one-way or two-way (repeated measures) ANOVA, as required, followed by the Tukey test, or using the Student's *t*-test, as required, using GraphPad Prism, version 8.00 for Windows (GraphPad Software, San Diego, CA, USA). A *p*-value lower than 0.05 was used as the limit for statistical significance. The sample sizes in the different experimental groups were always ≥ 5.

5. Conclusions

Therefore, our findings support the view that targeting the GPR55 with cannabinoids able to activate this receptor may afford neuroprotection in experimental PD, in particular, in models associated with mitochondrial dysfunction as in 6-OHDA-lesioned mice. Some beneficial effects were also found in LPS-lesioned mice, but with no effect against the intense glial activation occurring in this model. Future studies are projected to explore whether VCE-006.1 could also be active in mutant α-synuclein-based models of PD. Such a question is important to determine whether VCE-00.1 activity occurs exclusively in toxin-based models of PD or may also be found in models based on gene modifications. The need for this confirmation derives in part from the fact that VCE-006.1 was poorly active in experimental genetic models of ALS, although it is also possible that its development in this disease would require its combination with other cannabinoids active at additional endocannabinoid-related targets, in particular, anti-inflammatory targets. Collectively, these results demonstrate the specificities for the development of cannabinoid-based therapies for the different neurodegenerative disorders.

Author Contributions: Funding acquisition, E.M., E.d.L. and J.F.-R.; study design, coordination, and supervision, J.F.-R., C.G., N.J. and E.M.; studies of VCE-006.1: mechanisms of action, F.R.-P., J.D.U., P.M. and E.M.; studies in 6-OHDA-lesioned mice: design and methodology, S.B. and C.G.; studies in LPS-lesioned mice: design and methodology, S.B. and C.G.; studies in cultured cells, S.B. and M.G.-C.; development of the transgenic colonies of ALS mice, C.G.-C., M.G.-A. and C.R.-C.; studies in mSOD1 and TDP-43 transgenic mice: design and methodology, C.G.-C., M.G.-A., J.A., C.R.-C. and E.d.L.; statistical analysis of the data, S.B., C.G.-C., M.G.-A. and J.F.-R.; manuscript preparation, J.F.-R. with the revision and approval of all authors. All authors have read and agreed to the published version of the manuscript.

Funding: This work has been supported by grants from CIBERNED (CB06/05/0089), MICIU (RTI-2018-098885-B-100), ELA-Madrid-CM (B2017/BMD-3813), and Emerald Health Biotechnology-Spain. These agencies had no further role in study design, the collection, analysis, and interpretation of data, in the writing of the report, or in the decision to submit the paper for publication.

Institutional Review Board Statement: All experiments were conducted according to European guidelines (directive 2010/63/EU) and approved by the "Comité de Experimentación Animal" of our university (ref. PROEX 056/19).

Informed Consent Statement: Not applicable.

Data Availability Statement: Data supporting reported results may be supplied upon request to the authors.

Acknowledgments: Sonia Burgaz, Marta Gómez-Almería, and Claudia Gonzalo-Consuegra are predoctoral fellows supported by the FPI Programme-MICIU (MGA) and UCM-Predoctoral Programme (SB and CGC). Paula Morales is a postdoctoral fellow supported by the Juan de la Cierva Programme-MICIU (IJC 2019-042182-I).

Conflicts of Interest: The authors declare no conflict of interest.

Sample Availability: Samples of the compounds are available from the authors.

References

1. Fernández-Ruiz, J.; Moro, M.A.; Martinez-Orgado, J. Cannabinoids in Neurodegenerative Disorders and Stroke/Brain Trauma: From Preclinical Models to Clinical Applications. *Neurotherapeutics* **2015**, *12*, 793–806. [CrossRef]
2. Aymerich, M.S.; Aso, E.; Abellanas, M.A.; Tolon, R.M.; Ramos, J.A.; Ferrer, I.; Romero, J.; Fernández-Ruiz, J. Cannabinoid pharmacology/therapeutics in chronic degenerative disorders affecting the central nervous system. *Biochem. Pharmacol.* **2018**, *157*, 67–84. [CrossRef]
3. Fernández-Ruiz, J. The biomedical challenge of neurodegenerative disorders: An opportunity for cannabinoid-based therapies to improve on the poor current therapeutic outcomes. *Br. J. Pharmacol.* **2018**, *176*, 1370–1383. [CrossRef]
4. Chiarlone, A.; Bellocchio, L.; Blázquez, C.; Resel, E.; Soria-Gómez, E.; Cannich, A.; Ferrero, J.J.; Sagredo, O.; Benito, C.; Romero, J.; et al. A restricted population of CB1 cannabinoid receptors with neuroprotective activity. *Proc Natl Acad Sci USA* **2014**, *111*, 8257–8262. [CrossRef]

5. Hiebel, C.; Behl, C. The complex modulation of lysosomal degradation pathways by cannabinoid receptors 1 and 2. *Life Sci.* **2015**, *138*, 3–7. [CrossRef]

6. Aso, E.; Palomer, E.; Juvés, S.; Maldonado, R.; Muñoz, F.J.; Ferrer, I. CB1 Agonist ACEA Protects Neurons and Reduces the Cognitive Impairment of AβPP/PS1 Mice. *J. Alzheimer's Dis.* **2012**, *30*, 439–459. [CrossRef]

7. Navarro, G.; Borroto-Escuela, D.; Angelats, E.; Etayo, Í.; Reyes-Resina, I.; Pulido-Salgado, M.; Rodríguez-Pérez, A.I.; Canela, E.I.; Saura, J.; Lanciego, J.L.; et al. Receptor-heteromer mediated regulation of endocannabinoid signaling in activated microglia. Role of CB_1 and CB_2 receptors and relevance for Alzheimer's disease and levodopa-induced dyskinesia. *Brain Behav Immun.* **2018**, *67*, 139–151. [CrossRef]

8. Crunfli, F.; Vrechi, T.A.; Costa, A.P.; Torrão, A.S. Cannabinoid Receptor Type 1 Agonist ACEA Improves Cognitive Deficit on STZ-Induced Neurotoxicity Through Apoptosis Pathway and NO Modulation. *Neurotox. Res.* **2019**, *35*, 516–529. [CrossRef]

9. Chung, Y.C.; Bok, E.; Huh, S.H.; Park, J.Y.; Yoon, S.H.; Kim, S.R.; Kim, Y.S.; Maeng, S.; Park, S.H.; Jin, B.K. Cannabinoid receptor type 1 protects nigrostriatal dopaminergic neurons against MPTP neurotoxicity by inhibiting microglial activation. *J. Immunol.* **2011**, *187*, 6508–6517. [CrossRef]

10. Pérez-Rial, S.; García-Gutiérrez, M.S.; Molina, J.A.; Pérez-Nievas, B.G.; Ledent, C.; Leiva, C.; Leza, J.C.; Manzanares, J. Increased vulnerability to 6-hydroxydopamine lesion and reduced development of dyskinesias in mice lacking CB1 cannabinoid receptors. *Neurobiol. Aging* **2011**, *32*, 631–645. [CrossRef]

11. Abood, M.E.; Rizvi, G.; Sallapudi, N.; McAllister, S.D. Activation of the CB1 cannabinoid receptor protects cultured mouse spinal neurons against excitotoxicity. *Neurosci. Lett.* **2001**, *309*, 197–201. [CrossRef]

12. Zhao, P.; Ignacio, S.; Beattie, E.C.; Abood, M.E. Altered presymptomatic AMPA and cannabinoid receptor trafficking in motor neurons of ALS model mice: Implications for excitotoxicity. *Eur. J. Neurosci.* **2008**, *27*, 572–579. [CrossRef] [PubMed]

13. Rossi, S.; De Chiara, V.; Musella, A.; Cozzolino, M.; Bernardi, G.; Maccarrone, M.; Mercuri, N.B.; Carrì, M.T.; Centonze, D. Abnormal sensitivity of cannabinoid CB1 receptors in the striatum of mice with experimental amyotrophic lateral sclerosis. *Amyotroph. Lateral Scler.* **2010**, *11*, 83–90. [CrossRef]

14. Blázquez, C.; Chiarlone, A.; Sagredo, O.; Aguado, T.; Pazos, M.R.; Resel, E.; Palazuelos, J.; Julien, B.; Salazar, M.; Börner, C.; et al. Loss of striatal type 1 cannabinoid receptors is a key pathogenic factor in Huntington's disease. *Brain* **2011**, *134*, 119–136. [CrossRef]

15. Maya-López, M.; Colín-González, A.L.; Aguilera, G.; De Lima, M.E.; Colpo-Ceolin, A.; Rangel-López, E.; Villeda-Hernández, J.; Rembao-Bojórquez, D.; Túnez, I.; Luna-López, A.; et al. Neuroprotective effect of WIN55,212-2 against 3-nitropropionic acid-induced toxicity in the rat brain: Involvement of CB1 and NMDA receptors. *Am. J. Transl. Res.* **2017**, *9*, 261–274.

16. Ruiz-Calvo, A.; Maroto, I.B.; Bajo-Grañeras, R.; Chiarlone, A.; Gaudioso, Á.; Ferrero, J.J.; Resel, E.; Sánchez-Prieto, J.; Rodríguez-Navarro, J.A.; Marsicano, G.; et al. Pathway-specific control of striatal neuron vulnerability by corticostriatal cannabinoid CB1 receptors. *Cereb. Cortex* **2018**, *28*, 307–322. [CrossRef]

17. Rossi, S.; Furlan, R.; De Chiara, V.; Muzio, L.; Musella, A.; Motta, C.; Studer, V.; Cavasinni, F.; Bernardi, G.; Martino, G.; et al. Cannabinoid CB1 receptors regulate neuronal TNF-α effects in experimental autoimmune encephalomyelitis. *Brain Behav. Immun.* **2011**, *25*, 1242–1248. [CrossRef]

18. Moreno-Martet, M.; Feliú, A.; Espejo-Porras, F.; Mecha, M.; Carrillo-Salinas, F.J.; Fernández-Ruiz, J.; Guaza, C.; de Lago, E. The disease-modifying effects of a Sativex-like combination of phytocannabinoids in mice with experimental autoimmune encephalomyelitis are preferentially due to Δ9-tetrahydrocannabinol acting through CB1 receptors. *Mult. Scler. Relat. Disord.* **2015**, *4*, 505–511. [CrossRef] [PubMed]

19. Fernández-Ruiz, J.; Romero, J.; Velasco, G.; Tolón, R.M.; Ramos, J.A.; Guzmán, M. Cannabinoid CB2 receptor: A new target for controlling neural cell survival? *Trends Pharmacol. Sci.* **2007**, *28*, 39–45. [CrossRef]

20. Aso, E.; Ferrer, I. CB2 Cannabinoid Receptor As Potential Target against Alzheimer's Disease. *Front. Neurosci.* **2016**, *10*, 243. [CrossRef]

21. López, A.; Aparicio, N.; Pazos, M.R.; Grande, M.T.; Barreda-Manso, M.A.; Benito-Cuesta, I.; Vázquez, C.; Amores, M.; Ruiz-Pérez, G.; García-García, E.; et al. Cannabinoid CB_2 receptors in the mouse brain: Relevance for Alzheimer's disease. *J. Neuroinflamm.* **2018**, *15*, 158. [CrossRef]

22. Magham, S.V.; Krishnamurthy, P.T.; Shaji, N.; Mani, L.; Balasubramanian, S. Cannabinoid receptor 2 selective agonists and Alzheimer's disease: An insight into the therapeutic potentials. *J. Neurosci. Res.* **2021**, *99*, 2888–2905. [CrossRef] [PubMed]

23. Galán-Ganga, M.; Rodríguez-Cueto, C.; Merchán-Rubira, J.; Hernández, F.; Ávila, J.; Posada-Ayala, M.; Lanciego, J.L.; Luengo, E.; Lopez, M.G.; Rábano, A.; et al. Cannabinoid receptor CB2 ablation protects against TAU induced neurodegeneration. *Acta Neuropathol. Commun.* **2021**, *9*, 90. [CrossRef]

24. García, C.; Palomo-Garo, C.; García-Arencibia, M.; Ramos, J.; Pertwee, R.; Fernández-Ruiz, J. Symptom-relieving and neuroprotective effects of the phytocannabinoid Δ9-THCV in animal models of Parkinson's disease. *Br. J. Pharmacol.* **2011**, *163*, 1495–1506. [CrossRef]

25. Gómez-Gálvez, Y.; Palomo-Garo, C.; Fernández-Ruiz, J.; García, C. Potential of the cannabinoid CB2 receptor as a pharmacological target against inflammation in Parkinson's disease. *Prog. Neuropsychopharmacol. Biol. Psychiatry* **2016**, *64*, 200–208. [CrossRef] [PubMed]

26. Javed, H.; Azimullah, S.; Haque, M.E.; Ojha, S.K. Cannabinoid Type 2 (CB2) Receptors Activation Protects against Oxidative Stress and Neuroinflammation Associated Dopaminergic Neurodegeneration in Rotenone Model of Parkinson's Disease. *Front. Neurosci.* **2016**, *10*, 321. [CrossRef] [PubMed]

27. Shi, J.; Cai, Q.; Zhang, J.; He, X.; Liu, Y.; Zhu, R.; Jin, L. AM1241 alleviates MPTP-induced Parkinson's disease and promotes the regeneration of DA neurons in PD mice. *Oncotarget* **2017**, *8*, 67837–67850. [CrossRef]

28. Kim, K.; Moore, D.H.; Makriyannis, A.; Abood, M.E. AM1241, a cannabinoid CB2 receptor selective compound, delays disease progression in a mouse model of amyotrophic lateral sclerosis. *Eur. J. Pharmacol.* **2006**, *542*, 100–105. [CrossRef]

29. Shoemaker, J.L.; Seely, K.A.; Reed, R.L.; Crow, J.P.; Prather, P.L. The CB2 cannabinoid agonist AM-1241 prolongs survival in a transgenic mouse model of amyotrophic lateral sclerosis when initiated at symptom onset. *J. Neurochem.* **2006**, *101*, 87–98. [CrossRef] [PubMed]

30. Espejo-Porras, F.; García-Toscano, L.; Rodríguez-Cueto, C.; Santos-García, I.; de Lago, E.; Fernandez-Ruiz, J. Targeting glial cannabinoid CB_2 receptors to delay the progression of the pathological phenotype in TDP-43 (A315T) transgenic mice, a model of amyotrophic lateral sclerosis. *Br. J. Pharmacol.* **2019**, *176*, 1585–1600. [CrossRef]

31. Rodríguez-Cueto, C.; Gómez-Almería, M.; García Toscano, L.; Romero, J.; Hillard, C.J.; de Lago, E.; Fernández-Ruiz, J. Inactivation of the CB_2 receptor accelerated the neuropathological deterioration in TDP-43 transgenic mice, a model of amyotrophic lateral sclerosis. *Brain. Pathol.* **2021**, *31*, e12972. [CrossRef]

32. Rodríguez-Cueto, C.; García-Toscano, L.; Santos-García, I.; Gómez-Almería, M.; Gonzalo-Consuegra, C.; Espejo-Porras, F.; Fernández-Ruiz, J.; de Lago, E. Targeting the CB_2 receptor and other endocannabinoid elements to delay disease progression in amyotrophic lateral sclerosis. *Br. J. Pharmacol.* **2021**, *178*, 1373–1387. [CrossRef]

33. Sagredo, O.; González, S.; Aroyo, I.; Pazos, M.R.; Benito, C.; Lastres-Becker, I.; Romero, J.P.; Tolón, R.M.; Mechoulam, R.; Brouillet, E.; et al. Cannabinoid CB2 receptor agonists protect the striatum against malonate toxicity: Relevance for Huntington's disease. *Glia* **2009**, *57*, 1154–1167. [CrossRef]

34. Palazuelos, J.; Aguado, T.; Pazos, M.R.; Julien, B.; Carrasco, C.; Resel, E.; Sagredo, O.; Benito, C.; Romero, J.; Azcoitia, I.; et al. Microglial CB2 cannabinoid receptors are neuroprotective in Huntington's disease excitotoxicity. *Brain* **2009**, *132*, 3152–3164. [CrossRef]

35. Bouchard, J.; Truong, J.; Bouchard, K.; Dunkelberger, D.; Desrayaud, S.; Moussaoui, S.; Tabrizi, S.J.; Stella, N.; Muchowski, P.J. Cannabinoid receptor 2 signaling in peripheral immune cells modulates disease onset and severity in mouse models of Huntington's disease. *J. Neurosci.* **2012**, *32*, 18259–18268. [CrossRef] [PubMed]

36. Morales, P.; Gómez-Cañas, M.; Navarro, G.; Hurst, D.P.; Carrillo-Salinas, F.J.; Lagartera, L.; Pazos, R.; Goya, P.; Reggio, P.H.; Guaza, C.; et al. Chromenopyrazole, a versatile cannabinoid scaffold with in vivo activity in a model of multiple sclerosis. *J. Med. Chem.* **2016**, *59*, 6753–6771. [CrossRef]

37. Alberti, T.B.; Barbosa, W.L.; Vieira, J.L.; Raposo, N.R.; Dutra, R.C. (-)-β-Caryophyllene, a CB2 receptor-selective phytocannabinoid, suppresses motor paralysis and neuroinflammation in a murine model of multiple sclerosis. *Int. J. Mol. Sci.* **2017**, *18*, 691. [CrossRef]

38. Mecha, M.; Carrillo-Salinas, F.J.; Feliú, A.; Mestre, L.; Guaza, C. Perspectives on Cannabis-Based Therapy of Multiple Sclerosis: A Mini-Review. *Front. Cell. Neurosci.* **2020**, *14*, 34. [CrossRef]

39. O'Sullivan, S.E. An update on PPAR activation by cannabinoids. *Br. J. Pharmacol.* **2016**, *173*, 1899–1910. [CrossRef]

40. Iannotti, F.; Vitale, R. The Endocannabinoid System and PPARs: Focus on Their Signalling Crosstalk, Action and Transcriptional Regulation. *Cells* **2021**, *10*, 586. [CrossRef]

41. García, C.; Gómez-Cañas, M.; Burgaz, S.; Palomares, B.; Gómez-Gálvez, Y.; Palomo-Garo, C.; Campo, S.; Ferrer-Hernández, J.; Pavicic, C.; Navarrete, C.; et al. Benefits of VCE-003.2, a cannabigerol quinone derivative, against inflammation-driven neuronal deterioration in experimental Parkinson's disease: Possible involvement of different binding sites at the PPARγ receptor. *J. Neuroinflamm.* **2018**, *15*, 19. [CrossRef] [PubMed]

42. Junior, N.C.F.; dos-Santos-Pereira, M.; Guimarães, F.S.; Del Bel, E. Cannabidiol and Cannabinoid Compounds as Potential Strategies for Treating Parkinson's Disease and l-DOPA-Induced Dyskinesia. *Neurotox. Res.* **2020**, *37*, 12–29. [CrossRef]

43. Burgaz, S.; García, C.; Gómez-Cañas, M.; Rolland, A.; Muñoz, E.; Fernández-Ruiz, J. Neuroprotection with the Cannabidiol Quinone Derivative VCE-004.8 (EHP-101) against 6-Hydroxydopamine in Cell and Murine Models of Parkinson's Disease. *Molecules* **2021**, *26*, 3245. [CrossRef]

44. Burgaz, S.; García, C.; Gómez-Cañas, M.; Navarrete, C.; García-Martín, A.; Rolland, A.; Del Río, C.; Casarejos, M.J.; Muñoz, E.; Gonzalo-Consuegra, C.; et al. Neuroprotection with the cannabigerol quinone derivative VCE-003.2 and its analogs CBGA-Q and CBGA-Q-Salt in Parkinson's disease using 6-hydroxydopamine-lesioned mice. *Mol. Cell Neurosci.* **2021**, *110*, 103583. [CrossRef]

45. Cueto, C.R.; Santos-García, I.; García-Toscano, L.; Espejo-Porras, F.; Bellido, M.; Fernández-Ruiz, J.; Munoz, E.; de Lago, E. Neuroprotective effects of the cannabigerol quinone derivative VCE-003.2 in SOD1G93A transgenic mice, an experimental model of amyotrophic lateral sclerosis. *Biochem. Pharmacol.* **2018**, *157*, 217–226. [CrossRef] [PubMed]

46. Fakhfouri, G.; Ahmadiani, A.; Rahimian, R.; Grolla, A.A.; Moradi, F.; Haeri, A. WIN55212-2 attenuates amyloid-beta-induced neuroinflammation in rats through activation of cannabinoid receptors and PPAR-γ pathway. *Neuropharmacology* **2012**, *63*, 653–666. [CrossRef] [PubMed]

47. Cheng, Y.; Dong, Z.; Liu, S. β-Caryophyllene Ameliorates the Alzheimer-Like Phenotype in APP/PS1 Mice through CB2 Receptor Activation and the PPARγ Pathway. *Pharmacology* **2014**, *94*, 1–12. [CrossRef]

48. Kallendrusch, S.; Kremzow, S.; Nowicki, M.; Grabiec, U.; Winkelmann, R.; Benz, A.; Kraft, R.; Bechmann, I.; Dehghani, F.; Koch, M. The G protein-coupled receptor 55 ligand l-α-lysophosphatidylinositol exerts microglia-dependent neuroprotection after excitotoxic lesion. *Glia* **2013**, *61*, 1822–1831. [CrossRef] [PubMed]

49. Hill, J.D.; Zuluaga-Ramirez, V.; Gajghate, S.; Winfield, M.; Sriram, U.; Rom, S.; Persidsky, Y. Activation of GPR55 induces neuro-protection of hippocampal neurogenesis and immune responses of neural stem cells following chronic, systemic inflammation. *Brain Behav. Immun.* **2019**, *76*, 165–181. [CrossRef] [PubMed]

50. Minamihata, T.; Takano, K.; Moriyama, M.; Nakamura, Y. Lysophosphatidylinositol, an Endogenous Ligand for G Protein-Coupled Receptor 55, Has Anti-inflammatory Effects in Cultured Microglia. *Inflammation* **2020**, *43*, 1971–1987. [CrossRef]

51. Celorrio, M.; Rojo-Bustamante, E.; Fernández-Suárez, D.; Sáez, E.; Estella-Hermoso de Mendoza, A.; Müller, C.E.; Ramírez, M.J.; Oyarzábal, J.; Franco, R.; Aymerich, M.S. GPR55: A therapeutic target for Parkinson's disease? *Neuropharmacology* **2017**, *125*, 319–332. [CrossRef]

52. Martínez-Pinilla, E.; Aguinaga, D.; Navarro, G.; Rico, A.J.; Oyarzábal, J.; Sánchez-Arias, J.A.; Lanciego, J.L.; Franco, R. Targeting CB1 and GPR55 Endocannabinoid Receptors as a Potential Neuroprotective Approach for Parkinson's Disease. *Mol. Neurobiol.* **2019**, *56*, 5900–5910. [CrossRef]

53. Wu, C.S.; Chen, H.; Sun, H.; Zhu, J.; Jew, C.P.; Wager-Miller, J.; Straiker, A.; Spencer, C.; Bradshaw, H.; Mackie, K.; et al. GPR55, a G-Protein Coupled Receptor for Lysophosphatidylinositol, Plays a Role in Motor Coordination. *PLoS ONE* **2013**, *8*, e60314. [CrossRef] [PubMed]

54. Yang, H.; Zhou, J.; Lehmann, C. GPR55-a putative "type 3" cannabinoid receptor in inflammation. *J. Basic Clin. Physiol. Pharmacol.* **2016**, *27*, 297–302. [CrossRef]

55. Lauckner, J.E.; Jensen, J.; Chen, H.-Y.; Lu, H.-C.; Hille, B.; Mackie, K. GPR55 is a cannabinoid receptor that increases intracellular calcium and inhibits M current. *Proc. Natl. Acad. Sci. USA* **2008**, *105*, 2699–2704. [CrossRef] [PubMed]

56. Morales, P.; Reggio, P.H. An update on non-CB_1, non-CB_2 cannabinoid related G-protein-coupled receptors. *Cannabis Cannabinoid Res.* **2017**, *2*, 265–273. [CrossRef]

57. Khan, M.Z.; He, L. Neuro-psychopharmacological perspective of Orphan receptors of Rhodopsin (class A) family of G protein-coupled receptors. *Psychopharmacology* **2017**, *234*, 1181–1207. [CrossRef] [PubMed]

58. Sawzdargo, M.; Nguyen, T.; Lee, D.K.; Lynch, K.R.; Cheng, R.; Heng, H.H.Q.; George, S.R.; O'Dowd, B.F. Identification and cloning of three novel human G protein-coupled receptor genes GPR52, & Psi;GPR53 and GPR55: GPR55 is extensively expressed in human brain. *Mol. Brain Res.* **1999**, *64*, 193–198 . [CrossRef]

59. Alhouayek, M.; Masquelier, J.; Muccioli, G.G. Lysophosphatidylinositols, from Cell Membrane Constituents to GPR55 Ligands. *Trends Pharmacol. Sci.* **2018**, *39*, 586–604. [CrossRef]

60. Shore, D.M.; Reggio, P.H. The therapeutic potential of orphan GPCRs, GPR35 and GPR55. *Front. Pharmacol.* **2015**, *6*, 69. [CrossRef] [PubMed]

61. Ross, R.A. The enigmatic pharmacology of GPR55. *Trends Pharmacol. Sci.* **2009**, *30*, 156–163. [CrossRef] [PubMed]

62. Morales, P.; Jagerovic, N. Advances towards the Discovery of GPR55 Ligands. *Curr. Med. Chem.* **2016**, *23*, 2087–2100. [CrossRef]

63. Marichal-Cancino, B.A.; Fajardo-Valdez, A.; Ruiz-Contreras, A.E.; Mendez-Díaz, M.; Prospero-García, O. Advances in the physiology of GPR55 in the Central Nervous System. *Curr. Neuropharmacol.* **2017**, *15*, 771–778. [CrossRef] [PubMed]

64. Henstridge, C.M.; Balenga, N.A.; Kargl, J.; Andradas, C.; Brown, A.J.; Irving, A.; Sanchez, C.; Waldhoer, M. Minireview: Recent developments in the physiology and pathology of the lysophosphatidylinositol-sensitive receptor GPR55. *Mol. Endocrinol.* **2011**, *25*, 1835–1848. [CrossRef] [PubMed]

65. Morales, P.; Whyte, L.S.; Chicharro, R.; Gómez-Cañas, M.; Pazos, M.R.; Goya, P.; Irving, A.J.; Fernández-Ruiz, J.; Ross, R.A.; Jagerovic, N. Identification of novel GPR55 modulators using cell-impedance-based label-free technology. *J. Med. Chem* **2016**, *59*, 1840–1853. [CrossRef]

66. Jagerovic, N.; Morales, P.; Ross, R.; Whyte, L. Selective Modulators of the Activity of the gpr55 Receptor: Chromenopyrazole Derivatives. Patent WO2016177922A1, 27 April 2016.

67. Szliszka, E.; Czuba, Z.P.; Domino, M.; Mazur, B.; Zydowicz, G.; Krol, W. Ethanolic Extract of Propolis (EEP) Enhances the Apoptosis- Inducing Potential of TRAIL in Cancer Cells. *Molecules* **2009**, *14*, 738–754. [CrossRef]

68. Pietr, M.; Kozela, E.; Levy, R.; Rimmerman, N.; Lin, Y.H.; Stella, N.; Vogel, Z.; Juknat, A. Differential changes in GPR55 during microglial cell activation. *FEBS Lett.* **2009**, *583*, 2071–2076. [CrossRef]

69. Medina-Vera, D.; Rosell-Valle, C.; López-Gambero, A.; Navarro, J.; Zambrana-Infantes, E.; Rivera, P.; Santín, L.; Suarez, J.; De Fonseca, F.R. Imbalance of Endocannabinoid/Lysophosphatidylinositol Receptors Marks the Severity of Alzheimer's Disease in a Preclinical Model: A Therapeutic Opportunity. *Biology* **2020**, *9*, 377. [CrossRef] [PubMed]

70. Xiang, X.; Wang, X.; Jin, S.; Hu, J.; Wu, Y.; Li, Y.; Wu, X. Activation of GPR55 attenuates cognitive impairment and neurotoxicity in a mouse model of Alzheimer's disease induced by Aβ1–42 through inhibiting RhoA/ROCK2 pathway. *Prog. Neuro-Psychopharmacol. Biol. Psychiatry* **2021**, *112*, 110423. [CrossRef]

71. Martínez-Pinilla, E.; Rico, A.J.; Rivas-Santisteban, R.; Lillo, J.; Roda, E.; Navarro, G.; Lanciego, J.L.; Franco, R. Expression of GPR55 and either cannabinoid CB1 or CB2 heteroreceptor complexes in the caudate, putamen, and accumbens nuclei of control, parkinsonian, and dyskinetic non-human primates. *Brain Struct. Funct.* **2020**, *225*, 2153–2164. [CrossRef]

72. Fatemi, I.; Abdollahi, A.; Shamsizadeh, A.; Allahtavakoli, M.; Roohbakhsh, A. The effect of intra-striatal administration of GPR55 agonist (LPI) and antagonist (ML193) on sensorimotor and motor functions in a Parkinson's disease rat model. *Acta Neuropsychiatr.* **2020**, *33*, 15–21. [CrossRef]
73. Malek, N.; Popiolek-Barczyk, K.; Mika, J.; Przewlocka, B.; Starowicz, K. Anandamide, Acting viaCB2Receptors, Alleviates LPS-Induced Neuroinflammation in Rat Primary Microglial Cultures. *Neural Plast.* **2015**, *2015*, 1–10. [CrossRef]
74. Moreno-Martet, M.; Espejo-Porras, F.; Fernández-Ruiz, J.; de Lago, E. Changes in endocannabinoid receptors and enzymes in the spinal cord of SOD1(G93A) transgenic mice and evaluation of a Sativex®-like combination of phytocannabinoids: Interest for future therapies in amyotrophic lateral sclerosis. *CNS Neurosci.* **2014**, *20*, 809–815. [CrossRef]
75. Alvarez-Fischer, D.; Henze, C.; Strenzke, C.; Westrich, J.; Ferger, B.; Höglinger, G.U.; Oertel, W.H.; Hartmann, A. Characterization of the striatal 6-OHDA model of Parkinson's disease in wild type and α-synuclein-deleted mice. *Exp. Neurol.* **2008**, *210*, 182–193. [CrossRef]
76. Palkovits, M.; Browstein, J. *Maps and Guide to Microdissection of the Rat Brain*; Elsevier: Amsterdam, The Netherlands, 1988.
77. Hunter, R.L.; Cheng, B.; Choi, D.Y.; Liu, M.; Liu, S.; Cass, W.A.; Bing, G. Intrastriatal lipopolysaccharide injection induces parkinsonism in C57/B6 mice. *J. Neurosci Res.* **2009**, *87*, 1913–1921. [CrossRef]
78. Ko, Y.-H.; Kim, S.-K.; Kwon, S.-H.; Seo, J.-Y.; Lee, B.-R.; Kim, Y.-J.; Hur, K.-H.; Kim, S.Y.; Lee, S.-Y.; Jang, C.-G. 7,8,4′-Trihydroxyisoflavone, a Metabolized Product of Daidzein, Attenuates 6-Hydroxydopamine-Induced Neurotoxicity in SH-SY5Y Cells. *Biomol. Ther.* **2019**, *27*, 363–372. [CrossRef]
79. Coughlan, K.S.; Halang, L.; Woods, I.; Prehn, J.H. A high-fat jelly diet restores bioenergetic balance and extends lifespan in the presence of motor dysfunction and lumbar spinal cord motor neuron loss in TDP-43A315T mutant C57BL6/J mice. *Dis. Model Mech.* **2016**, *9*, 1029–1037. [CrossRef]
80. Fleming, S.M.; Ekhator, O.R.; Ghisays, V. Assessment of Sensorimotor Function in Mouse Models of Parkinson's Disease. *J. Vis. Exp.* **2013**, *76*, e50303. [CrossRef]
81. Guyenet, S.J.; Furrer, S.A.; Damian, V.M.; Baughan, T.D.; La Spada, A.R.; Garden, G.A. A Simple Composite Phenotype Scoring System for Evaluating Mouse Models of Cerebellar Ataxia. *J. Vis. Exp.* **2010**, *21*, e1787. [CrossRef]
82. Alvarez, F.J.; Lafuente, H.; Rey-Santano, M.C.; Mielgo, V.E.; Gastiasoro, E.; Rueda, M.; Pertwee, R.G.; Castillo, A.I.; Romero, J.; Martínez-Orgado, J. Neuroprotective effects of the nonpsychoactive cannabinoid cannabidiol in hypoxic-ischemic newborn piglets. *Pediatr. Res.* **2008**, *64*, 653–658. [CrossRef]

MDPI
St. Alban-Anlage 66
4052 Basel
Switzerland
Tel. +41 61 683 77 34
Fax +41 61 302 89 18
www.mdpi.com

Molecules Editorial Office
E-mail: molecules@mdpi.com
www.mdpi.com/journal/molecules